Mechanisms in Pigmentation

Pigment Cell

Vol. 1

Editor: V. Riley, Seattle, Wash.

S. Karger · Basel · München · Paris · London · New York · Sydney

Proceedings of the 8th international Pigment Cell Conference
Sydney, March 13–17, 1972

Mechanisms in Pigmentation

Editors:
V. J. McGovern and P. Russell, Sydney, N.S.W.

166 figures, 89 tables

RC 280
S5
I57
1972

S. Karger · Basel · München · Paris · London · New York · Sydney · 1973

The Biology of Melanomas
Edited by MYRON GORDON. Proceedings of the Conference on the Biology of Normal and Atypical Cell Growth
(Ist International Pigment Cell Conference), New York City, N.Y., November 1946. XII + 466 p., 1948 (The New York Academy of Sciences, New York).

Zoologica, Vol. 35, Abstracts of Papers
Proceedings of the Second Conference on the Biology of Normal and Atypical Cell Growth (IInd International Pigment Cell Conference), New York City, N.Y., November 1949. 32 p., 1950 (The New York Zoological Society).

Pigment Cell Growth
Edited by MYRON GORDON. Proceedings of the IIIrd International Pigment Cell Conference New York City, N.Y., November 1951. XIII + 365 p., 1953 (Academic Press, Inc., New York).

Pigment Cell Biology
Edited by MYRON GORDON. Proceedings of the IVth International Pigment Cell Conference Houston, Tex., November 1957. XIV + 647 p., 1959 (Academic Press, Inc., New York).

The Pigment Cell: Molecular, Biological, and Clinical Aspects
Edited by VERNON RILEY and JOSEPH G. FORTNER. Proceedings of the Vth International Pigment Cell Conference
New York City, N.Y., October 1961. XII + 1123 p., 1963 (The New York Academy of Sciences, New York).

Structure and Control of the Melanocyte
Edited by G. DELLA PORTA and O. MÜHLBOCK. Proceedings of the VIth International Pigment Cell Conference
Sofia, Bulgaria, May 1965. XIV + 374 p., 1966 (Springer, New York).

Pigmentation: Its Genesis and Biologic Control
Edited by VERNON RILEY. Proceedings of the VIIth International Pigment Cell Conference Seattle, Wash., September 1969. XX + 682 p., 1972 (Appleton-Century-Crofts, New York).

S. Karger · Basel · München · Paris · London · New York · Sydney
Arnold-Böcklin-Strasse 25, CH-4011 Basel (Switzerland)

All rights, including that of translation into other languages, reserved. Photomechanic reproduction (photocopy, microcopy) of this book or parts thereof without special permission of the publishers is prohibited.

© Copyright 1973 by S. Karger AG, Verlag für Medizin und Naturwissenschaften, Basel
Printed in Switzerland by City-Druck, Glattbrugg
ISBN 3-8055-1480-8

Contents

Editorial IX
Foreword XI
Myron Gordon Award XIII

Melanocyte Morphology

CHEN, SHAN-TE; TCHEN, T. T., and TAYLOR, J. (Detroit, Mich.): The Role of c-AMP in Hormone (MSH or ACTH)-Induced Melanocyte Development in Organ Cultures of Caudal Fins of the Xanthic Goldfish 1

NIXON, P. F. (Canberra): The 'Red-skinned' New Guinean: Distinctive Melanocytes . 6

SCHAIBLE, R. H. (Indianapolis, Ind.): Identification of Variegated and Piebald-Spotted Effects in Dominant Autosomal Mutants 14

SATO, S.; KUKITA, A., and JIMBOW, K. (Sapporo): Electron Microscopic Studies of Dendritic Cells in Human Gray and White Hair Matrix during Anagen .. 20

WIKSWO, M. A. and SZABO, G. (Boston, Mass.): Studies on the Interaction between Melanocytes and Keratinocytes with Special Reference to the Role of Microfilaments 27

SEIJI, M.; FUKUZAWA, H.; SHIMAO, K., and ITAKURA, H. (Sendai): On the Melanization Process of Melanosomes 39

BRUMBAUGH, J. A.; BOWERS, R. R., and CHATTERJEE, G. E. (Lincoln, Nebr.): Genotype-Substrate Interactions Altering Golgi Development during Melanogenesis 47

ROST, F. W. D.; POLAK, J. M., and PEARSE, A. G. E. (London): The Cytochemistry of Normal and Malignant Melanocytes, and their Relationship to Cells of the Endocrine Polypeptide (APUD) Series 55

TODA, K.; PATHAK, M. A.; FITZPATRICK, T. B.; QUEVEDO, W. C., Jr.; MORIKAWA, F., and NAKAYAMA, Y. (Tokyo): Skin Color: Its Ultrastructure and its Determining Mechanism 66

Chemistry of Melanogenesis

CHEN, YU MIN and HUO, ANNE (Detroit, Mich.): Biochemical Characterization of Tyrosinase in Vertebrates 82

OIKAWA, A.; NAKAYASU, M., and NOHARA, M. (Tokyo): Metabolism of Tyrosine and its Control in Cultured Melanoma Cells 90

OKUN, M. R.; PATEL, R. P.; DONNELLAN, B., and EDELSTEIN, L. M. (Boston, Mass.): Subcellular Localization of Peroxidase-Mediated Oxidation of Tyrosine to Melanin 98

ROMSDAHL, M. M. and O'NEILL, P. A. (Houston, Tex.): Tyrosinase Inhibition Studies in Human Malignant Melanoma Grown *in vitro* 111

COOPER, M. L. (Allen Park, Mich.): Melanogenic Inhibition by Protein Derivatives .. 118

VAN WOERT, M. H. (New Haven, Conn.): Some Properties of the Outer Melanosomal Membrane 125

HIRAGA, M.; NAKAJIMA, K., and ANAN, F. K. (Tokyo): Microsomal Cytochrome b_5 and Electron-Transfer System of Mouse Melanoma 134

Chemistry of Melanin

NICHOLLS, E. M. and RIENITS, K. G. (Kensington, N.S.W.): Marsupial Pigments 142

SWAN, G. A. (Newcastle upon Tyne): Current Knowledge of Melanin Structure .. 151

THATHACHARI, Y. T. (Stanford, Calif.): Structure of Melanins 158

DUCHON, J.; BOROVANSKY, J., and HACH, P. (Prague): Chemical Composition of Ten Kinds of Various Melanosomes 165

RORSMAN, H.; ROSENGREN, A.-M., and ROSENGREN, E. (Lund): Fluorimetry of a Dopa Peptide and Dopa Thioethers 171

Control of Pigmentation

FINNIN, B. C. and REED, B. L. (Parkville, Vic.): A Quantal Bioassay for Melatonin 180

NOVALES, R. and NOVALES, B. J. (Evanston, Ill.): The Effect of Various Drugs on the Response of Isolated Frog Skin Melanophores to Melanocyte-Stimulating Hormone (MSH) and Adenosine 3',5'-Monophosphate (Cyclic AMP) 188

FUJII, R.; NAKAZAWA, T., and FUJII, Y. (Sapporo): Effects of Ultraviolet Radiation on Melanophore System of Fish 195

Experimental Pharmacology

BLEEHEN, S. S. (Sheffield): The Effect of 4-Isopropylcatechol on the Harding-Passey Melanoma 202

BART, R. S. and KOPF, A. W. (New York, N.Y.): Studies on the Mechanism of the Anti-Melanoma Effect of Polyinosinic-Polycytidylic Acid (PIC). Does Exogenous Interferon Mimic the Anti-Melanoma Effect of PIC? Does Thymectomy Plus Irradiation Abrogate the Effect of PIC? 208

Mishima, Y. (Wakayama-City): Neutron Capture Treatment of Malignant Melanoma Using ^{10}B-Chlorpromazine Compound 215

Epidemiology of Melanoma

Macdonald, E. J.; McGuffee, V., and White, E. (Houston, Tex.): Status of Epidemiology of Melanona 1971 222
Lane-Brown, M. M. and Melia, D. F. (Boston, Mass.): 'Celticity' and Cutaneous Malignant Melanoma in Massachusetts 229
Rose, E. F. (East London, South Africa): Pigment Variation in Relation to Protection and Susceptibility to Cancer 236
Mori, W. (Tokyo): Geographic Pathology of Malignant Melanoma in Japan 246
Fletcher, W. S. (Portland, Ore.): The Incidence of other Primary Tumors in Patients with Malignant Melanoma 255

Melanoma Biology

Clark, R. L. (Houston, Tex.): The Research and Clinical Approach to Melanoma at The University of Texas M.D. Anderson Hospital and Tumor Institute ... 261
Cochran, A. J. and Cochran, K. M. (Glasgow): Naevi of Childhood 276
Thompson, P. G. (Bristol): Relationship of Lymphocytic Infiltration to Prognosis in Primary Malignant Melanoma of Skin 285
Mishima, Y. and Matsunaka, M. (Wakayama-shi): Macromolecular Pathology of Pagetoid Melanoma 292
Gilbert, E. F.; McCord, R. G.; Skibba, J. L.; Fallon. J. F.; Croft, W. A., and Jaeschke, W. F. (Madison, Wis.): Local Recurrence and Diffuse Multiple Melanoma Following Chronic Oral Administration of L-Dopa 300

Biochemistry of Melanoma

Goodall, McC. (Galveston, Tex.): Metabolism of L-Dopa-3-^{14}C (3,4-Dihydroxyphenylalanine) in Human Subjects 308
Hinterberger, H.; Freedman, A., and Bartholomew, R. J. (Little Bay, N.S.W.): Precursors of Melanin in the Urine and 3,4-Dihydroxyphenylalanine in the Blood of Patients with Malignant Melanoma 312
Duchon, J. and Matous, B. (Prague): Dopa and Its Metabolites in Melanoma Urine 317
Taskovich, L.; Banda, P. W., and Blois, M. S. (Palo Alto, Calif.): Chemical Studies on Urinary Melanogens 323
Riley, V.; Spackman, D., and Fitzmaurice, M. A. (Seattle, Wash.): Plasma and Urine Amino Acid Changes Associated with Melanoma 331
Bourgoin, J. J.; Manuel, Y.; Sonneck, J. M.; Gronneberg, K., and Defontaine, M. C. (Lyon): Serum Haemopexin in Human Malignant Melanomas 346

Immunology of Melanoma

CLARK, D. A. and NATHANSON, L. (Boston, Mass.): Cellular Immunity in Malignant Melanoma ... 350

COCHRAN, A. J.; JEHN, U. W., and GOTHOSKAR, B. P. (Glasgow): Cell-Mediated Immunity to Malignant Melanoma ... 360

BOURGOIN, J. J. and BOURGOIN, A. (Lyon): Cytoplasmic Antigens in Human Malignant Melanoma Cells ... 366

COX, I. S. and ROMSDAHL, M. M. (Houston, Tex.): Immunoglobulins Associated with Human Malignant Melanoma Tumors ... 372

WHITEHEAD, R. H. and LITTLE, J. H. (Brisbane): Tissue Culture Studies on Human Malignant Melanoma ... 382

FOSTER, M.; HERMAN, J., and THOMSON, L. (Ann Arbor, Mich.): Genetic and Immunologic Approaches to Transplantable Mouse Melanomas ... 390

JADIN, J. M. and VAN DER SCHUEREN, G. (Leuven): Direct Heterologous Transplantation of a Malignant Human Melanoma and its Ultrastructure. Preliminary Report ... 399

IKONOPISOV, R. L. (Sofia): Morphologic Patterns of Spontaneously Regressing Malignant Melanoma in Relation to Host Immune Reactions ... 402

Subject Index ... 411

Editorial

Since the first Pigment Cell Conference, held in 1948, the number of participating scientists and physicians has increased and the scope of their scientific and medical activities has been substantially extended. A rather remarkable social phenomenon exhibited by this group has been the persisting, long-term maintenance of scientific activity and communication in the absence of any formal organization, officers, treasury, or Journal, all of which seems to testify to the natural organizational inclinations and needs of this heterogenous assembly.

A glance at the list of the Proceedings of the previous seven Conferences, on an adjoining page, provides not only a useful guide to the historical literature dealing with the pigment cell and melanoma field, but it also indicates the lack of continuity in publisher and format, a matter perhaps of serious concern only to the librarians and Editor charged with responsibility for publishing the Conference Proceedings without subvention.

This Volume 1, of Pigment Cell, constitutes the Proceedings of the 8th International Pigment Cell Conference, *Mechanisms in Pigmentation,* and represents the first stage of a new, long-term plan intended to meet the expanding needs of the International Pigment Cell Group, and to provide a future continuity and acceptable scientific form for accomodating the increasing research productivity in this field, which embraces fundamental biochemical and biological inquiries on the cell as well as more mission-oriented research in cancer and other pathological aspects of pigment-producing tissues.

As a consequence of discussions with Mr. THOMAS KARGER, the Publishing House of S. Karger AG, Basel, Switzerland, has agreed to publish the future Proceedings of these conferences, as well as appropriate interim volumes on related subject matter. These monographs shall constitute a numbered series, starting with these Proceedings as Volume 1. If an interim monograph is published prior to the next International Conference, it shall designated Pigment Cell, Volume 2, and will carry an appropriate sub-title indicating its essential subject matter.

The Series Editor invites proposals for such interim monographs from investigators or physicians working in the field who would be interested in serving as authors, co-authors, or editors of such volumes.

As a further logical step toward integrating the diverse and increasingly voluminous literature expected from current and future investigations on the pigment cell, it has been suggested that the International Pigment Cell Group be formalized into a Society, and that a Journal be initiated in order to provide a vehicle for publishing, in one place, papers of wide diversity in terms of scientific discipline but having a common denominator in terms of pigmentation. A factor which makes this proposal realistic, and offers an opportunity for early initiation of such a Journal, is the willingness of the S. Karger Publishers to join in a meaningful way in the undertaking.

In order to obtain the widest possible participation in future conferences of the International Pigment Cell Group, and to make more fully available the newly developing publication opportunities for meritorious papers concerned with pigmentation or the pigment cell, communications are invited from new workers in the field, and from others who have not previously participated in the International Conferences.

VERNON RILEY
Series Editor

Foreword

When the International Union Against Cancer decided to hold its 1972 Interim Meeting in Sydney with the themes of Leukaemia and Skin Cancer-Melanoma it was thought appropriate that the VIIIth International Pigment Cell Conference be held concurrently. Accordingly an invitation was issued on behalf of the Australian Cancer Society at the VIIth International Pigment Cell Conference held in Seattle in September 1969.

The Australian Cancer Society appointed Sir WILLIAM KILPATRICK as the Chairman of the Organizing Committee of the Cancer Conference and Dr. VINCENT J. MCGOVERN as the Chairman of the Pigment Cell Subcommittee. The expenses of mounting the Pigment Cell Conference were borne by the Australian Cancer Society with funds raised personally by Sir WILLIAM.

A great deal of the success of the Pigment Cell Conference was due to the Secretary General of the Organizing Committee of the Cancer Conference Professor KENNETH COX, and to his executive secretary, Mr. PETER SCHRADER and to Mrs. KAREN FORBES.

Myron Gordon Award

At the conjoined banquet of the Cancer Conference and the VIIIth International Pigment Cell Conference, in the presence of Sir WILLIAM KILPATRICK, President of the Cancer Conference, Sir GARFIELD BARWICK, Chief Justice of the Commonwealth, and numerous distinguished guests, Dr. VERNON RILEY, Chairman of the International Pigment Cell Steering Committee, announced the recipients of the Myron Gordon Award. Amongst those present were two of the three previous recipients of the Award,

ELEANOR J. MACDONALD   VINCENT J. MCGOVERN

Dr. G. A. SWAN of the University of Newcastle-upon-Tyne and Dr. T. B. FITZPATRICK of Harvard.

As Dr. RILEY said, the Award was instituted to perpetuate the memory of a beloved and respected scientist, MYRON GORDON, who together with GEORGE MILTON SMITH was responsible for inaugurating the Pigment Cell Conferences.

On this occasion a double award was made, the recipients being ELEANOR J. MACDONALD, Professor of Epidemiology, M. D. Anderson Hospital and Tumor Institute, University of Texas, Houston, and VINCENT J. McGOVERN, Director of the Fairfax Institute of Pathology, Royal Prince Alfred Hospital, Sydney.

Miss MACDONALD received her award for her meticulous epidemiological and ethnic survey over a 23-year period of the regional patterns of melanoma incidence and mortality in Texas, and their relationship with race, sex, and solar exposure, thereby establishing a parallelism with the Australian experience in this disease.

Dr. McGOVERN received his award for his studies of melanoma in which he drew attention 20 years ago to the aetiological factors of solar exposure and Celtic inheritance. Arising out of his work, partly in conjunction with Dr. WALLACE CLARK of Philadelphia and fellow workers in Australia, there has evolved a valuable new classification of melanoma.

Melanocyte Morphology

The Role of c-AMP in Hormone (MSH or ACTH)-Induced Melanocyte Development in Organ Cultures of Caudal Fins of the Xanthic Goldfish[1]

SHAN-TE CHEN, T. T. TCHEN and J. TAYLOR

Departments of Chemistry and Biology, Wayne State University, Detroit, Mich.

We have previously reported that melanocyte stimulating hormone (MSH) and other structurally related polypeptides (including ACTH) induce the differentiation of melanoblasts (or pro-melanocytes) into melanocytes in organ cultures of caudal fin from xanthic goldfish. We have developed the theory that a) this involves the differential mitosis of stem cells; b) the hormone acts during or shortly preceding mitosis; c) the newly-formed melanocyte is small but capable of melanin synthesis in the absence of protein synthesis, presumably by inactivating an existing tyrosinase inhibitor and thus activating existing tyrosinase (both the enzyme and an inhibitor have been shown to be present in completely xanthic skin); d) the newly-formed melanocyte normally (without addition of any inhibitor of RNA or protein synthesis) undergoes cytoplasmic growth and morphological development to become a large dendritic pigmented cell, and e) this hormonal effect can also be produced by dibutyryl-c-AMP, a derivative of the second messenger c-AMP.

On the other hand, other workers have reported that a) the hormone produces melanosome-like structures in xanthophores, and b) purified pterinosomes, the characteristic pigment organelles of xanthophores, contain tyrosinase and protein inhibitor of tyrosinase. These results suggest the possibilities that a) xanthophores may be induced to synthesize melanin, or b) the newly-formed melanocytes may be capable of synthesizing both kinds of pigments. (See CHEN and TCHEN [4] for references.)

---

[1] Supported by grants from the USPHS AM 12713 and AM 05384, and National Science Foundation grant GB-16392.

We wish to present some newer findings, particularly with regard to the size of the newly-formed melanocytes and the dual role of dibutyryl-c-AMP, and by extrapolation also of MSH and c-AMP, in the development of melanocytes.

With organ cultures, dibutyryl-c-AMP at optimal concentrations (approximately 3×10^{-4}M) can replace the hormone (ACTH was used instead of MSH in these studies) to induce the formation of melanocytes. However, not all fish whose skin explants responded to the hormone responded to dibutyryl-c-AMP. Conversely, not all the fish whose skin explants responded to dibutyryl-c-AMP responded to the hormone either. In the explants from fish that responded to both the hormone and dibutyryl-c-AMP, the response to the two inducers was comparable. c-AMP is essentially ineffective by itself at all concentrations tested. However, 3×10^{-4}M c-AMP in the presence of 1mM theophyllin can induce melanocyte formation although the response obtained was considerably poorer than that obtained with hormone both in the number of melanocytes formed per explant and in the percentage of fish whose skin explants gave a positive response.

The morphology of the melanocytes formed depends upon the duration of incubation with dibutyryl-c-AMP. Using pre-synchronized preparations, incubation with dibutyryl-c-AMP for two days produced mainly (>90%) large dendritic melanocytes whereas incubation with dibutyryl-c-AMP for shorter periods of time (20 h or less), followed by wash and re-incubation with basal organ culture medium, produced more (50–80%) small melanocytes without dendrites. If the re-incubation after wash was in the presence of another dose of dibutyryl-c-AMP, mainly large dendritic melanocytes were formed. These results indicate that dibutyryl-c-AMP (and by extrapolation also c-AMP and MSH) is required not only for the formation of melanocytes (small, but capable of melanin synthesis) from melanoblasts, but also for the subsequent cytoplasmic growth and morphological development of the small melanocytes into the large dendritic melanocytes.

Due to the structural complexity of the explants (small pieces of caudal fin), it is difficult to see the outlines of individual cells. Consequently, the 'small' melanocytes could be rather large dendritic cells with concentration of melanin granules in a small area (like perinuclear). However, examination of 1μ thick epon sections showed clearly that these small melanocytes are indeed small cells (fig. 1).

Addition of cytochalasin B (1 μg/ml) produced melanocytes of abnormal morphology (fig.2 and 3). Many of the melanocytes have the shape of a dumb-bell with a single long dendrite at each end. These cells are binucleated.

Since cytochalasin B is known to inhibit cytoplasmic cleavage without inhibiting nuclear division in several other systems [1–3], the formation of these

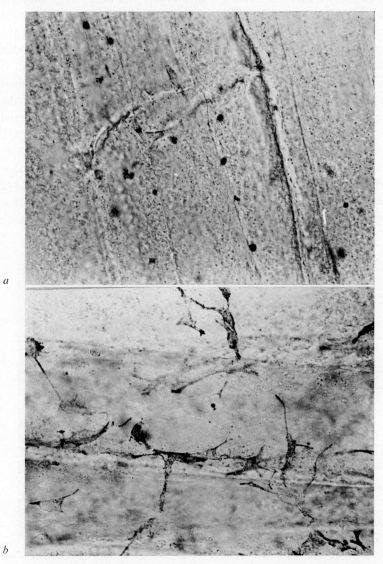

Fig. 1. Comparison of small and large melanocytes. *a* Explant was treated with dibutyryl-c-AMP for a relatively short period (see text). The melanocytes are small and without dendrites; *b* Explant was treated with dibutyryl-c-AMP for 48 h. The melanocytes are large and dendritic x 300.

binucleated cells is in agreement with our theory that hormone-induced melanocyte formation involves an obligatory mitotic event. Our theory further postulates that this mitosis is asymmetric and normally produces a melanocyte and a melanoblast. We are currently investigating by electron microscopy the possibility that the two nuclei in these cells formed in the presence of cytochalasin B may not be identical, with perhaps only one nucleus which has an active Golgi complex and produces melanosomes.

In summary, these recent results (a) offer more direct experimental support for our theory that hormone-induced melanocytogenesis in the xanthic goldfish involves the mitosis, probably an asymmetric or differential mitosis, of a small pro-melanocyte (or melanoblast, or stem cell) to form a small melanocyte; (b) the hormone acts via the second messenger c-AMP, and (c) the hormone or its second messenger is also required for the further cytodifferentiation (cytoplasmic growth and morphological development) of the small melanocyte. These results do not support the hypothesis that existing xanthophores can be induced to synthesize melanin. The possibility

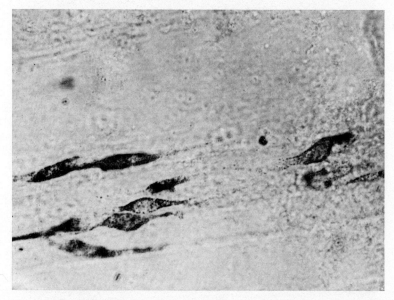

Fig. 2. Melanocytes formed in the presence of cytochalasin B (1 μg/ml): most of the melanocytes formed in the presence of cytochalasin B, with ACTH as inducer, are of abnormal morphology. Many of these cells are shaped like dumb-bells with a single dendrite at each end. These cells are binucleated as shown in figure 3. This photograph was taken by a modified phase-contrast microscopy. x 450.

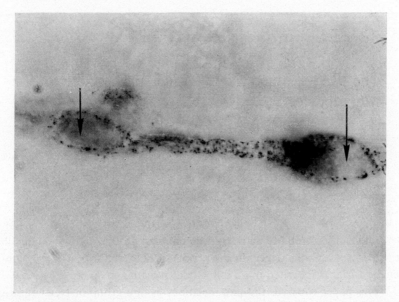

Fig. 3. Higher magnification of a dumb-bell-shaped melanocyte: same as figure 2 with higher magnification (x 950). The arrows point to the nuclei in this cell.

that the newly-formed melanocytes may be pluripotential and can synthesize also the pteridine pigment characteristic of xanthophores cannot be eliminated. Since pterinosomes cannot be resolved in thick sections examined by light microscopy, we must await the results of current electron microscopic studies before we can arrive at a definitive conclusion.

References

1 CARTER, S. B.: 'Effects of cytochalasins on mammalian cells'. Nature, Lond. *213:* 261–264 (1967).
2 RIDLER, M.A.C. and SMITH, G. F.: 'The response of human cultured lymphocytes to cytochalasin B. J. Cell Sci. *3:* 595–602 (1968).
3 SCHROEDER, T. E.: 'The role of 'contractile ring' filaments in dividing Arbacia egg'. Biol. Bull. *137:* 413–414 (1969).
4 CHEN, S-t. and TCHEN, T. T.: Induction of melanocytogenesis in explants of *Currassius Auratus L.* the xanthic goldfish by dibutyryl-c AMP Biochem. Biophys. Res. Comm. *41:* 964–968 (1970).

Author's address: Dr. SHAN-TE CHEN, Department of Chemistry, Wayne State University, *Detroit, MI 48202* (USA)

The 'Red-skinned' New Guinean: Distinctive Melanocytes

P. F. NIXON

Department of Biochemistry, John Curtin School of Medical Research, Australian National University, Canberra, and Institute of Human Biology, Goroka

The lack of mobility between different clan groups in New Guinea has resulted in a high incidence, in the people of that country, of expression of genetic variants inherited in an autosomal recessive manner.

This paper concerns the skin melanocytes of a group of New Guineans whose skin is a red colour, similar to the red colour of northern European skin about 24 hours after severe sunburning. Their hair is fair but not red. Each one of 12 subjects who were adequately examined had a tigroid distribution of retinal pigment. Many, but not all, suffered an apparent defect of gaze fixation, manifest as a constant oscillating nystagmus. WALSH [2] and HARVEY [1] have shown that the inheritance of this phenotype is consistent with an autosomal recessive mechanism and that among the populations which they studied its incidence was between 1 % and 2 %. In pidgin English the New Guineans use the word 'red skin' to refer to both this phenotype and another, characterized by an unremarkable pale brown 'milk coffee' skin colour but otherwise similar to the red skin phenotype described above.

As the initial step of an investigation of the site and nature of the mutation which results in the red-skin phenotype, light microscopic examination has been carried out on preparations of biopsied skin from the following numbers of New Guineans of characteristic skin colour: red skin: 19; pale brown skin with nystagmus: 4; pale-to mid-brown skin normal controls: 9; dark brown skin normal controls: 19. The 9 normal subjects whose skin colour was pale- to mid-brown were of about the same intensity of pigmentation as the red-skin subjects.

Skin was obtained under local anaesthesia from the upper arm in the anterior axillary line by means of a 2 mm punch or by freehand cutting of a small, very thin 'split skin graft'. Portion of each of the latter biopsies was trypsinized for preparation of epidermal sheets free of dermis. All biopsied skin was fixed in glutaraldehyde or combined glutaraldehyde/formaldehyde fixative. Following fixation, small pieces from each skin biopsy were incubated either in buffer or in one of the melanin precursor substrates: 5mM L-dopa or 2.5 mM L-tyrosine plus 0.1 mM L-dopa.

Sections were examined with a view to determining if there were any morphological features which distinguished the melanocytes of red skin, and if these melanocytes differed from those of control subjects in histochemical reactions for melanin synthesis. The overall objective at this stage was to provide a guide to the direction of future biochemical research into the site of action of the red-skin mutation.

The distribution of melanin granules throughout the epidermis differed with skin colour. Dark control skin had many granules throughout the epidermis, but the highest density of granules was in the apices of the basal keratinocytes (fig. 1). Pale brown control skin had few granules, even in the basal keratinocytes (fig. 2). The overall amount of melanin granules in red skin (fig. 3–6) appeared to be intermediate between the extremes illustrated by the dark and pale controls, but there were generally disproportionately fewer granules in the basal keratinocytes (fig. 3, 5, 6).

In sections of skin which had not been incubated with any melanin substrate the melanocytes were quite difficult to see and contained few granules whatever the skin colour. However, the melanocytes were darkened and easily seen in all skin which had been incubated in L-dopa (fig. 1, 2, 3, 5, 6). The melanocytes of red skin (fig. 3, 5, 6) darkened at least as much as, or even more than, controls. Incubation of skin in L-tyrosine containing catalytic amounts of L-dopa also resulted in darkening of all melanocytes, including those of red skin (fig. 4). Irrespective of the skin colour the darkening of melanocytes by tyrosine was never as intense as that by dopa. Unstained sections showed that the darkening of melanocytes by dopa or tyrosine was not dependent on counterstaining, and preliminary electron microscopy of one red and two control skins has confirmed that the melanocyte darkening was associated in these cases, with specific darkening of melanosomes.

In both dark and pale control skins the melanocytes appeared relatively rounded and discrete (fig. 1, 2). In red skin the melanocytes appeared more variable in morphology, tending to be spread out over the basal surface of the epidermis, to be more numerous and larger, and to have more prominent dendrites (fig. 3–6). The dendrites of melanocytes of red skin

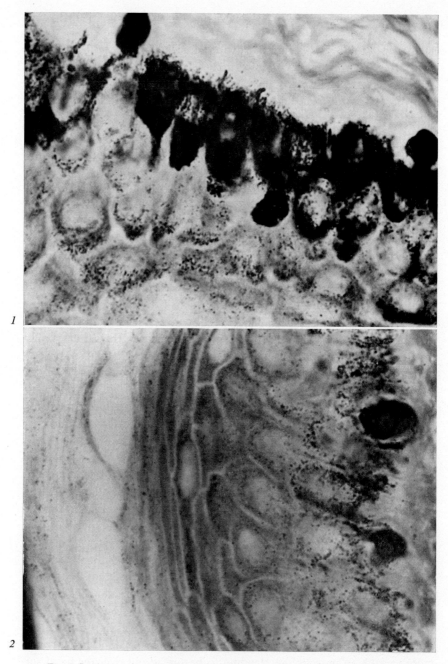

Fig. 1. Section, stained by Masson's ammoniacal silver stain for melanin, of control dark skin incubated in L-dopa. Figures 1–5 are at the same magnification.

Fig. 2. Section of control mid-brown skin treated as in figure 1.

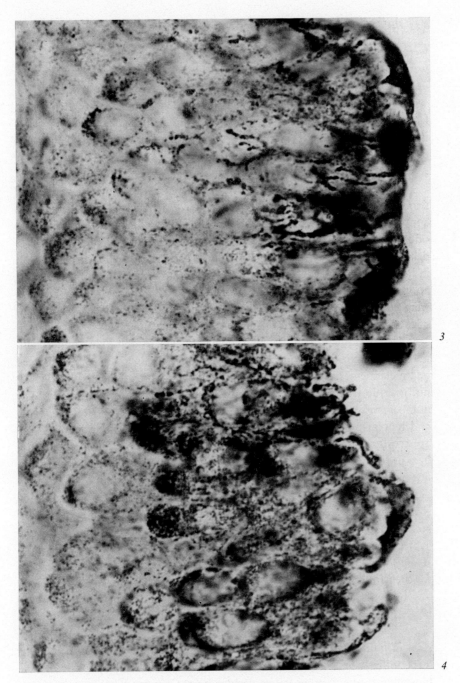

Fig. 3. Section of red skin treated as in figure 1.
Fig. 4. Section stained for melanin, of red skin incubated in L-tyrosine.

were apparently thicker and longer than those of control skin; also they were more spread over the epidermal basal surface and the contacts between dendrites of adjacent melanocytes appeared more prominent (fig. 5, 6, 8).

The observation that the dendrites of red skin melanocytes appear larger than those of controls, together with the paucity of melanin granules in the basal keratinocytes of red skin, suggests the possibility that the process of transfer of pigment granules from melanocytes to keratinocytes may be relatively ineffective in red-skin subjects.

Preparations of epidermal sheets incubated with L-dopa and observed from the basal surface showed the melanocytes quite well in both red and control skin (fig. 7, 8). Such preparations showed that the melanocytes of control skin (fig. 7) were usually bipolar in shape and were oriented into distinct whorls about the dermal papillae, or into pallisade formations. In red skin (fig. 8) the melanocytes were much more frequently polydendritic and lacking in obvious polarity or orientation. The melanocytes of red skin were also more numerous and their dendrites more prominent. The extent of inter-melanocyte dendritic contact can be seen in figure 8.

The prominent features of the melanocyte of red skin, under light microscopy, can be summarized as follows. Melanin granules are formed and melanocytes are well darkened by melanin substrate incubation. Melanin granule transfer to basal keratinocytes is apparently diminished. The number and size of melanocytes is apparently increased. The orientation and bipolarity of melanocytes is apparently diminished. The dendrites are apparently more numerous, thicker, longer and darker and the inter-melanocyte dendritic contact is more readily apparent. Although each feature was not necessarily obvious in every section of red skin, or confined to red skin alone, when taken together these features consistently enabled one to assign sections correctly to the skin colour of the subject, with the exception of sections from 2 red-skin children aged 5 and 8. Sections from these two subjects were occasionally incorrectly assigned to the pale control group. It may well be that some of the features of red-skin melanocytes depend, for their full expression, upon age or period of exposure to solar irradiation.

Since the red-skin variation is inherited in an autosomal recessive manner, it may be assumed that the mechanism of its expression involves the alteration of a single protein concerned with melanin biosynthesis. The results of light microscopy suggest that the mutation to red skin in humans does not involve the tyrosine reaction directly and is unlikely to involve the biosynthesis of any other low molecular weight intermediates in the melanin pathway. The site of action of the mutation which results in red skin appears

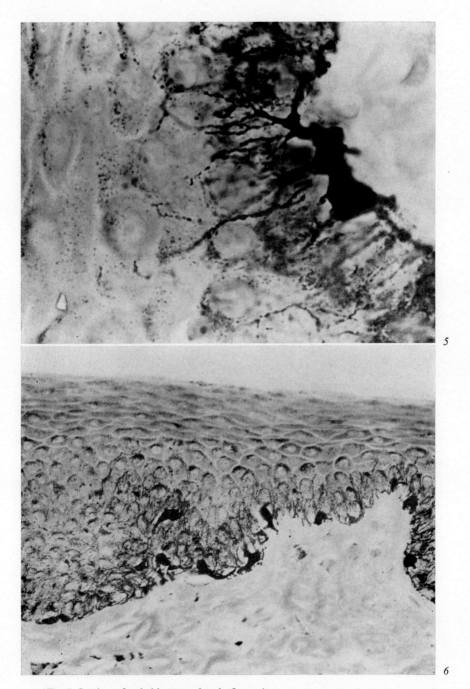

Fig. 5. Section of red skin treated as in figure 1.
Fig. 6. Section of red skin treated as in figure 1. Low magnification.

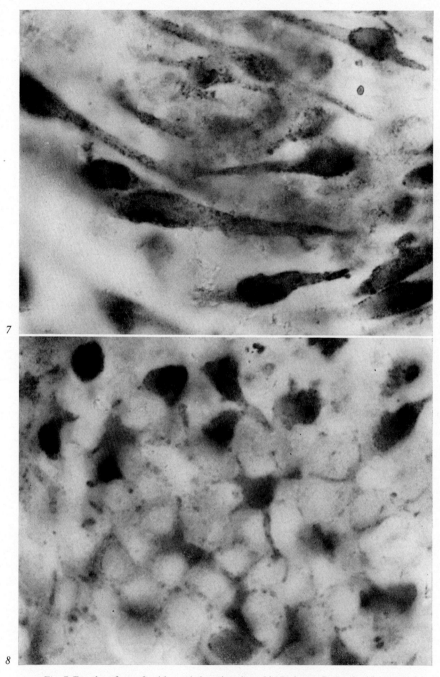

Fig. 7. Basal surface of epidermal sheet incubated in L-dopa. Control mid-brown skin.
Fig. 8. Red skin prepared as in figure 7, and at the same magnification as figure 7.

more likely, therefore, to involve either one or more of the following: a) the biosynthesis of the premelanosome, and, hence, the physical character of the melanin granule; or b) the process of polymerization of the immediate melanin subunit; or c) the process of transfer of melanin granules to keratinocytes.

References

1 HARVEY, R. G.: Hum. Biol. Oceania *1:* 123 (1971).
2 WALSH, R. J.: Ann. hum. Genet. *34:* 379 (1971).

Author's address: Dr. P. F. NIXON, Department of Biochemistry, John Curtin School of Medical Research, Australian National University, *Canberra* (Australia)

Identification of Variegated and Piebald-Spotted Effects in Dominant Autosomal Mutants

R. H. Schaible

Biology Department, Indiana University – Purdue University, Indianapolis, Ind.

Most investigators have found no pigment cells in the white regions of piebald-spotted mammals [16, 3] and birds [12]. Therefore, Schaible [14] used the congenital absence of pigment cells in the white regions as the basis for the definition of piebald spotting, or piebaldism as it was termed by Fitzpatrick and Quevedo [4].

Some investigators have reported the occurrence of amelanotic pigment cells in the white regions of piebald-spotted human subjects [6]. Breathnach et al. [2] reported one case which lacked melanocytes in the white forelock but had abnormal pigment cells in a white region of the forearm. Such results appear to be contradictory but may be expected if the various types of piebald-spotted mutants in other mammals are compared with those in man.

Piebald spotting is transmitted as an autosomal dominant in nearly all of the 25 human families that have been reported [6]. Searle [15] listed 7 loci that could carry at least one allele for dominant spotting in the mouse. Heterozygotes for dominant mutant alleles at three of those loci (Mi^{wh}, W^a and V^a) show variegation of the pigmented regions in addition to the usual white breaks which occur between primary pigment areas [13]. Variegation may be defined as the occurrence of hyperpigmented and/or hypopigmented spots against the shade of background color expected for the mutant.

If there is a light shade of pigment in the hypopigmented regions of variegated mutants, pigment cells are certainly expected to be present and possibly abnormal in morphology. A more severe abnormality in the pigment cells could result in a hypopigmented spot that is grossly white. Such spots will be referred to as 'white hypopigmented spots' in contrast to the characteristic white breaks between primary pigment areas in piebald-spotted mutants.

McClintock's [8] analyses of variegation in maize serve as a precedent for expecting complete suppression of pigment production in some spots in variegated mutants. She has provided evidence for mutation to several 'states' (shades of pigmentation) including complete suppression of pigment production in variegated mutants.

The best evidence for mutation to a different state in variegated mutants is the transmission of the new state to the next generation, i. e., germinal mutation. Most of the somatic mutation in $Mi^{wh}/+$, $W^a/+Va/+$ variegated mice is presumably to wild-type since hyperpigmented spots are more frequent than hypopigmented spots [11]. However, there was no evidence of germinal mutation for these dominant, variegated, piebald-spotted, mutants. In crosses of Mi^{wh}/Mi^{wh} and Va/Va mutants to non-mutant mates, there were no non-mutant types among 5165 and 25 progeny respectively. Very few progenies were obtained from Va homozygotes which are generally poor breeders. Homozygotes of W^a cannot be tested for they die of anemia shortly after birth.

Although no germinal mutations have been obtained for dominant, variegated, piebald-spotted mutants of the mouse, repeated occurrences of germinal mutations have been reported for recessive, variegated, non-piebald mutants [10]. Nicholls [9] presented evidence that freckling of homozygotes and heterozygotes for red hair in man is a form of variegation due to somatic mutation.

Variegation in Piebald-Spotted Dogs

The merle mutant in dogs seems to be the best model for studies of variegation in dominant piebald-spotted mutants in mammals. Ford [5] has obtained a low percentage of non-merle pups from a homozygous merle Collie dog. Table I summarizes data on the progenies of a Collie male and a Great Dane male which are presumed merle homozygotes. Since their progenies from non-merle mates resulted in a significant deviation (P < .001 in both cases) from the ratio of 1 merle:1 non-merle, it can be concluded that the non-merle pups were the result of germinal mutation rather than heterozygosity of the sires. Blindness due to incomplete development of the eyes, a characteristic of merle homozygotes in the Collie breed, was additional evidence that the Collie sire in table I was indeed homozygous. The Great Dane appears to have completely normal eyes. The merle progeny from the Great Dane included 20 pups which breeders of Great Danes call

Table I. Progeny of homozygous merle dogs bred to non-merle bitches

Sires	Dams	Pups	
		Merle	Non-merle
Collie	Tri-color Collie[1]	8	0
	Liver-spotted Dalmatian	12	1
Great Dane	5 black Great Danes[2]	58	1

1 Data supplied by Mrs. GERHARD E. KOEHLINGER, Rt. 3, Box 179, Bremen, Indiana 46506.
2 Data supplied by PATTI MULLINS, 6471 North Parker Avenue, Indianapolis, Indiana 46220.

'harlequin'. LITTLE [7] stated that all harlequins carry the mutant gene for merle.

There is evidence in merle dogs for the existence of two kinds of white regions. Primary breaks between some of the pigment areas are evident as white regions on the puppy in figure 1. Presumably, no pigment cells exist within the primary breaks. The primary pigment areas are basically gray, the usual effect of the mutant gene for merle. Within the gray areas are black patches which are presumed to be the result of back mutation to non-merle in some of the stem melanoblasts. Note that the margins of the primary pigment areas are rounded convexly as though populations of pigment cells were migrating into unpopulated territory. In contrast, the margins of black patches located within the gray areas are jagged as though the melanocytes which produce black pigment are competing for territory with those melanocytes which produce gray pigment. In the harlequin Great Dane, the gray color is apparently suppressed to white by a second genetic factor. Harlequin Danes have jagged black patches against a white background. Presumably part of that white background contains melanocytes in which the production of pigment has been suppressed.

Classification of Piebald-Spotted Mutants

It is clear from studies on mice and dogs that the dominant piebald-spotting mutants in mammals can be separated into two groups, those

Fig. 1. Merle puppy produced by the cross of the homozygous merle Collie dog and Dalmatian bitch given in table I.

which are variegated and those which are not. Electron microscopy should reveal no melanocytes within the white regions of the latter group. In the former group some white regions should be void of pigment cells and others should contain amelanotic pigment cells.

The occurrence of hyperpigmented spots may possibly be of aid in classifying particular cases. The patient reported by COMINGS and ODLAND [3] had no melanocytes in the white regions and no hyperpigmented spots within the primary pigment areas. In contrast, KUGELMAN and LERNER [6] observed amelanotic pigment cells in the white regions and hyperpigmented spots within the primary pigment areas in two piebald-spotted individuals from one family. If these correlations are always complete, the occurrence of both primary breaks and white hypopigmented spots would be expected in the Waardenburg syndrome since hyperpigmented spots do occur in affected individuals [1]. SEARLE [15] proposed the merle mutant in dogs as the animal model for the Waardenburg syndrome on the basis of other symptoms.

In practice, knowledge of the location of primary pigment areas is essential for selection of the proper locations on the body for skin biopsies. Otherwise, samples may not be obtained from both primary breaks and white hypopigmented areas if they both exist. SCHAIBLE [14] reviewed most of what is known regarding the location of pigment areas in mammals and birds. Less is known about man than about mice, cattle, dogs and chickens.

A Request

During the next few years I plan to map the primary pigment areas of man. It will not be possible to use the experimental approaches that were available in the study on mice [13]. However, much useful information could be obtained by comparing locations of pigmented areas on a large number of human piebalds and pigment mosaics. I would be grateful if clinicians having piebald or mosaic patients would send me photographs or carefully prepared drawings of their subjects.

Conclusions

Comparisons of the microscopic observations reported for piebaldism in man with the effects of the variety of piebald-spotted mutants known in mice and dogs indicate that piebald-spotted mutants should be divided into two categories, variegated piebaldism and non-variegated piebaldism. The non-variegated type is void of pigment cells in the white regions. The variegated type would be expected to have amelanotic pigment cells in white hypopigmented areas but complete absence of pigment cells in the white regions which separate primary pigment areas.

References

1 ARIAS, S.: Genetic heterogeneity in the Waardenburg syndrome. Birth Defects: Original Article Series 7 (No. 4): 87–101 (1971).
2 BREATHNACH, A. S.; FITZPATRICK, T. B. and WYLLIE, L. M.: Electron microscopy of melanocytes in a case of human piebaldism. J. Invest. Derm. *45:* 28–37 (1965).
3 COMINGS, D. E. and ODLAND, G. F.: Partial albinism. J. Amer. med. Ass. *195:* 519–523 (1966).
4 FITZPATRICK, T. B. and QUEVEDO, W. C.: Albinism; in: STANBURG, WYNGAARDEN and FREDRICKSON, The metabolic basis of inherited disease. 3rd ed., pp. 326–337 (McGraw-Hill, New York 1972).
5 FORD, L.: Personal comm. Address: Genetics Consultant Service, Butler, Ind. 46721 (1971).

6 KUGELMAN, T. P. and LERNER, A. B.: Albinism, partial albinism and vitiligo. Yale J, Biol. Med. *33:* 407–414 (1961).
7 LITTLE, G. C.: The inheritance of coat color in dogs (Comstock Pub. Ass., Ithaca N.Y. 1957).
8 MCCLINTOCK, B.: Intranuclear systems controlling gene action and mutation. Brookhaven Symp. Biol. *8:* 58–74 (1956).
9 NICHOLLS, E. M.: Phacomatoses, the inheritance of cancer, and somatic mutation. Clin. Genet. *1:* 245–257 (1970).
9a NICHOLLS, E. M. and RIENITS, K. G.: Marsupial pigments. Pigment Cell, vol. 1, pp. 142–150 (Karger, Basel 1972).
10 RUSSELL, L. B.: Genetic and functional mosaicism in the mouse; in: LOCKE, The role of chromosomes in development, pp. 153–181 (Academic Press, New York 1964).
11 SCHAIBLE, R. H.: Developmental genetics of spotting patterns in the mouse; Ph.D. thesis, Iowa State University, Ames (1963).
12 SCHAIBLE, R. H.: Development of transitory piebald spotted and completely pigmented patterns in the chicken. Devel. Biol. *18:* 117–148 (1968).
13 SCHAIBLE, R. H.: Clonal distribution of melanocytes in piebald-spotted and variegated mice. J. exp. Zool. *172:* 181–199 (1969).
14 SCHAIBLE, R. H.: Comparative effects of piebald spotting genes on clones of melanocytes in different vertebrate species; in V. RILEY, Pigmentation: its genesis and control (Appleton-Century-Crofts, New York, pp. 343–357, 1972).
15 SEARLE, A. G.: Comparative genetics of coat colour in mammals (Logos Press/Academic Press, London/New York 1968).
16 SILVERS, W. K.: Pigment cells: occurrence in hair follicles. J. Morph. *99:* 41–56 (1956).

Author's address: Dr. ROBERT H. SCHAIBLE, Biology Department, Indiana University-Purdue University, 925 W. Michigan St., *Indianapolis, IN 46202* (USA)

Electron Microscopic Studies of Dendritic Cells in the Human Gray and White Hair Matrix During Anagen[1]

S. Sato, A. Kukita and K. Jimbow

Department of Dermatology, Sapporo Medical College, Sapporo

Introduction

The graying and whitening of the human scalp hair are seen not only in the physiological process of aging but also as a symptom of various diseases. The genesis of gray and white hairs is generally attributed to the gradual diminution of the enzymic activity responsible for melanogenesis [2, 7, 8]. But the presence or absence of dendritic cells in the hair matrix has not been conclusively determined. A descriptive term, dendritic cells is customarily used as a synonym for epidermal melanocytes and Langerhans cells. During the last few years it became clear that the third type of dendritic cells (indeterminate type) occurs in the epidermis. These dendritic cells are reported to reveal in many respects similar configuration to that of epidermal melanocytes as well as of Langerhans cells, but they lack their specific and characteristic organelles [5, 9, 10, 11, 12, 13, 14].

Little data is at present available to compare these three types of dendritic cells in the hair matrix with those of the epidermis. The purpose of this paper is to demonstrate dendritic cells in gray and white hair matrix in anagen.

Materials and Methods

The scalp hair bulbs during anagen of senile gray and white hairs, of gray hairs in incomplete albinism, and of white hairs in vitiligo vulgaris were used for an electron

---

1   This investigation was supported by grants from the Ministry of Education, Japan; Japan O'Leary Inc. Fund for Pigment Research, and the Far East Basic Research Fund of Sears, Roebuck and Co., Inc., Chicago, Ill.

microscopic study. As a control study, black hairs from normal subjects were obtained. The materials were fixed in 4% glutaraldehyde and 2% osmium tetroxide adjusted to pH 7.4 with cacodylate buffer. They were dehydrated in ethanol, and embedded in Epon 812. Thin sections cut with a Porter-Blum ultramicrotome MT-2 were stained with uranyl acetate and lead citrate, and examined in a Hitachi HS-8E electron microscope.

Results

a) Black Hair Matrix

A vast number of melanocytes were seen in the melanocytic zone of the hair bulb. Their cytoplasmic organelles were less abundant than those found in the epidermal melanocytes and fine filaments were seldom observed. In the matrix below the critical level, melanocytes were also noted few in number. Langerhans cells were identified infrequently in the hair matrix, and they were usually located in the amelanotic outer root sheath of the hair bulb and in the matrix below the critical level. The ultrastructural characteristics of the cells were almost identical to those seen in the epidermis. The well developed cytoplasmic organelles suggested that the cells were healthy. Indeterminate type of cells were not observed in the melanocytic zone.

b) Gray Hair Matrix of Albinism

Numerous melanocytes were present in the melanocytic zone. Melanin granules in their pericarya and dendrites were decreased in size and number, and revealed low degrees of melanization. It was not possible to identify the indeterminate type of cells within the melanocytic zone. Langerhans cells were observed in fewer numbers in the hair matrix, both above and below the critical level. It is interesting to note that a mesenchymal type of cell containing both melanosome complexes and Langerhans granules was present in the hair dermal papilla.

c) Senile Gray Hair

Melanocytes were present in normal numbers in the melanocytic zone. Their cytoplasmic organelles were scanty and melanin granules were small. The simultaneous occurrence of melanocytes and indeterminate dendritic cells was seldom observed in the melanocytic zone and Langerhans cells were infrequently identified in the hair matrix. Melanosomes, if present, were seen always in the form of membrane-bounded aggregations (fig. 1).

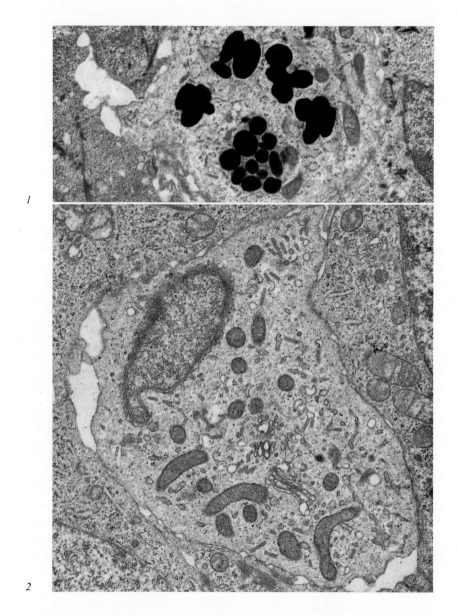

Fig. 1. Langerhans cell in senile gray hair matrix. Note melanosomes in form of membrane-bounded aggregations. × 12,000.

Fig. 2. Langerhans cell in senile white hair matrix. A large number of Langerhans granules and cyptoplasmic organelles are seen. × 13,000.

Fig. 3. Indeterminate dendritic cell in senile white hair matrix. Note well developed cytoplasmic organelles. × 14,000.

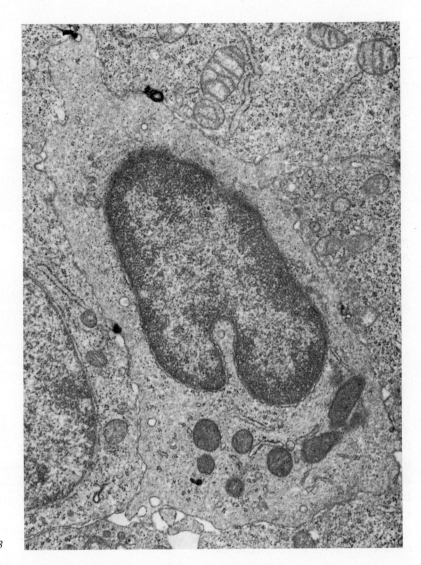

3

d) Senile White Hair

Melanocytes engaging actively in melanogenesis were no longer identified in the melanocytic zone. However, indeterminate dendritic cells, often with convoluted nuclei, were observed in the melanocytic zone in the second or third row above the basal lamina. The fine structure of the cells resembled that of the indeterminate dendritic cells in the epidermis. Relatively prominent cytoplasmic organelles was a characteristic of these cells. Langerhans cells, small in number, were observed in the hair matrix several rows above

Table I. Presence of dendritic cells in human hair matrix

	Melanocyte	I.D.C.[1]	Langerhans cell
Normal black hair Albino gray hair Senile gray hair	identified	not identified	identified
Senile white hair Vitiligo white hair	not identified	identified	

1 I.D.C. = Indeterminate dendritic cell

the basal lamina. In their cytoplasm numerous Langerhans granules and prominent cytoplasmic organelles were present (fig. 2 and 3).

e) White Hair Matrix of Vitiligo vulgaris

No melanocytes were observed in the hair matrix. Indeterminate dendritic cells were noted in lesser number. Langerhans cells were also identified in the hair matrix.

The results of the present qualitative study are summarized in table I.

Discussion

The present qualitative survey of gray and white hair matrices in different disorders revealed: a) gray hair matrix undergoes a reduction in the quantity of melanin granules and in the degree of melanization without a decrease in the number of melanocytes, while white hair matrix has no longer any types of cells capable of melanogenesis; b) reciprocal occurrence of melanocytes and indeterminate dendritic cells were observed in gray and white hair matrix; c) Langerhans cells in fewer numbers were identified consistently in the hair matrix examined.

These results may suggest that the reduced number of melanocytes and final disappearance from the hair matrix signify one of the pathological mechanisms of graying and whitening of the human scalp hair. The ultrastructural changes recognized in this study appear to confirm previous electron microscopic investigations [1, 3, 4, 12]. If there are immature melanocytes in the hair matrix [3], the method employed here failed to reveal them.

Contrary to the current view, Langerhans cells were identified consistently in the hair matrix. There were also dendritic cells identical to the epidermal Langerhans cells in the outer root sheath of the hair follicle between the infundibulum and in the excretory duct of the sebaceous gland [6]. The functional significance of the cells in the pilosebaceous system is obscure because of their infrequent occurrence. From the fact that the cells are present not only within the hair matrix but in the hair dermal papilla, it might at least be considered that Langerhans cells are mesenchymal in origin.

The presence of indeterminate dendritic cells in the white hair matrix is another fascinating subject to be discussed. Indeterminate dendritic cells [14] or alpha-dendritic cells [10] have been reported in human epidermis where both melanocytes and Langerhans cells are normally present. ZELICKSON et al. [14] reported in human adult epidermis, whether normal or vitiliginous, indeterminate dendritic cells, constantly numbered between 0.7 to 1.2%. MISHIMA et al. [10] stated that the presence of these cells is reciprocal with those of epidermal melanocytes. As regards the significance of these cells, it has been said that they may be a) a form of melanocytes in which melanin synthesis could be induced [10, 14]; b) effete melanocytes in which melanogenesis has ceased [5, 10, 14]; c) undifferentiated cells which could give rise to either melanocytes or Langerhans cells [13, 14]; d) melanocytes and Langerhans cells which could transform one into the other through indeterminate dendritic cells [13]; and e) completely unrelated cells [14].

In the present study, the impressive feature is the occurrence of indeterminate dendritic cells in the absence of melanocytes, and in locations similar to those of melanocytes. This fact suggests a possible relationship between melanocytes and indeterminate dendritic cells. According to this hypothesis, the cells must be presumed to be worn-out melanocytes, having lost the capability to produce specific organelles and replace those that had been expended. This would represent melanocytes as dynamic structures which may undergo transformation in response to the requirements for viability or to metabolic changes of the hair matrix. Reduced or negative tyrosinase activity in gray and white hair matrix might be partly based on the transformation of melanocytes into indeterminate dendritic cells. While there may be a qualitative reciprocity between indeterminate dendritic cells and melanocytes, a quantitatively reciprocal relationship between the two types of cells remains to be investigated.

Conclusions

The present qualitative observations on dendritic cells in the human hair matrix revealed the following results. Gray hair matrix undergoes a reduction in the amount of melanin granules without a diminution of the number of melanocytes, but white hair matrix has no melanocytes in the hair matrix. Indeterminate dendritic cells are observed in reciprocal relationship with melanocytes. Langerhans cells in lesser number were consistently identified in the hair matrix.

References

1 BIRBECK, M. S. C. and BARNICOT, N. A.: Electron microscope studies on pigment formation in human hair follicle; in GORDON, Pigment cell biology (Academic Press, New York 1959).
2 BLOCH, B.: Über die Entwicklung des Haut- und Haarpigments beim menschlichen Embryo und über das Erlöschen der Pigmentbildung im ergrauenden Haar (Ursache der Canities). Arch. Derm. Syph. *135:* 77–108 (1921).
3 FITZPATRICK, T. B.; BRUNET, T. B. and KUKITA, A.: The nature of hair pigment; in Biology of the hair growth (Academic Press, New York 1958).
4 HERZBERG, J. und GUSEK, W.: Das Ergrauen des Kopfhaares. Arch. klin. exp. Derm. *236:* 368–384 (1970).
5 HORIKI, M.: Electron microscopic study of dendritic cells of human epidermis. Skin Res. *8:* 233–250 (1966).
6 JIMBOW, K.; SATO, S. and KUKITA, A.: Langerhans cells of the normal human pilosebaceous system. J. Invest. Derm. *52:* 177–180 (1969).
7 KUKITA, A. and FITZPATRICK, T. B.: Demonstration of tyrosinase in melanocytes of the human hair matrix by autoradiography. Science *121:* 893–894 (1955).
8 KUKITA, A.: Changes in tyrosinase of melanocytes during the hair growth cycle. J. Invest. Derm. *28:* 273–274 (1957).
9 KUKITA, A.; SATO, S. and JIMBOW, K.: The electron microscopic study on dendritic cells in the hair matrix of human white and gray hair. Jap. J. Derm. Ser. A *81:* 777–787 (1971).
10 MISHIMA, Y. and KAWASAKI, H.: Role of alpha-dendritic cell in the pathogenesis of depigmentation. Jap. J. Derm. Ser. A *81:* 802–809 (1971).
11 SNELL, R. S.: An electron microscopic study of the dendritic cells in the basal layer of guinea-pig epidermis. Z. Zellforsch. *66:* 457–470 (1965).
12 SZABÓ, G.: Current state of pigment research with special reference to macromolecular aspects; in Biology of the skin and hair growth (Angus and Robertson, Sydney 1965).
13 TSUJI, T.; SUGAI, T. and SAITO, T.: Ultrastructure of three types of epidermal dendritic cells in hairless mice. J. Invest. Derm. *53:* 332–346 (1969).
14 ZELICKSON, A. S. and MOTTAZ, J. H.: Epidermal dendritic cells. A quantitative study. Arch. Derm. *98:* 625–659 (1968).

Author's address: Dr. SYOZO SATO, Department of Dermatology, Sapporo Medica College, *Sapporo* (Japan)

Studies on the Interaction Between Melanocytes and Keratinocytes With Special Reference to the Role of Microfilaments[1]

Muriel A. Wikswo and G. Szabo

Harvard School of Dental Medicine, Boston, Mass.

Introduction

Pigmentation in mammals involves an interaction between pigment cells or melanocytes and the surrounding epidermal cells known as keratinocytes. Together the melanocytes and keratinocytes form a structural and functional unit known as the epidermal melanin unit [9]. This unit is involved in at least four fundamental biological processes. These are the manufacture and melanization of melanosomes within the melanocytes and their movement out into the melanocyte dendrites, the attachment of the melanocytes to the keratinocytes, the transfer of melanosomes from the melanocytes to the keratinocytes, and the movement and breakdown of melanosome complexes within the keratinocytes.

Little is known concerning the manner of attachment between the melanocytes and keratinocytes or the mechanism of pigment transfer. *In vitro*, at least, it appears that a portion of the melanocyte dendrite is pinched off and transferred into the keratinocyte. *In vitro* observations also show that the cell membrane of the keratinocyte undulates very actively in the area of contact with a melanocyte dendrite [5, 15]. Membrane ruffling or undulation in a number of cells has been attributed to the presence of intracellular microfilaments [1, 3, 10, 13, 14, 25]. Ultrastructural studies demonstrate that filaments are present in mammalian keratinocytes and melanocytes both *in vivo* [4, 7, 19] and *in vitro* [11, 20], however, their role in cell movement is unclear.

---

[1] This work was supported by grants DE 01766-09 from the NIDR and 1 FO 2 Am 47, 747-01 from the NIH.

Recently a fungal metabolite, cytochalasin B, has been found which disrupts microfilaments in a variety of cells [2, 12, 24–27]. The effect of cytochalasin on microfilaments in these cells correlates with changes in cellular activity such as membrane undulation and cell movement. Cytochalasin has also recently been found to have an effect on microfilaments in melanophores [17, 18], melanocytes [18, 28, 29] and keratinocytes [28, 29]. It is the purpose of this paper to review the effects of cytochalasin B on microfilaments in mammalian melanocytes and keratinocytes, and to show the possible relationship that microfilaments bear to cell membrane activity and pigment transfer in these cells.

Materials and Methods

Cell cultures of adult guinea pig epidermis were established in Cruickshank chambers and plastic Cooper dishes according to previously described methods [6, 16, 29]. Small grafts of ear skin were placed at 37 °C in a 0.04% EDTA solution for 25 min followed by a 0.135% trypsin solution for 10 min. The skin was split and the epidermal cells were dispersed and suspended in Eagle's Minimal Essential Media containing 10% calf serum to which 20% fetal calf serum was added. The cell suspension was diluted and injected into Cruickshank chambers and plastic Cooper dishes and incubated at 37 °C.

The cells were examined by phase contrast microscopy and their behavior recorded by time-lapse cinemicrography. Cultures ranging in age from 7 to 15 days were treated with 1 to 10 µg/ml cytochalasin B (Imperial Chemical Industries, Ltd.). Stock solutions of 1 mg cytochalasin per ml dimethyl sulfoxide (DMSO) were prepared. Culture media containing 1 to 10 µg/ml cytochalasin was introduced to the cultures. Controls received from 0.1 to 1% DMSO in the culture media.

Cells were then fixed for electron microscopy in 3% glutaraldehyde followed by 1% osmium tetroxide [29]. They were then treated *en bloc* with 1% uranyl acetate [8, 29], rapidly dehydrated in a graded series of alcohols and embedded. Thin sections were cut parallel to the surface of the dish, stained with uranyl acetate and lead citrate, and examined with an RCA-3G electron microscope.

Results

The cultures have become well established by the end of one week (fig. 1a). The keratinocytes spread out and begin to form sheets, melanocytes become associated with keratinocytes and pigment donation is observed.

Addition of cytochalasin B to the cultures produces marked morphological changes in the cells. As early as 2 min after addition of 10 µg/ml cytochalasin, the edges of the keratinocytes begin to retract and the melanocytes shrink slightly in diameter. By 5 min, there is a decrease in undulation of the keratinocyte membrane as well as further retraction of the edges of the keratinocytes and further shrinkage of the melanocytes. After 20 min cell movement is inhibited, the ruffling or undulations of the keratinocyte membrane cease and only slight surface activity is observed (fig. 1b). The melanocytes have shrunk further in diameter and the dendritic processes have become beaded in appearance (fig. 1b). The melanocytes remain attached to the keratinocytes. Particle movement within both melanocytes and keratinocytes is still observed but pigment donation is not.

The effects of cytochalasin are reversible. By 5 min after removal of cytochalasin (10 μg/ml) from the medium the melanocytes and keratinocytes begin to re-expand and some ruffling of keratinocyte membranes is observed. Within 30 min to 1 h after drug removal the cells appear to have recovered. Keratinocytes and melanocytes have re-expanded, keratinocyte membranes undulate actively, pigment donation between melanocytes and keratinocytes is noted and in many instances both cell types change shape and migrate (fig. 1c).

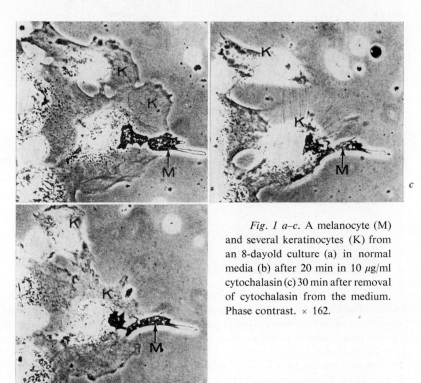

Fig. 1 a–c. A melanocyte (M) and several keratinocytes (K) from an 8-dayold culture (a) in normal media (b) after 20 min in 10 μg/ml cytochalasin (c) 30 min after removal of cytochalasin from the medium. Phase contrast. × 162.

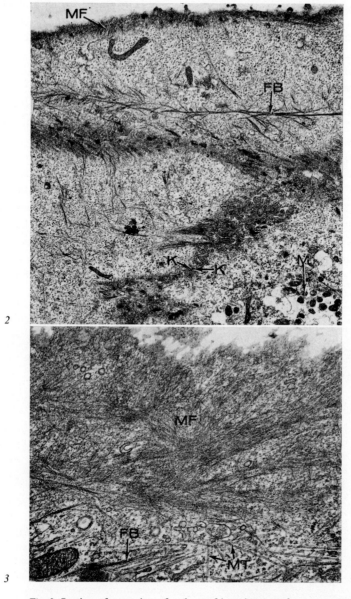

Fig. 2. Section of a portion of a sheet of keratinocytes from an untreated 8-day old culture. The keratinocytes (K) are closely attached to each other (arrows). A band of microfilaments (MF) is observed just beneath the cell membrane. Bundles of filaments (FB) are also present as well as melanosomes (M). × 4,000.

Fig. 3. High magnification of a section through a keratinocyte treated for 20 min with 1 % DMSO. Note the band of microfilaments (MF) beneath the cell membrane. Bundles of filaments (FB) and microtubules (MT) are also present. × 23,500.

No change in activity or morphology of cells in control cultures is observed when medium with DMSO is added.

Lower dosages of cytochalasin (5 and 1 µg/ml) were also tested. A longer treatment time (1 to 4½ h respectively) is required in order to produce an effect. The results are similar to those obtained with a higher dosage of cytochalasin.

Ultrastructural studies were next carried out on these cultured cells. Both untreated and DMSO-treated cells appear the same. Cells in sheets of keratinocytes are closely attached to one another (fig. 2). In keratinocytes, a band of 30–70 Å microfilaments is present just beneath the cell membrane (fig. 2 and 3). Bundles of 60–110 Å filaments as well as microtubules are also observed within these cells. In melanocytes a band of 30–70 Å microfilaments is likewise observed just beneath the cell membrane (fig. 6 and 7). Microtubules are present and are orientated for the most part, parallel to the longitudinal axis of the cell process; 40–110 Å filaments are also scattered throughout the cell (fig. 7).

After a 20 min exposure to cytochalasin, the edges of the keratinocytes have retracted somewhat and in many areas the cells have pulled apart (fig. 4). The band of microfilaments just beneath the cell membrane is replaced by masses of granular and short filamentous material (fig. 5). The bundles of 60–110 Å filaments and microtubules are unaffected. Melanocytes treated with cytochalasin also show a decrease in number of microfilaments just beneath the cell membrane; this change is accompanied by the presence of granular and filamentous material (fig. 9). Microtubules and scattered filaments are still present.

In cytochalasin-treated cells many of the areas of contact between the melanocytes and keratinocytes remain (fig. 8). All other organelles, other than microfilaments, appear normal in the presence of cytochalasin.

The effects of cytochalasin on microfilaments are reversible. Microfilaments begin to reappear by 5–10 min after cytochalasin (10 µg/ml) is removed from the medium (fig. 10). By 1 h after cytochalasin is removed, the band of microfilaments just beneath the cell membrane in both keratinocytes and melanocytes is again observed.

Discussion

This work provides new and additional information concerning the relationship between melanocytes and keratinocytes. It gives information

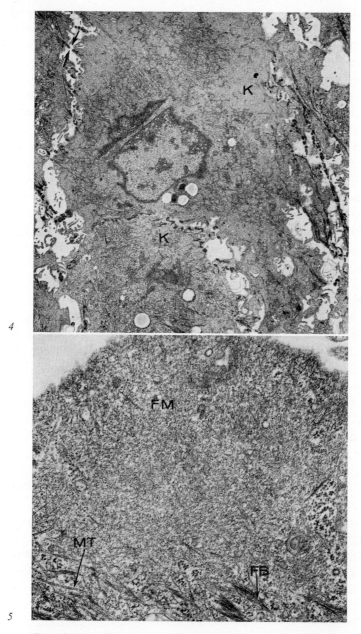

Fig. 4. Section of a portion of a sheet of keratinocytes treated for 20 min with 10 µg/ml cytochalasin. The keratinocytes (K) are pulling apart (arrows). × 2,700.

Fig. 5. Higher magnification of a portion of a keratinocyte treated for 20 min with 10 µg/ml cytochalasin. Note the mass of fine filamentous material (FM) beneath the cell membrane. Bundles of filaments (FB) and microtubules (MT) are present. × 29,000.

concerning the nature of attachment of the melanocyte to the keratinocyte. In the presence of cytochalasin the keratinocytes retract and pull apart from each other, however, the melanocytes remain attached to the keratinocytes.

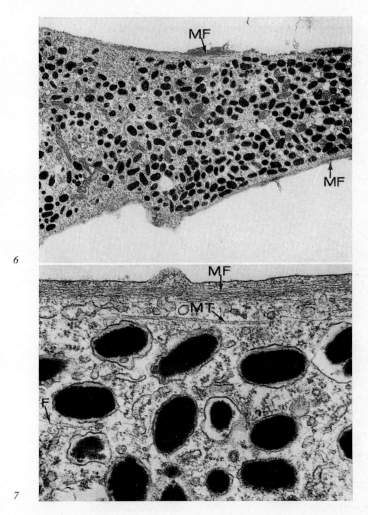

Fig. 6. Longitudinal section through a portion of a melanocyte process treated for 20 min with 1% DMSO. A band of microfilaments (MF) is observed just beneath the cell membrane. × 6,000.

Fig. 7. Higher magnification of a portion of a 1% DMSO-treated melanocyte. Note the band of microfilaments (MF) beneath the cell membrane, microtubules (MT) and scattered filaments (F). × 35,000.

Little is known about the manner of attachment of the melanocytes to the keratinocytes. It does not appear to be desmosomal as does the keratinocyte-keratinocyte attachment. When treated with cytochalasin, many of the desmosomal attachments between keratinocytes break apart whereas the

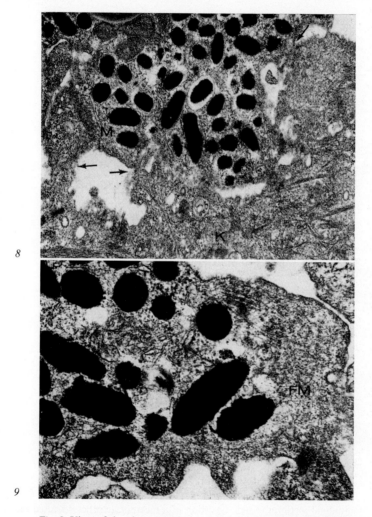

Fig. 8. View of the close association (arrow) between a melanocyte (M) and a keratinocyte (K) treated for 20 min with 10 μg/ml cytochalasin. × 20,000.

Fig. 9. Longitudinal section through a melanocyte process treated for 20 min with 10 μg/ml cytochalasin. A mass of granular and filamentous material (FM) is observed just beneath the cell membrane. × 37,500.

attachments between melanocytes and keratinocytes remain. This does not mean that cytochalasin is having a direct effect on cell junctions. It does indicate, though, that the nature of the melanocyte-keratinocyte attachment is such that it can withstand the tension created when the keratinocytes retract whereas the keratinocyte-keratinocyte desmosomal attachments cannot. Recently ROTH *et al.* [21–23] have indicated that in embryonic cells there are specific moieties on apposing cell surfaces which are responsible for the intercellular adhesion of these cells. It is possible then, though as yet there is no evidence to support the idea, that there are sites on the keratinocyte and melanocyte cell surfaces which are involved in the adhesion of these two cell types to each other.

This work also indicates that cytochalasin inhibits ruffling or undulation of keratinocyte membranes and inhibits cell movement in both melanocytes and keratinocytes. This change in cell activity in turn correlates with the breakdown of a peripheral band of microfilaments in these cells. In a number of different cell types a similar correlation has been found between cessation

Fig. 10. Section through a keratinocyte 10 min after 10 μg/ml cytochalasin (for 20 min) was removed from the medium. Note reappearance of microfilaments (MF) beneath the cell membrane. × 19,000.

of cell movement and breakdown of microfilaments [2, 12, 25–27]. It is possible, as has been speculated [1, 3, 10], that the microfilaments are contractile and play a role in membrane movements.

Pigment donation is also not observed in cytochalasin-treated cultures. As to whether pigment donation bears any relationship with undulation of cell membranes and the presence of microfilaments is unclear. However, an increase in membrane ruffling of keratinocytes is observed in the area of contact with a melanocyte process [5, 15] and this membrane ruffling is inhibited by cytochalasin. Recent work with liver cells indicates that microfilaments just under the plasma membrane may be necessary for membrane movements such as ruffling and undulation which are involved in endocytosis [26]. This work shows that treatment of these cells with cytochalasin B leads to cessation of membrane movements, an inhibition of endocytosis and a simultaneous disappearance of microfilaments subjacent to the plasma membrane. Though endocytosis may not be basically the same process as pigment donation, it is not unreasonable to suggest the possibility of a similarity between endocytosis and the uptake of a melanosome packet by a keratinocyte. Thus, membrane activity could be important for pigment donation and this membrane activity in turn could be related to the presence of microfilaments.

Conclusions

Cytochalasin B produces marked changes in the morphology of cultured mammalian melanocytes and keratinocytes. After addition of cytochalasin, the edges of the keratinocytes retract and ruffling and undulation of the cell membrane ceases. Melanocytes shrink in diameter but maintain contact with the keratinocytes. Ultrastructurally in keratinocytes, the band of 30–70 Å microfilaments just beneath the cell membrane is replaced by masses of granular and short filamentous material. Microtubules and bundles of 60–110 Å filaments dispersed throughout the keratinocyte are unaffected. In cytochalasin-treated melanocytes, a band of 30–70 Å microfilaments just beneath the cell membrane is also replaced by granular and filamentous material. Pigment donation from melanocytes to keratinocytes is not observed during cytochalasin treatment. The effects of cytochalasin are reversible. These results show that there is a correlation between the integrity of the microfilament systems and the maintenance of morphology and movement of the melanocytes and keratinocytes.

References

1   AMBROSE, E. J.: The movements of fibrocytes. Exp. Cell Res., Suppl. *8:* 54–73 (1961).
2   AUERSPERG, N.: Microfilaments in epithelial morphogenesis. J. Cell Biol. *52:* 206–211 (1972).
3   BUCKLEY, I. K. and PORTER, K. R.: Cytoplasmic fibrils in living cultured cells. A light and electron microscope study. Protoplasma *64:* 349–380 (1967).
4   CLARK, W. H. and HIBBS, R. G.: Electron microscope studies of the human epidermis. The clear cell of Masson (dendritic cell or melanocyte). J. biophys. biochem. Cytol. *4:* 679–684 (1958).
5   COHEN, J. and SZABO, G.: Study of pigment donation in vitro. Exp. Cell Res. *50:* 418–434 (1968).
6   CRUICKSHANK, C. N. D. and HARCOURT, S. A.: Pigment donation in vitro. J. Invest. Derm. *42:* 183–184 (1964).
7   DROCHMANS, P.: Electron microscope studies of epidermal melanocytes and the fine structure of melanin granules. J. biophys. biochem. Cytol. *8:* 165–180 (1960).
8   FARQUHAR, M. G. and PALADE, G. E.: Cell junctions in amphibian skin. J. Cell Biol. *26:* 263–291 (1965).
9   FITZPATRICK, T. B. und BREATHNACK, A. S.: Das epidermale Melanin-Einheit-System. Derm. Wschr. *147:* 481–489 (1963).
10 FRANKS, L. M.; RIDDLE, P. N. and SEAL, P.: Actin-like filaments and cell movement in human ascites tumor cells. Exp. Cell Res. *54:* 157–162 (1969).
11 GAZZOLO, L. et PRUNIERAS, M.: Culture de longue durée de cellules issues de l'épiderme de cobaye adulte. III. Essai de caractérisation ultrastructurale. Path. Biol. *17:* 251–259 (1969).
12 GOLDMAN, R. D.: The effects of cytochalasin B on the microfilaments of baby hamster kidney (BHK-21) cells. J. Cell Biol. *52:* 246–254 (1972).
13 GOLDMAN, R. D. and FOLLETT, E. A. C.: The structure of the major cell processes of isolated BHK-21 fibroblasts. Exp. Cell Res. *57:* 263–276 (1969).
14 INGRAM, V. M.: A side view of moving fibroblasts. Nature, Lond. *222:* 641 (1969).
15 KLAUS, S. N.: Pigment transfer in mammalian epidermis. Arch. Derm. *100:* 756–762 (1969).
16 KLAUS, S. N. and SNELL, R. S.: The response of mammalian epidermal melanocytes in culture to hormones. J. Invest. Derm, *48:* 352–358 (1967).
17 MALAWISTA, S. E.: Reversible inhibition of melanin granule movement by cytochalasin B. Nature, Lond. *234:* 354 (1971).
18 MCGUIRE, J. and MOELLMANN, G.: Cytochalasin B: effects on microfilaments and movement of melanin granules within melanocytes. Science *175:* 642–644 (1972).
19 ODLAND, G. F.: The fine structure of the interrelationship of cells in the human epidermis. J. biochem. biophys. Cytol. *4:* 529–533 (1958).
20 PRUNIERAS, M.: Interactions between keratinocytes and dendritic cells. J. Invest. Derm. *52:* 1–17 (1969).
21 ROTH, S.: Studies on intercellular adhesive selectivity. Develop. Biol. *18:* 602–631 (1968).
22 ROTH, S.; MCGUIRE, E. J. and ROSEMAN, S.: An assay for intercellular adhesive specificity. J. Cell Biol. *51:* 525–535 (1971).

23 ROTH, S.; MCGUIRE, E. J. and ROSEMAN, S.: Evidence for cell-surface glycosyltransferases. Their potential role in cellular recognition. J. Cell Biol. *51:* 536–547 (1971).
24 SCHROEDER, T. E.: The contractile ring. I. Fine structure of dividing mammalian (HeLa) cells and the effects of cytochalasin B. Z. Zellforsch. *109:* 431–449 (1970).
25 SPOONER, B. S.; YAMADA, K. M. and WESSELLS, N. K.: Microfilaments and cell locomotion. J. Cell Biol. *49:* 595–613 (1971).
26 WAGNER, R.; ROSENBERG, M. and ESTENSEN, R.: Endocytosis in Chang liver cells. Quantitation by sucrose-^{3}H uptake and inhibition by cytochalasin B. J. Cell Biol. *50:* 804–817 (1971).
27 WESSELLS, N. K.; SPOONER, B. S.; ASH, J. F.; BRADLEY, M. O.; LUDUENA, M. A.; TAYLOR, E. L.; WRENN, J. T. and YAMADA, K. M.: Microfilaments in cellular and developmental processes. Science *171:* 135–143 (1971).
28 WIKSWO, M. A. and SZABO, G.: In vitro effects of cytochalasin on melanocytes and keratinocytes. Proc. Amer. Soc. Cell Biol., 11th Ann. Meet. p. 323 (1971).
29 WIKSWO, M. A. and SZABO, G.: Effects of cytochalasin B on mammalian melanocytes and keratinocytes. J. Invest. Derm. (59 [in press] 1972).

Author's address: Dr. MURIEL A. WIKSWO, Department of Dermatology, Yale University School of Medicine, *New Haven, CT 06510* (USA)

On the Melanization Process of Melanosomes

M. Seiji, H. Fukuzawa, K. Shimao and H. Itakura

Department of Dermatology, Tohoku University School of Medicine, Sendai; Department of Physics, Faculty of General Education, Tokyo Medical and Dental University, Ichikawa, and Central Biochemistry Laboratory Juntendo University School of Medicine, Tokyo

Introduction

In mammals, melanin formation and deposition occur on cytoplasmic particles of the melanocyte because the enzyme, tyrosinase, is bound to specialized organelles, the melanosomes [5, 6]. Since tyrosinase is present on the melanosomes, which themselves undergo melanization, the relationship between melanization and tyrosinase activity of the melanosome constitutes an interesting problem.

The present experiments were performed in an effort to clarify the relationship between the tyrosinase activity and the amount of melanin formed. Kinetic studies were carried out in order to further understand the mechanisms of the oxidation of dopa by melanosome tyrosinase.

Materials and Methods

Preparation of Melanosomes

An entire, actively growing Harding-Passey mouse melanoma was excised and was promptly homogenized in 0.25 M sucrose. The large granule fraction was prepared by Seiji's [4] method and melanosomes were isolated. Density gradients were prepared in the tubes of the Hitachi swinging-bucket rotor (RPS 40A) by layering 1 ml of 4 different concentrations of sucrose solution in serial order (2.0, 1.8, 1.6, and 1.55 M). They were then allowed to stand for 16 h. At the end of this interval, 1 ml of freshly prepared large granule fraction was layered carefully over the top of each tube. The tubes thus prepared were centrifuged at 103,000 r.p.m. for 1 h. At the end of centrifugation, the fractions No. 1 and No. 2 were separated from the middle and the bottom of the centrifuge tubes, respectively. Melanosomes used for the other experiment were prepared according to the method of Seiji et al. [7].

Determination of Enzyme Activity

Tyrosinase activity was estimated by the colorimetric method of SHIMAO [8]. The optical density was determined with the Klett-Summerson photoelectric colorimeter (ΔE = scale reading).

Analytical Procedure

Protein was determined according to the method of LOWRY et al. [1].

Incorporation of ^{14}C-Dopa in vitro by Melanosome No. 1 and No. 2 Fractions

Melanosome No. 1 and No. 2 fractions were incubated at 37 °C with 16 mμCi ^{14}C-DL-dopa (5.45 mCi/mm), 0.1 M phosphate buffer, pH 6.8 and L-dopa, 3 mg in the total volume of 3.5 ml. Equal amounts of the two melanosome fractions were incubated in two different experiments. In one the protein content was the same in each fraction while in the other, the tyrosinase content was the same. At the end of various incubation times, 0.5 ml of perchloric acid solution (PCA) was added to the reaction mixtures, then filtered through the millipore filter (HAW P 025 00 H.A. 0.45 μ), washed twice with 0.5 ml of 5% PCA and the millipore filters with melanosomes were dried.

In vitro Incorporation of ^{14}C-Dopa by Melanosomes for the Kinetic Study

Various amounts of melanosome suspensions were incubated at 37 °C with 3.6×10^{-3} M ^{14}C-dopa (8 mμCi ^{14}C-DL-dopa + 2.5 mg L-dopa), 1000 units of penicillin G potassium, M/15 phosphate buffer pH 6.8 in a total volume of 3.5 ml. At the end of various incubation times, 5% trichloroacetic acid solution was added to the reaction mixture, then centrifuged at $20,000 \times g$ for 20 min. The radioactivity of an aliquot of the supernatant thus obtained was determined.

Measurement of Radioactivity

The radioactivity of the melanosomes on the millipore filter was determined by means of the Nihon Musen gas flow counter. The radioactivity of an aliquot of the supernatant was measured in glass vials in a Packard TRI-CARB model 3380 spectrometer.

Results

A Comparison Between Melanosome No. 1 and No. 2 Fractions

As shown in table I, the optical densities of the melanosome No. 2 are higher than those of No. 1 at both wave lengths. On the other hand, the tyrosinase level of melanosome No. 1 is twice that of No. 2. From these observations, it can be said that melanosome No. 1 is less melanized or less mature than No. 2. Comparison of the rate of incorporation of ^{14}C-dopa into the melanosome No. 1 and No. 2 are given in figure 1. In this particular experiment, the amount of both No. 1 and No. 2 fraction is the same in mg protein. The immediate rapid labeling of No. 1 fraction was observed

Table I. Comparison of optical densities and tyrosinase activities of melanosome No. 1 fraction and melanosome No. 2 fraction

No.	O.D.[1]		Tyrosinase activity
	280 mμ/mg protein	500 mμ/mg protein	ΔE 10 min/mg protein
1	12.3	4.05	142.3
2	16.5	7.12	74.0

1 O.D. = optical density.

and at the end of 7 h, the incorporation of ^{14}C-dopa into both fractions appears to be almost complete. Melanosome No. 1, which possesses higher tyrosinase activity in terms of mg protein and is less melanized than No. 2, incorporates more ^{14}C-dopa than melanosome No. 2 when *in vitro* melanization is complete.

Figure 2 also shows the time courses of the incorporation of ^{14}C-dopa in these two fractions. In this experiment, the tyrosinase activities of both No. 1 and No. 2 fractions are the same. The time courses of incorporation of No. 1 and No. 2 are quite similar, and at the end of 4 h, when the in-

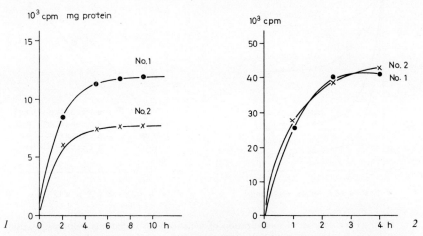

Fig. 1. Comparison of the rate of incorporation of ^{14}C-dopa into the melanosome No. 1 fraction and melanosome No. 2 fraction. The amounts (mg protein) of both melanosome fractions incubated are the same.

Fig. 2. Comparison of the rate of incorporation of ^{14}C-dopa into the melanosome No. 1 fraction and melanosome No. 2 fraction. The tyrosinase activity of both melanosome fractions incubated are the same.

corporation appears to be complete, the amounts of ^{14}C-dopa incorporated are nearly the same.

Melanin Formation From ^{14}C-Dopa and Melanosomes *in vitro*

Figure 3 shows the time courses of the melanization of melanosomes. The radioactivity of the ^{14}C-dopa left in the supernatant decreased as melanin formation proceeded. The rate of decrease in the radioactivity of ^{14}C-dopa appears to depend on the amount of melanosomes incubated. The larger the amount of melanosomes incubated, the faster and greater the decrease in radioactivity of ^{14}C-dopa. At the end of 8 h of incubation, the decrease in ^{14}C-dopa appears to reach a final value in all experimental groups. In other words, although the substrate, L-dopa, still remains in the reaction medium, the tyrosinase activity of melanosomes ceases and is no longer active. Figure 4 shows the relationship between tyrosinase activity and the amount of melanin formed. The amount of melanin formed at the

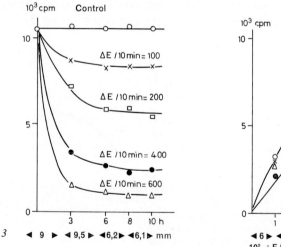
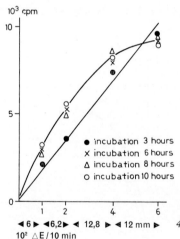

Fig. 3. Time courses of melanization of melanosomes in various amounts. Various amounts of melanosome suspensions (ΔE = 100, 200, 400, and 600 as tyrosinase activity) were incubated.

Fig. 4. Relationship between the amount of melanosomes and the amount of melanin synthesized. A linear relationship was obtained at 3 h incubation when the melanin formation was still in progress, but after the melanin formation ceased, the relationship was not linear but an upward convex curve.

end of 3 h incubation appears to be related linearly to the amount of melanosomes incubated. At the end of 6, 8, and 10 h incubation, the relationship between these are not linear but upward convex curves.

Discussion

The observation of the relationship between tyrosinase activity and the melanization of the melanosome of the retinal pigmented epithelium of chick embryo [2] and the experimental results of the *in vitro* melanization of melanosomes of mouse melanomas led us to assume that there is an inverse relationship between melanin content and tyrosinase activity in the melanosome [3]. The reduction in tyrosinase activity that occured with an increase in melanization of the melanosome appeared to result from a blocking of the active centers on the enzyme, probably by the quinoid intermediates. Recently, *in vitro* melanin formation with ^{14}C labeled soluble tyrosinase and L-dopa was studied [7]. Time courses of the disappearance of labeled tyrosinase from the reaction system clearly indicated that soluble tyrosinase combined with the products of the dopa-tyrosinase reaction and precipitated out with them. It was also shown that the tyrosinase molecules thus precipitated out with melanin no longer possessed any enzyme activity.

Two kinds of melanosomes in different developmental stages were prepared. Melanosome No. 1 was less melanized and possessed more tyrosinase activity and melanosome No. 2 more melanized with less tyrosinase activity. The following *in vitro* experiments were carried out to prove the hypothesis that the less melanized melanosomes will synthesize and deposit more melanin than the more melanized melanosomes. As figure 1 shows (the amount of ^{14}C-dopa incorporated), the amount of melanin synthesized was larger in melanosome No. 1 than in melanosome No. 2 at the end of 9 h incubation when the incorporation appeared to be complete. In order to clarify the quantitative relationship between tyrosinase activity and the amount of melanin synthesized in the melanosomes, the following *in vitro* experiments and kinetic studies were carried out.

In this experiment, varying amounts of melanosome suspensions were incubated with ^{14}C-dopa in phosphate buffer solution at fixed concentration, and time courses of the decrease in ^{14}C-dopa were observed. Figure 3 shows that the rate of reduction of ^{14}C-dopa in the supernatants appears to depend on the amount of melanosomes incubated. The larger the amount of melanosomes, the faster the rate of reduction, and the more the melanin formation.

After 8 h incubation, melanin formation no longer seems to take place even though the substrate, L-dopa, still remains in the reaction medium. In other words, there seems to exist some kind of mechanism which inactivates the enzyme, because, if the enzyme remained active, the final ^{14}C-dopa concentration should approach zero.[1] From the experimental results obtained previously [2, 3, 7], it is quite accurate, at the present time, to assume that the reaction product, P, reacts with the enzyme, E, to form an inactive E-P complex. On the basis of this assumption, kinetic studies were carried out in order to interpret the experimental results obtained.

The following three reaction mechanisms were considered;
Case I: Enzyme-product complex EP.
$D + E \rightleftharpoons DE \rightarrow E + P$, $E + P \rightarrow EP$,
Case II: Enzyme-product complex E_2P.
$D + E \rightleftharpoons DE \rightarrow E + P$, $2E + P \rightarrow E_2P$.
Case III: Reversibility of the enzyme-product complex formation in Case I.
$D + E \rightleftharpoons DE \rightarrow E + P$, $E + P \rightleftharpoons EP$.
Where D = dopa, E = tyrosinase, DE = enzyme-substra tecomplex, P = product, which, for the sake of simplicity, is assumed to be in only one form, EP = inactive enzyme-product complex.

However, the experimental results could not be explained by these reaction mechanisms. When simultaneous and irreversible formation of two rypes of inactive enzyme-product complexes, EP and E_2P, is assumed, the teaction mechanism and the corresponding rate equations are;

Case IV:

$$D + E \underset{k_2}{\overset{k_1}{\rightleftharpoons}} DE \overset{k_3}{\longrightarrow} E + P,$$
$$u \quad v \quad\quad w \quad\quad v \quad x$$

$$E + P \overset{k_4}{\longrightarrow} EP,$$
$$v \quad x \quad\quad y$$

$$2E + P \overset{k_5}{\longrightarrow} E_2P$$
$$2v \quad x \quad\quad z$$

and,

---

1   Irreversibility of step 2) in the following scheme is postulated:
   1)   2)
$D + E \rightleftharpoons DE \rightarrow E + P$, where D = dopa, E = tyrosinase, P = the reaction product (dopa quinone).

$$u' = -k_1uv + k_2w,$$
$$v' = -k_1uv + (k_2+k_3)w - k_4vx - 2k_5v^2x,$$
$$w' = k_1uv - (k_2+k_3)w,$$
$$x' = \qquad\qquad k_3w - k_4vx - k_5v^2x,$$
$$y' = \qquad\qquad k_4vx,$$
$$z' = \qquad\qquad k_5v^2x.$$

As the integration of these equations in the same way as carried out in Cases I to III was unfortunately impossible in this case, numerical integration of the rate equations was performed by the following equations:

$$u_{i+1} = u_i - k_1 u_i v_i \Delta t + k_2 w_i \Delta t,$$
$$v_{i+1} = v_i - k_1 u_i v_i \Delta t + (k_2+k_3)w_i \Delta t - k_4 v_i x_i \Delta t - 2k_5 v_i^2 x_i \Delta t,$$
$$w_{i+1} = w_i + k_1 u_i v_i \Delta t - (k_2+k_3)w_i \Delta t,$$
$$x_{i+1} = x_i + k_3 w_i \Delta t - k_4 v_i x_i \Delta t - k_5 v_i^2 x_i \Delta t,$$
$$y_{i+1} = y_i + k_4 v_i x_i \Delta t,$$
$$z_{i+1} = z_i + k_5 v_i^2 x_i \Delta t,$$

where values of u, etc., at time $t + \Delta t$, u_{i+1}, etc., are calculated from those at time t, u_i, etc.

The results for $u_0 = 1$, $v_0 = 0.2$, 0.4, 0.7 and 1.0 are shown in figure 5. Rate constants were arbitrarily chosen as $k_1 = k_2 = k_3 = 1$ and $k_4 = k_5 = 10$. In spite of the arbitrary selection of the initial conditions and values of rate constants, figure 5 is of the same type as the corresponding figures in the experiment (fig. 3). Moreover, the plot of u_0 minus u at large values of t,

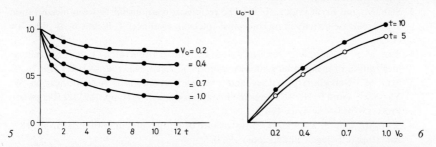

Fig. 5. Change in dopa concentration with time in the reaction mechanism discussed under Case IV in the text. See text for the reaction mechanism and rate equations. Plots were obtained by numerical integration of the rate equations, putting $u_0 = 1$; $v_0 = 0.2, 0.4, 0.7,$ and 1.0; $w_0 = 0$, $x_0 = 0$, $y_0 = 0$, $z_0 = 0$; $k_1 = k_2 = k_3 = 1$, and $k_4 = k_5 = 10$.

Fig. 6. Relation between the initial enzyme concentration, v_0, and the amount of dopa reached at 5 and 10 units of incubation time as plotted from figure 5.

which can be taken as the approximate value of D_∞ versus $v_0(=E_0)$, gives upward convex curves, as shown in figure 6, which correspond to the experimental results shown in figure 4.

From the above results, it might be possible to explain the results of the experiment by the mechanism treated in Case IV. Of course, the reaction scheme discussed in Case IV is not the only mechanism which can explain the results of the experiments. Further investigation, both experimental and theoretical, is necessary in order to deduce clear conclusions concerning the mechanism of the reactions involved. It would be safe to assume, however, that formation of several types of inactive enzyme-product complex might be involved in the inactivation of tyrosinase during melanization of melanosomes.

Acknowledgments

This work was supported by a research grant from the Ministry of Education of Japan and in part by grants from the Takeda Science Foundation, Osaka, and the Japan O'Leary Pigment Research Fund, Osaka.

References

1 LOWRY, O. H.; ROSEBROUGH, N. U.; FARR, A. L. and RANDALL, R. J.: Protein measurement with the Folin phenol reagent. J. biol. Chem. *193:* 265–275 (1951).
2 MIYAMOTO, M. and FITZPATRICK, T. B.: On the nature of the pigment in retinal pigment epithelium. Science *126:* 449–450 (1957).
3 SEIJI, M. and FITZPATRICK, T. B.: The reciprocal relationship between melanization and tyrosinase activity in melanosomes (melanin granules). J. Biochem. *49:* 700–706 (1960).
4 SEIJI, M.; SHIMAO, K.; BIRBECK, M. S. C. and FITZPATRICK, T. B.: Subcellular localization of melanin biosynthesis. Ann. N.Y. Acad. Sci. *100:* 497–533 (1963).
5 SEIJI, M.: Subcellular particles and melanin formation to melanocytes; in W. MONTAGNA und F. HU Advances in biology of skin, vol. VIII, pp. 189–222 (Pergamon Press, Oxford/New York 1967).
6 SEIJI, M.: in S. S. ZELICKSON Ultrastructure of normal and abnormal skin, pp. 183–201. (Lea & Febiger, Philadelphia 1967).
7 SEIJI, M. and MIYAZAKI, K.: Melanization and tyrosinase activity. J. Invest. Derm. *57:* 316–322 (1971).
8 SHIMAO, K.: Partial purification and kinetic studies of mammalian tyrosinase. Biochem. biophys. Acta *62:* 205–215 (1962).

Author's address: Dr. MAKOTO SEIJI, Department of Dermatology, Tohoku University School of Medicine, *Sendai* (Japan)

Genotype-Substrate Interactions Altering Golgi Development During Melanogenesis

J. A. Brumbaugh, R. R. Bowers and G. E. Chatterjee

Department of Zoology, University of Nebraska, Lincoln, Nebr.

Introduction

Ultrastructural observations and cytochemistry have shown that the Golgi system is associated with melanogenesis [2, 3, 5, 9, 13, 14, 16]. In this study two pigment mutants in the fowl which affect Golgi development are compared to the standard genotype and their responses to increased tyrosine levels examined. The results suggest the specific function of the Golgi complex during melanogenesis and the gene action of the two mutants.

Materials and Methods

Melanocytes from the regenerating breast feathers of adult males of three genotypes were examined. Standard males ($+^E/+^E, +^C/+^C, +^{Pk}/+^{Pk}$) have black breast feathers and dark eyes and are normal pigment producers. Albino-like males ($e^y/e^y, c/c, +^{Pk}/+^{Pk}$) have white feathers and pink eyes. Pink-eye males ($E/E, +^C/+^C, pk/pk$) have gray feathers and pink eyes. The two types of mutant males would both have black breast feathers in the absence of their respective mutations (c and pk) even though they differ at the E locus. Thus this study examined differences controlled only by the c and pk loci.

Stocks are maintained in the Department of Zoology of the University of Nebraska-Lincoln. The standard stock is of junglefowl origin [4]. The albino-like effect is due to the presence of the recessive white [c] mutation [10] and the pink-eye stock has previously been described [2].

Breast feather tracts were plucked and allowed to regenerate for 2 weeks. The breast areas were then disinfected and several feathers carefully removed. The portion of each feather cylinder which had been below the follicle mouth was cut off with sterile scissors and placed in dissecting medium where it was split in half longitudinally and the central pulp teased away. One half of each feather was placed in control medium and the other half in high-tyrosine medium. At least 3 feathers of each genotype were sampled. The explants were incubated at 38 °C for 20 h in 95 % O_2 and 5 % CO_2.

Control medium consisted of Waymouth's formula (752/1; GIBCO) with 5% chick serum plus antibiotics (100 u penicillin, 80 μg streptomycin, and 2.5 μg Fungizone/ml). The concentration of L-tyrosine in control medium was 0.22 mM. Dissecting medium was like control medium but contained 2× the concentration of antibiotics. The high tyrosine medium was also like control medium but had a final concentration of L-tyrosine hydrochloride of 4.0 mM.

After incubation, the feather pieces were fixed in 3% glutaraldehyde in 0.1 M phosphate buffer (pH 7.2; 2 h), osmicated for 1 h (2% OsO_4) in phosphate buffer, dehydrated, and embedded in Epon 812. Ultrathin sections were mounted on formvar coated grids and viewed with an RCA EMU 3-B (modified) electron microscope after being stained with uranyl acetate and lead citrate. Thin sections were always taken from the barb-ridge forming region of the feather to ensure that corresponding stages of development were compared [3]. Six to thirteen photographs of melanocytes were taken for each genotype and treatment and always included a nuclear portion to ensure uniform comparisons.

Each negative was quantitatively evaluated by placing it over a grid of regularly placed dots 1 cm apart. The dictyosomes ('stacks' of Golgi cisternae) were counted and the area of cytoplasm determined by counting the number of squares filled with cytoplasm. The number of dictyosomes per μ^2 of cytoplasm was calculated. A 't' test [17] was used to determine whether any statistical significance could be assigned to the differences between control and tyrosine-treated means.

Results

Melanocytes incubated in control medium resembled melanocytes which had not been subjected to dissection and incubation. Incubation, then, did not seem to harm the cells and apparently allowed melanogenesis to proceed.

Incubation in high-tyrosine medium did not alter standard melanocytes when compared to control cells (fig. 1 and 2). Both types of melanocytes were copious melanosome producers and possessed Golgi systems (arrows, fig. 1 and 2), indicating that melanogenesis was in progress. Dictyosome number was not altered by high-tyrosine since the average was $0.046/\mu^2$ in control tissue and $0.047/\mu^2$ in treated tissue (table I). No obvious increase in deposited melanin was noted in the tyrosine-treated melanocytes.

Albino-like melanocytes from control cultures were strikingly different from standard melanocytes (fig. 3). They produced no melanin and pos-

---

Fig. 1. Standard melanocyte incubated in control medium. × 20,500.

Fig. 2. Standard melanocyte incubated in high-tyrosine medium. Note the similarity to figure 1. × 20,500.

Fig. 3. Albino-like melanocyte incubated in control medium. Note the hypertrophy of the Golgi system and the absence of definitive premelanosomes. × 20,500.

Fig. 4. Albino-like melanocyte incubated in high-tyrosine medium. Note the presence of well-formed premelanosomes (pm). × 20,500.

Golgi Development During Melanogenesis

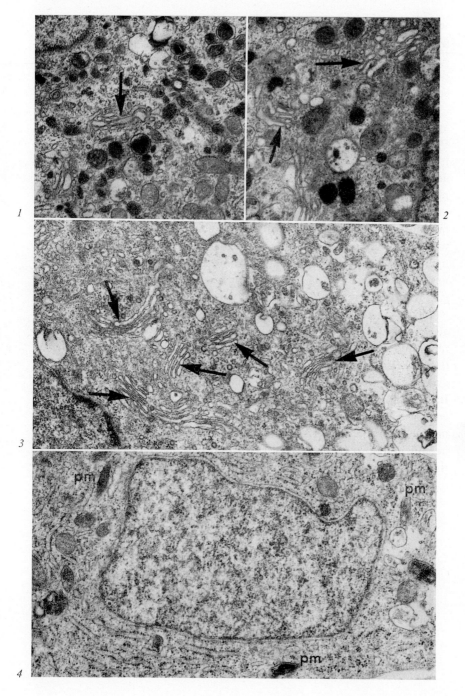

sessed numerous vacuoles and only an occasional premelanosome. Their Golgi system was hypertrophied and the number of dictyosomes (arrows, fig. 3) was almost three times that of standard control cells (table I). Incubation does not cause Golgi hypertrophy because unincubated albino-like cells have the same appearance.

Incubation in high-tyrosine medium definitely caused albino-like melanocytes to assume a more normal conformation and metabolism (fig. 4). Premelanosomes were formed (fig. 4) some of which contained obvious electron-opaque areas of melanin deposition. Golgi hypertrophy was not evident and dictyosome number was reduced from $0.132/\mu^2$ in control tissue to a normal value of $0.046/\mu^2$. This three-fold reduction is statistically significant.

Pink-eye melanocytes are characterized by incompletely melanized premelanosomes [2]. In control cultures most of the melanogenic organelles were premelanosomes (fig. 5). Although a Golgi complex is not seen in figure 5, they were present and exhibiting some hypertrophy. Dictyosome number was twice that of standard control cells (table I).

Incubation in high-tyrosine medium seemed to increase Golgi hypertrophy in pink-eye melanocytes rather than reduce it as was noted for the albino-like cells (arrows, fig. 6). Dictyosome number increased from 0.098 to $0.157/\mu^2$ (table I) but a comparison of these values is statistically equivocal (table I). Tyrosine incubation definitely had an effect upon pink-eye melanocytes, however, since a selective, premature death of melanocytes occurred. In many instances, the areas normally occupied by melanocytes contained cellular debris like that shown in figure 7. The outline of each dead or dying melanocyte was clearly visible since the surrounding keratinocytes appeared healthy and normal. There was no obvious increase in melanization due to the high-tyrosine treatment.

Table I. The number of dictyosomes per μ^2 of cytoplasm

Genotype	Control	High-tyrosine	Probability[1]
Standard	0.046 (8/174)	0.047 (10/213)	$p > .5$
Albino-like	0.132 (36/273)	0.046 (11/239)	$p < .001$
Pink-eye	0.098 (45/459)	0.157[2] (37/236)	$.10 > p > .05$

1 The probability that the 2 means are from the same population.
2 Selective melanocyte death also occurs.
The numerator of each parenthesis is the total number of dictyosomes observed, while the denominator is the number of square microns surveyed.

Discussion

The frequent appearance of a well-developed Golgi system in differentiating melanocytes led to the conclusion that the Golgi system was involved in melanogenesis [2]. More convincing proof was provided when electron-opaque deposits were found in the Golgi-associated membrane systems of dopa-reacted melanoma cells [14, 16], fowl melanocytes [3, 5, 13], and human epidermal melanocytes [9]. These findings coupled with the fluctuations in the Golgi system observed in this investigation suggest how the Golgi apparatus is involved in melanogenesis.

Experimental conditions have produced a hypertrophied Golgi system in plant cells [7, 20], amebae [6], mammalian pancreas [8, 11], and thyroid [1]. In most of these cases secretion was impeded. It has been suggested that the Golgi system may be responsible for assembling protein subunits for delivery to other cellular components or for secretion [19]. Experimental manipulations may interfere with the assembly processes causing precursor substances to remain in the dictyosomes producing hypertrophy. On this basis, MORRÉ et al, [15] predicted that certain pigment cell mutants would alter Golgi expression.

Several investigations suggest that premelanosome components and tyrosinase molecules come from separate intracellular sources [9, 12, 13, 14, 18]. Premelanosome components seem to be synthesized in the rough endoplasmic reticulum and partially assembled in dilatations of the smooth endoplasmic reticulum [12] (fig. 8). Tyrosinase molecules are transferred to the Golgi system after synthesis and are then channeled to the premelanosomes probably through a system of coated vesicles and tubules [13, 18] (fig. 8.)

The results of this study concur with the suggested melanogenic scheme described above. Because increased tyrosine levels *reduced* Golgi development in albino-like cells but *increased* its development in pink-eye melanocytes, it is logical to conclude that each mutation affects a different function. Since the mutants also show genetic complementation ($+^c/c, +^{Pk}/pk$ individuals are normally pigmented) they are probably affecting different protein molecules. Albino-like melanocytes are cytochemically dopa negative [unpublished results] so they probably have altered tyrosinase molecules. On the other hand, pink-eye melanocytes are dopa positive [5], suggesting the presence of normal tyrosinase molecules and possibly altered premelanosome components.

The c mutation could reduce the affinity of tyrosinase for tyrosine (fig. 8); the main function of the Golgi system being to form this substrate-enzyme complex. Failure of this complex to form could produce a 'back-up' of unused molecules in the dictyosomes typical of albino-like melanocytes. Increased tyrosine possibly shifts equilibrium in the direction of complex formation reducing the profusion of the Golgi system, allowing premelanosomes to assemble, and melanin deposition to proceed. In the fowl, normal premelanosome structures are apparently not completed until after premelanosome components are joined by tyrosinase molecules.

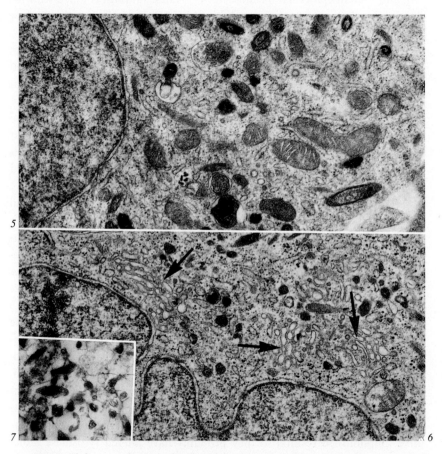

Fig. 5. Pink-eye melanocyte incubated in control medium. Note the presence of unmelanized premelanosomes. × 20,500.

Fig. 6. Pink-eye melanocyte incubated in high-tyrosine medium. Note the hypertrophy of the Golgi system. × 20,500.

Fig. 7. Portion of a dead pink-eye melanocyte incubated in high-tyrosine medium. × 19,000.

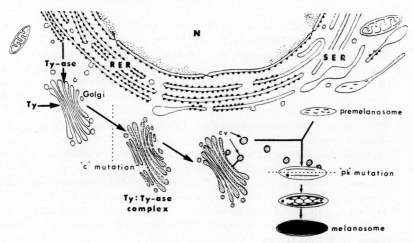

Fig. 8. Proposed schematic of melanogenesis in the fowl with suggested positions of *c* and *pk loci* functions indicated. cv = coated vesicles, RER = rough endoplasmic reticulum, SER = smooth endoplasmic reticulum, Ty-ase = tyrosinase, Tyr = tyrosine.

It also seems plausible to suggest that the *pk* mutation alters premelanosome components so that the function of premelanosome-tyrosinase complexes is impaired (fig. 8). This might also cause a 'back-up' of unused molecules in the Golgi system which would only be made worse by adding excess tyrosine since the genetic block occurs *after* the enzyme-substrate complex is formed. The unused melanin precursors might become toxic to the cells and could explain the cell death observed in our high-tyrosine cultures.

Acknowledgments

This work was supported by NSF Research Grant GB 12429, PHS Research Grant GM 18969 from NIGMS, and by PHS Research Career Development Award GM 42355 from NIGMS to the senior author. The excellent technical assistance of Dr. K.W. LEE is gratefully acknowledged.

References

1 BEAMS, H. and KESSEL, R.: The Golgi apparatus: structure and function. Int. Rev. Cytol. *23:* 209–276 (1968).
2 BRUMBAUGH, J.: Ultrastructural differences between forming eumelanin and pheomelanin as revealed by the pink-eye mutation in the fowl. Develop. Biol. *18:* 375–390 (1968).

3 BRUMBAUGH, J.: The ultrastructural effects of the *I* and *S loci* upon black-red melanin differentiation in the fowl. Develop. Biol. *24:* 392–412 (1971).
4 BRUMBAUGH, J. and HOLLANDER, W.: A further study of the *E* pattern *locus* in the fowl. Iowa State J. Sci. *40:* 51–64 (1965).
5 BRUMBAUGH, J. and ZIEG, R.: The ultrastructural effects of the dopa reaction upon developing retinal and epidermal melanocytes in the fowl. J. Invest. Derm. *54:* 84 (1970).
6 FLICKINGER, C.: Alternations in the Golgi apparatus of amebae in the presence of an inhibitor of protein synthesis. Exp. Cell Res. *68:* 381–387 (1971).
7 HALL, W. and WITKUS, E.: Some effects on the ultrastructure of the root meristem of *Allium cepa* by 6-aza uracil. Exp. Cell Res. *36:* 494–501 (1964).
8 HRUBAN, Z.; SWIFT, H. and WISSLER, R.: Effect of β-3-thienyalalanine on the formation of zymogen granules of exocrine pancreas. J. Ultrastruct. Res. *7:* 359–372 (1962).
9 HUNTER, J.; MOTTAZ, J. and ZELICKSON, A.: Melanogenesis: ultrastructural histochemical observations on ultraviolet irradiated human melanocytes. J. Invest. Derm. *54:* 213–221 (1970).
10 HUTT, F.: Genetics of the fowl (McGraw-Hill, New York 1949).
11 JAMIESON, J. and PALADE, G.: Synthesis, intracellular transport, and discharge of secretory proteins in stimulated pancreatic exocrine cells. J. Cell Biol. *50:* 135–158 (1971).
12 MAUL, G.: Golgi-melanosome relationship in human melanoma *in vitro*. J. ultrastruct. Res. *26:* 163–176 (1969).
13 MAUL, G. and BRUMBAUGH, J.: On the possible function of coated vesicles in melanogenesis of the regenerating fowl feather. J. Cell Biol. *48:* 41–48 (1971).
14 MAUL, G. and ROMSDAHL, M.: Ultrastructural comparison of two human malignant melanoma cell lines. Cancer Res. *30:* 2782–2790 (1970).
15 MORRÉ, D.; MOLLENHAUER, H. and BRACKER, C.: Origin and continuity of Golgi apparatus. Results and problems in cell differentiation, Vol. 2, pp. 82–126 (1971).
16 NOVIKOFF, A.; ALBALA, A. and BIEMPCA, L.: Ultrastructural and cytochemical observations on B-16 and Harding-Passey mouse melanomas. The origin of premelanosomes and compound melanosomes. J. Histochem. Cytochem. *16:* 299–319 (1968).
17 SNEDECOR, G.: Statistical methods; 5th ed. (Iowa State Univ. Press, Ames, Ia. 1956).
18 STANKA, P.: Elektronenmikroskopische Untersuchung über die Pramelanosomenentstehung im retinalen Pigmentepithel von Hühnerembryonen. Z. Zellforsch. *112:* 120–128 (1971).
19 WHALEY, W.: The Golgi apparatus. Biol. Basis Med. *1:* 179–208 (1968).
20 WHALEY, W.; KEPHART, J. and MOLLENHAUER, H.: The dynamics of cytoplasmic membranes during development. Proc. Soc. Study Growth and Develop. *22:* 135–173 (1964).

Author's address: Dr. J. A. BRUMBAUGH, Department of Zoology, University of Nebraska, *Lincoln, NB 68508* (USA)

The Cytochemistry of Normal and Malignant Melanocytes, and Their Relationship to Cells of the Endocrine Polypeptide (APUD) Series

F. W. D. Rost, Julia M. Polak and A. G. E. Pearse

Department of Histochemistry, Royal Postgraduate Medical School, Hammersmith Hospital, London

Introduction

The term 'APUD' has been given to a group of cell types sharing a number of endocrine, cytochemical, and ultrastructural characteristics, and probably of a common embryological origin [34]. These cell types are listed in table I.

The common functional characteristic of APUD cells is the production of a polypeptide hormone, e.g. ACTH, calcitonin. Investigation of a number of polypeptide-producing cell types revealed that these cells shared a number of cytochemical characteristics, listed in table II. A number of other cell types were noted to possess some or all of these cytochemical characteristics, and these were included in the APUD series.

A brief description of the cytochemical properties listed may be necessary. α-Glycerophosphate menadione reductase, commonly known as α-glycerophosphate dehydrogenase (α-GPD) is provisionally regarded as the mitochondrial component of the α-glycerophosphate shuttle. No functional significance can be attached to high levels of histochemically demonstrable activity of α-GPD, non-specific esterases (ns-E) and cholinesterases (Ch-E), although these may all be related to phospholipid synthesis.

5-hydroxytryptamine (5-HT), noradrenaline, adrenaline, dopamine and other arylethylamines can be demonstrated histochemically by condensation reactions with formaldehyde, acetaldehyde, or acetic acid, leading to fluorescent products. A formaldehyde-induced fluorescence (FIF) due to noradrenaline was discovered in the adrenal medulla by Eränkö [16, 17, 18]. Another FIF was reported in enterochromaffin cells [46]; the nature of the

Table I. Cells of the APUD series

a) Known polypeptide-secreting endocrine cells

Tissue	Cell type	Hormone	Amine stored
1. Pituitary		ACTH, MSH	(5-HT)
2. Pancreatic islet	α	Glucagon	
3. Pancreatic islet	β	Insulin	
4. Pancreatic islet	delta	Gastrin	
5. Thyroid, etc.	C	Calcitonin	
6. Stomach	G	Gastrin	

b) Possible polypeptide-secreting cells

Tissue	Cell type	Postulated hormone	Amine stored
7. Stomach	Enterochromaffin (EC)		5-HT
8. Stomach	A	Enteroglucagon	
9. Stomach	D		
10. Stomach	EC-like		
11. Intestine	EC		5-HT
12. Intestine	'Large-granule' (L)	Enteroglucagon	
13. Intestine	'Small-granule' (S)	Secretin	
14. Intestine	X	CK-PZ	
15. Carotid body	Type I	(Glomin)[1]	5-HT
16. Lung	Feyrter	Vasoactive lung peptide	
17. Adrenal medulla	Adrenaline (A)	(Medullarin)[1]	A
18. Adrenal medulla	Noradrenaline (NA)	Neuraleistin	NA
19. Skin, etc.	Melanocyte	(Nigrin)[1]	?

[1] Hypothetical names in parentheses.

Table II. Common cytochemical properties of cells producing polypeptide hormones

1. Arylethylamine content (usually 5-HT)
2. Uptake and decarboxylation of arylethylaminoacids
3. Masked metachromasia
4. Staining with lead haematoxylin
5. High α-GPD (α-glycerophosphate menadione reductase)
6. High non-specific esterase or cholinesterase or both

β-carboline derivative of 5-HT which was produced in freeze-dried sections exposed to formaldehyde vapour was described by BARTER and PEARSE [3, 4]. Further development of the latter technique for the demonstration of FIF led to standard procedures [26, 19] of greater sensitivity. The application of microspectrofluorometry to FIF enabled individual arylethylamines to be identified [13, 28, 41, 6, 7, 9, 8]. Similar reactions are possible using acetaldehyde or acetic acid instead of formaldehyde [42, 20]. The freeze-drying and hot formaldehyde vapour technique (FDFV) can thus be used to demonstrate and identify endogenous 5-HT or another arylethylamine, or to demonstrate the uptake of the corresponding amino acids (5-hydroxytryptophan, dopa) after intravenous or intraperitoneal administration.

The administration of the amino-acid precursors of fluorogenic monoamines, particularly dopa and 5-HTP, gives rise to fluorescence in a number of different cell types after the FDFV technique. Since both the precursor amino acids (dopa, 5-HTP) and the corresponding amines (dopamine, 5-HT) are demonstrated by the FDFV technique, there is normally no direct evidence that a cell shown to contain a fluorogenic amine, after injection of the precursor amino acid, has carried out the decarboxylation itself. However, HÅKANSON and OWMAN [27] have produced strong evidence that decarboxylation does occur in certain cells, which they described as 'enterochromaffin-like cells', situated in the gastric mucosa of the rat and rabbit. It seems probable that similar decarboxylation does take place in other cells [34].

The acronym APUD refers primarily to this process of 'Amine and Precursor Uptake and Decarboxylation', but is applied to refer to the whole series of cells and therefore implies also the other cytochemical and ultrastructural characteristics.

Ability to take up amine precursors is present at an early stage in the embryological development of these cells, and is a valuable property for tracing the migration of these cells in the embryo [36, 40, 37, 38].

Masked Metachromasia

Masked metachromasia is demonstrated by staining with a metachromatic basic dye after mineral acid treatment of suitably fixed tissue [47], and is believed to be due to polypeptides with a high concentration of side-chain acidic groups and a random-coil conformation [12].

Specific Immunofluorescence

As soon as pure hormones become available, anti-sera can be prepared and immunofluorescence techniques can be applied to confirm the presence of the specific hormone within cells of the type responsible for its production. This has already been carried out in respect of some cell types (see table I). In so far as the demonstration of a specific immunofluorescence reaction is merely confirmation of the specific function of the cell, this is not a characteristic as such. Even with 'pure' anti-sera to pure hormones, however, there will be some degree of cross-reactivity with other polypeptides. Thus anti-(human gastrin) has been shown to react with caerulein in the skin of the frogs *Hyla caerulea* [29], and *Hyla crepitans* [39], and with an unidentified polypeptide in malignant melanoma cells [44].

Embryology

The concept of the APUD cell series carried with it the implication that the common cytochemical and ultrastructural characteristics were either significant in terms of function, or of embryological development, or both. It is now believed that these cells have a common embryological origin and have retained, usefully or otherwise, a distinct set of ancestral functions. If this hypothesis is correct, there is only one possible ancestor: the neuro-endocrine cell derived from the neural crest [33]. This cell is regarded as a direct ancestor of melanoblasts [49], and of the adrenal medullary amine-producing cells. APUD cells in embryos have been shown to originate from the neural crest, and to colonize the chick adrenal, and the developing foregut and its derivatives, including pharynx, stomach, duodenum, ultimo-branchial body, and pancreas in mice [37, 38]. Complete proof that these early APUD cells, which demonstrably rise from the neural crest, are the precursors of all the endocrine polypeptide cells of the adult pancreas, stomach, duodenum, and of the small and large intestine, is not at present available. However, the hypothesis seems likely to be true.

Melanocytes

Because of the well-known ability of melanocytes to take up dopa [10], and the known neural crest origin of melanoblasts it seemed logical to

Table III. Melanocyte cytochemistry: summary of results (From Rost et al. [44], with additions)

	Normal melanocytes			Melanomas		Naevus
	man	mouse	frog	man	hamster	man
α-GPD	+	+(strong)	+	+(strong)	+	+
Ch-E	+(weak)	+	+	0	0	variable
NsE	+	+(weak)	+	0	0	0
FIF	+	+(v. weak)	+?	+(usually)	0	+
Uptake:						
dopa *in vitro*		+			+	
dopa *in vitro*		+	+		+	
Masked metachromasia	+	+	+		+	
Lead haematoxylin	+	+	+	+	+	+
Immunofluorescence:						
anti-gastrin	0	0		+(1 case)	0	0

consider whether melanocytes might belong to the APUD series [44]. We therefore investigated normal and neoplastic melanocytes in man, mouse and hamster, and later in the frogs *Hyla crepitans* and *Hyla caerulea*, to find out which, if any, of the characteristics listed in table I were possessed by these cells. Our findings are summarized in table III.

Enzymes

Normal melanocytes in man, mouse and frog were found to have the histochemical enzyme pattern characteristic of APUD cells. High α-GPD levels were noted in malignant melanomas and naevi; this may be due to the fact that their cells are metabolically very active. The absence of histochemically-demonstrable esterase activity may be related to a loss of normal physiological function.

Arylethylamine Content

BARONI [2] applied fluorescence microscopy to the study of 3 melanomas of the skin, and reported a yellow fluorescence in the tumour cells and in cells of the basal layer of the epidermis. Unfortunately he did not report the fixative used, but it was probably formalin. The yellow colour observed by BARONI, rather than the bluish-green reported by later workers, was probably due to the use of a yellow barrier filter.

FALCK et al. [23, 24, 25] studied malignant melanomas, naevi and normal skin using the FDFV method [21], and fluorescence microscopy using a yellow barrier filter. Under these conditions, a green to yellow fluorescence was noted in melanoma cells, in naevus cells showing junctional activity, and in normal melanocytes. This fluorescence appeared to be specific for normal and neoplastic melanocytes. EHINGER et al. [15] found a similar fluorescence in some but not all of 11 ocular malignant melanomas. OLIVECRONA and RORSMAN [32] found an increase in melanocyte fluorescence after Röntgen irradiation. The same authors [31] reported absence of such fluorescence in malignant melanomas in Syrian golden hamsters. ROST and POLAK [43] reported findings similar to those of FALCK et al. [22, 23, 24] in human melanomas, naevi and normal melanocytes: however, melanomas containing a large amount of melanin were non-fluorescent. Fluorescence was also noted in two experimental hamster melanomas [44]. An increase in fluorescence in otherwise normal melanocytes was noted after irritation of the skin with 20-methylcholanthrene [44].

Since dopa is known to be present in melanocytes [22, 24, 25, 48, 14]; it seemed reasonable to presume that this substance was responsible for the fluorescence reaction, partly if not wholly [22, 23, 24, 25]. In addition, a second substance with a fluorescence emission at about 500 nm was found [22]; the fluorophore was stable to hydrochloric acid.

EHINGER et al. [15] performed microspectrofluorometry and found that the fluorescence did not appear to be typical of any catechol substance hitherto investigated. ROST and POLAK [43] carried out microspectrofluorometry of human malignant melanomas, naevi and normal melanocytes, using a Leitz microspectrograph modified for fluorescence giving corrected excitation and emission spectra. The FIF of dopa was found to have excitation/emission peaks at about 420/490 nm, similar to those found for dopamine by BJÖRKLUND et al. [7]. In contrast, all melanocytes gave excitation/emission peaks at about 440/490 nm; the fluorescence was not affected by hydrochloric acid applied according to the technique of BJÖRKLUND et al. [6, 7]. This obviously corresponds to the fluorescence of the second substance found by FALCK et al. [22]. CEGRELL et al. [14], using the microspectrofluorometer of BJÖRKLUND et al. [6], obtained similar results: melanocytes showed excitation/emission peaks at 430/480-520 nm, and dopa 410/480 nm.

The nature of the substance responsible for the FIF is still unknown. From the nature of the Bischler-Napieralski reactions [5], upon which these fluorescence reactions are based [42], it is indicated that the substance responsible is some form of arylethylamine.

Uptake of Amine Precursors

The ability of melanocytes to take up tyrosine and dopa is now so well known as to hardly merit specific documentation. The uptake of dopa is the basis of the dopa-oxidase reaction [10], and uptake has also been demonstrated by autoradiography [11, 30], biochemically by C^{14} label [45] and by the FDFV technique [44].

From the above it may be concluded unequivocably that melanocytes, both normal and malignant, have the ability to take up dopa and to store it; decarboxylation of dopa to dopamine, however, does not appear to have been reported.

Masked Metachromasia

Masked metachromasia has been reported in normal melanocytes in man and mouse [44], and is also found in epidermal melanocytes in the frog *Hyla crepitans* [ROST, POLAK and PEARSE, unpublished; 39].

Specific Immunofluorescence

Following the isolation of a polypeptide, caerulein, from the skin of the tree frog *Hyla caerulea*, it was recognized that the amino-acid sequence of a portion of the polypeptide molecule was identical with part of the human gastrin molecule [1]. The finding that anti-gastrin may bind caerulein [29] prompted us to investigate the possibility of a cross-reaction between antigastrin and a presumptive polypeptide in melanocytes. No reaction could be demonstrated in normal melanocytes, but one malignant melanoma gave a strongly positive reaction [44]. It was thought that this might represent a polypeptide not normally produced by melanocytes, or not normally accessible to immunofluorescent reactions.

It has been established that melanocytes in several species possess many of the characteristics common to APUD cells: histochemically demonstrable α-GPD, nsE, and ChE; uptake of dopa; and masked metachromasia. The known origin of melanoblasts from the neural crest is also in accordance with the overall concept.

The Significance of the APUD Concept, in Relation to Melanocytes

Probably the greatest significance of the APUD concept in relation to melanocytes is to be found in diagnostic pathology. The histochemical characteristics of APUD cells can be used in the differential diagnosis of malignant melanomata, provided that suitably fixed tissue is available. Examination for formaldehyde-induced fluorescence, although best carried out on FDFV treated material, is possible on formalin-fixed material. Lead haematoxylin staining is a simple technique, requiring no special apparatus.

References

1 Anastasi,A.; Erspamer,V. and Endean, R.: Isolation and structure of caerulein, an active decapeptide from the skin of *Hyla caerulea*. Experientia *23:* 699–700 (1967).
2 Baroni,B.: Contributo allo studio dei melanomi cutanei al lume di un moderno mezzo d'indagine: del microscopio a fluorescenza. Arch. ital. Derm. *9:* 543–586 (1933).
3 Barter, R. and Pearse, A. G. E.: Detection of 5-hydroxytryptamine in mammalian enterochromaffin cells. Nature, Lond. *172:* 810 (1953).
4 Barter, R. and Pearse, A. G. E.: Mammalian enterochromaffin cells as the source of seritonin (5-hydroxytryptamine). J. Path. Bact. *65:* 25–31 (1955).
5 Bischler, A. and Napieralski, B.: Zur Kenntnis einer neuen Isochinolinsynthese. Ber. Dtsch. chem. Ges. *26:* 1903–1908 (1893).
6 Björklund, A.; Ehinger, B. and Falck, B.: A method for differentiating dopamine from noradrenaline in tissue sections by microspectrofluorometry. J. Histochem. Cytochem. *16:* 263–270 (1968).
7 Björklund, A.; Falck, B. and Håkanson, R.: Histochemical demonstration of tryptamine. Properties of the formaldehyde-induced fluorophores of tryptamine and related indole compounds in models. Acta physiol. scand. Suppl. *318* (1968).
8 Björklund, A.; Nobin, A. and Stenevi, U.: Acid catalysis of the formaldehyde condensation reaction for a sensitive histochemical demonstration of tryptamines and 3-methoxylated phenylethylamines. 2. Characteristics of amine fluorophores and application to tissues. J. Histochem. Cytochem. *19:* 286–298 (1971).
9 Björklund, A. and Stenevi, U.: Acid catalysis of the formaldehyde condensation reaction for sensitive histochemical demonstration of tryptamines and 3-methoxylated phenylethylamines. 1. Model experiments. J. Histochem. Cytochem. *18:* 794–802 (1970).

10 BLOCH, B.: Das Problem der Pigmentbildung in der Haut. Arch. Derm. Syph., Berl. *124:* 129–208 (1917–1918).
11 BLOIS, M. S. and KALLMAN, R. F.: Incorporation of C^{14} from 3,4-dihydroxyphenyl-alanine-2'-C^{14} into the melanin of mouse melanomas. Cancer Res. *24:* 863–868 (1964).
12 BUSSOLATI, G.; ROST, F.W.D. and PEARSE, A.G.E.: Fluorescence metachromasia in polypeptide hormone-producing cells of the APUD series, and its significance in relation to the structure of the precursor protein. Histochem. J. *1:* 517–530 (1969).
13 CASPERSSON, T.; HILLARP, N.-Å. and RITZÉN, M.: Fluorescence microspectrophotometry of cellular catecholamines and 5-hydroxytryptamine. Exp. Cell Res. *42:* 415–428 (1966).
14 CEGRELL, L.; FALCK, B. and ROSENGREN, A.M.: Extraction of dopa from the integument of pigmented animals. Acta physiol. scand. *78:* 76–69 (1970).
15 EHINGER, B.; FALCK, B.; JACOBSSON, S. and RORSMAN, H.: Formaldehyde induced fluorescence of intranuclear bodies in melanoma cells. Brit. J. Derm. *81:* 115–118 (1969).
16 ERÄNKÖ, O.: Histochemical evidence of the presence of acid-phosphatase-positive and -negative cell islets in the adrenal medulla of the rat. Nature, Lond. *168:* 250–251 (1951).
17 ERÄNKÖ, O.: On the histochemistry of the rat adrenal medulla. Acta physiol. scand. *25:* Suppl. 89: 22–23 (1951).
18 ERÄNKÖ, O.: Distribution of adrenaline and noradrenaline in the adrenal medulla. Nature, Lond. *175:* 88–89 (1955).
19 ERÄNKÖ, O.: The practical histochemical demonstration of catecholamines by formaldehyde-induced fluorescence. J. roy. micr. Soc. *87:* 259–276 (1967).
20 EWEN, S.W.B. and ROST, F.W.D.: New methods for the fluorescence histochemical demonstration of catecholamines and tryptamines microspectrofluorimetric characterization of the fluorophores in models. Histochem. J. *4:* 59–69 (1972).
21 FALCK, B.; HILLARP, N.-Å.; THIEME, G. and TORP, A.: Fluorescence of catecholamines and related compounds condensed with formaldehyde. J. Histochem. Cytochem. *10:* 348–354 (1962).
22 FALCK, B.; JACOBSSON, S.; OLIVECRONA, H.; ROSENGREN, A.M. and ROSENGREN, E.: On the occurrence of catechol derivatives in malignant melanomas. Comm. Dept. Anatomy, Univ. of Lund, Sweden, No. 5 (1966).
23 FALCK, B.; JACOBSSON, S.; OLIVECRONA, H. and RORSMAN, H.: Pigmented nevi and malignant melanomas as studied with a specific fluorescence method. Science *149:* 439–440 (1965).
24 FALCK, B.; JACOBSSON, S.; OLIVECRONA, H. and RORSMAN, H.: Specific fluorescence in pigmented nevi and malignant melanomas. The Swedish Cancer Soc. Yearbook 1963-5, No. 4, pp. 95–98 (1966).
25 FALCK, B.; JACOBSSON, S.; OLIVECRONA, H. and RORSMAN, H.: Fluorescent dopa reaction of naevi and melanomas. Arch. Derm. *94:* 363–369 (1966).
26 FALCK, B. and OWMAN, C.: A detailed methodological description of the fluorescence method for the cellular demonstration of biogenic amines. Acta univ. Lund. Sect. II, No. 7 (1965).
27 HÅKANSON, R. and OWMAN, C.: Distribution and properties of amino acid decarboxylases in gastric mucosa. Biochem. Pharmacol. *15:* 489–499 (1966).

28 JONSSON, G. and RITZÉN, M.: Microspectrofluorometric identification of metaraminol in sympathetic adrenergic neurones. Acta physiol. scand. *67:* 505–513 (1966).
29 MCGUIGAN, J. E.: Binding of caerulein by antibodies to human gastrin. I. Gastroenterology *56:* 858–861 (1969).
30 MODEL, P. G. and DALTON, H. C.: The uptake and localization of radioactive DOPA by amphibian melanoblasts *in vitro*. Develop. Biol. *17:* 245–271 (1968).
31 OLIVECRONA, H. and RORSMAN, H.: Fluorescence microscopy of malignant melanomas in the Syrian golden hamster. Acta derm. venereol., Stockh. *46:* 401–402 (1966).
32 OLIVECRONA, H. and RORSMAN, H.: The effect of Roentgen irradiation on the specific fluorescence of epidermal melanocytes. Acta derm. venereol., Stockh. *46:* 403–405 (1966).
33 PEARSE, A. G. E.: Common cytochemical properties of cells producing polypeptide hormones, with particular reference to calcitonin and the C cells. Vet. Rec. *79:* 587–590 (1966).
34 PEARSE, A. G. E.: Common cytochemical and ultrastructural characteristics of cells producing polypeptide hormones (The APUD Series) and their relevance to thyroid and ultimobranchial C cells and calcitonin. Proc. roy. Soc. B. *170:* 71–80 (1968).
35 PEARSE, A. G. E.: Histochemistry. Theoretical and applied, 3rd ed., vol. 1 (Churchill, London 1968).
36 PEARSE, A. G. E. and CARVALHEIRA, A. F.: Cytochemical evidence for an ultimobranchial origin of rodent thyroid C cells. Nature, Lond. *214:* 929–930 (1967).
37 PEARSE, A. G. E. and POLAK, Julia M.: Cytochemical evidence for the neural crest origin of mammalian ultimobranchial C cells. Histochemie *27:* 96–102 (1971).
38 PEARSE, A. G. E. and POLAK, Julia M.: Neural crest origin of the endocrine polypeptide (APUD) cells of the gastrointestinal tract and pancreas. Gut *12:* 783–788 (1971).
39 POLAK, Julia M. and PEARSE, A. G. E.: Anti-gastrin immunofluorescence in the skin of *Hyla crepitans* and the cytochemistry of the cells involved. Experientia *26:* 288–289 (1970).
40 POLAK, Julia M.; ROST, F. W. D. and PEARSE, A. G. E.: Fluorogenic amine tracing of neural crest derivatives forming the adrenal medulla. Gener. comp. Endocr. *16:* 132–136 (1971).
41 RITZÉN, M.: Cytochemical identification and quantitation of biogenic amines; M. D. thesis, Stockholm (1967).
42 ROST, F. W. D. and EWEN, S. W. B.: New methods for the histochemical demonstration of catecholamines, tryptamines, histamine and other arylethylamines by acid- and aldehyde-induced fluorescence. Histochem. J. *3:* 207–212 (1970).
43 ROST, F. W. D. and POLAK, Julia M.: Fluorescence microscopy and microspectrofluorimetry of malignant melanomas, naevi and normal melanocytes. Virchows Arch. path. Anat. *347:* 321–326 (1969).
44 ROST, F. W. D.; POLAK, Julia M. and PEARSE, A. G. E.: The melanocyte: its cytochemical and immunological relationship to cells of the endocrine polypeptide (APUD) series. Virchows Arch. Abt. B Zellpath. *4:* 93–101 (1969).
45 SEIJI, MAKOTO: Subcellular particles and melanin formation in melanocytes; in W. MONTAGNA and FUNAN HU, Advances in biology of skin, vol. 8: The pigmentary system, pp. 189–222 (Pergamon Press, Oxford 1966).

46 SHEPHERD, D. M.; WEST, G. B. and ERSPAMER, V.: Detection of 5-hydroxytryptamine by paper chromatography. Nature, Lond., *172:* 357 (1953).
47 SOLCIA, E.; VASSALLO, G. and CAPELLA, C.: Selective staining of endocrine cells by basic dyes after acid hydrolysis. Stain Technol. *43:* 257–263 (1968).
48 TAKAHASHI, H. and FITZPATRICK, T. B.: Large amounts of dioxyphenylalanine in the hydrolysate of melanosomes from Harding-Passey mouse melanoma. Nature, Lond. *209:* 888–890 (1966).
49 TEILLET, Marie-Aimée et LE DOUARIN, Nicole.: La migration des cellules pigmentaires étudiée par la méthode des greffes hétérospécifiques de tube nerveux chez l'embryon d'Oiseau. C. R. Acad. Sci., Sér. D *270:* 3095–3098 (1970).

Author's address: Dr. F. W. D. ROST, Department of Histochemistry, Royal Postgraduate Medical School, Hammersmith Hospital, *London W12 OHS* (England)

Skin Color:
Its Ultrastructure and Its Determining Mechanism

K. Toda, M. A. Pathak, T. B. Fitzpatrick, W. C. Quevedo, Jr.,
F. Morikawa and Y. Nakayama

Department of Dermatology, Tokyo Teishin Hospital, Tokyo, Department of Dermatology, Harvard Medical School, Massachusetts General Hospital, Boston, Mass., Division of Biological and Medical Sciences, Brown University, Providence, R.I., Shiseido Chemical Research Laboratory, Yokohama

Introduction

The most obvious difference between the various human races is the variation in the color of the skin. The factors determining the skin color of normal skin include: a) reflection co-efficient of skin surface; b) absorption co-efficient of epidermal-cell and dermal-cell constituents; c) scattering co-efficient of various cell layers; d) thickness of the individual cell layers (stratum corneum, epidermis, and dermis); e) biochemical content of ultraviolet- and visible-light-absorbing components, such as proteins (keratins, elastin, collagen, and lipoproteins), melanin in melanosomes, nucleic acids, urocanic acid, carotenoids, hemoglobin (reduced and oxidized), and lipids; f) number and spatial arrangement of melanosomes, melanocytes, and blood vessels; and g) the relative quantity of blood and red cells flowing through the vessels.

Pigmentation of the skin, as viewed clinically, is largely related to the content of melanin in the keratinocytes. According to present concepts, the normal skin color is primarily influenced by the number, size, type, and distribution pattern of melanosomes in the epidermis [3–5]. Careful studies of human skin, particularly of the unexposed regions of the body, have revealed that there is no significant difference between the number of melanocytes in various racial groups, although there are definite regional differences in the population density of dopa-positive melanocytes in various areas of the body. Recent data based on light and electron microscopic observation [4, 5, 9] indicate that the color of the skin is due to variations in the number of melanin granules or melanosomes in the epidermal melanocytes and epidermal keratinocytes, as well as to melanosomes. In Caucasoids, American Indians, and Mongoloids, usually groups of two or more melanosomes are found in the epidermal keratinocytes, and such aggregates of melanosomes are surrounded by a unit membrane. In Negroids and Australian Aborigines, usually non-aggregated (single) melanosomes are found in the epidermal keratinocytes. Until just recently, it was believed that this normal racial distribution of melanosomes in the non-follicular keratinocytes was controlled by genetic factors and was not affected by non-genetic factors, such as ultraviolet-light irradiation [9]. Our recent data, however, revealed that this distribution pattern of melanosomes in the epidermal keratinocytes of various races can be affected by non-genetic factors, in as much as the distribution of melanosomes in non-follicular keratinocytes can be changed by exposure to ultraviolet light [10, 11].

In this paper, we are concerned with: a) some additional pigmentary factors that determine the skin color; b) pigmentary changes that occur and influence the human skin color after topical application of 4,5′,8-trimethyl psoralen (TMP) and exposure to long-wave ultraviolet radiation (UVL; 320–400 nm); and c) changes in acid-phosphatase activity in guinea-pig skin after application of TMP and exposure to UVL or after the horny cells (stratum corneum) are stripped off with an adhesive tape ('Scotch').

Skin Color and Melanosomes

Melanin Pigment in the Epidermis

Investigations concerning the relationship between the visible skin color and the melanosome number in the keratinocytes were carried out by light and electron microscopic observations of the unexposed buttock skin of fourteen American Negroids. These subjects represented a varying pattern of skin coloration ranging from very heavily pigmented to moderately pigmented. Buttock-skin reflectance values ranged from 8 to 27% when measured against 100% reflectivity of pure white magnesium oxide. Based on visual observations of skin color and reflectance values, three 'skin color classes' were arbitrarily assigned: Class 1 represented very-dark or heavily-pigmented skin with reflectance values ranging from 8 to 10%; Class 2 represented dark to medium-dark skin with skin reflectance values of 12 to 16%; Class 3 represented medium-light skin with reflectance values of 22–27%.

Table I. Skin color and number of melanocytes in 14 subjects with Negroid skin

Code	Class	Visual grade	Reflectance	Melanocytes/mm^2
S5	1	very dark	8	1216
W7	1	very dark	8	–[1]
R7	1	very dark	9	–[1]
M1	1	very dark	10	1498.4
P9	2	medium dark	12	1173.5
N3	2	dark	13	1118.0
Mc4	2	dark	13.5	1084.2
S8	2	medium dark	15	1225.8
W8	2	dark	15	–[1]
R9	2	medium	16	1513.8
T6	2	medium	16	1215.1
J2	3	medium	22	1135.0
T56	3	medium-light	24	1027.2
M9	3	medium-light	26	1175.0

1 Could not be counted.

In these three skin color classes, the population of melanocytes per square millimeter of skin surface did not vary according to the age or color of the skin and appeared to be equal to the population of melanocytes of Caucasoid buttock skin (table I).

The number of melanosomes per basal cell (the mean value of melanosomes obtained by a count of the number of melanosomes in 500 basal cells) varied in different classes and correlated well with the reflectance value and the visual color grade (fig. 1).

The number of melanosomes per malpighian cell also showed good correlation with visual skin color grade. Interestingly enough, the number of melanosomes per cell from the lowest layer of the malpighian cells to the highest layer did not show much variation. The diminution in the number of melanosomes from basal cells to malpighian cells is related to the degradation of melanosomes and appears to take place probably in the basal layer (fig. 2).

The number of melanosomes in the horny layer did not show good correlation with skin reflectance values. Likewise, the count of pigment granules in the horny layer obtained after skin-stripping by the adhesive tape in these fourteen subjects also showed no satisfactory correlation with the visual skin color grade. Although skin reflectivity is inversely related to the pigment content of the horny cells, the skin reflectance value is evidently more dependent on the pigment content and the number of the melanosomes in the lower layer of the epidermis.

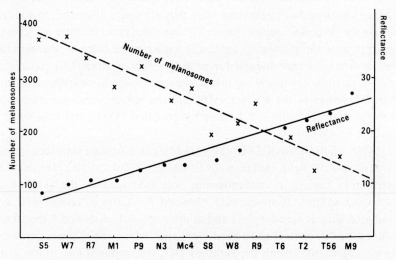

Fig. 1. Shows relationship between the number of melanosomes per basal cell nad skin reflectance values in 14 Negroid subjects.

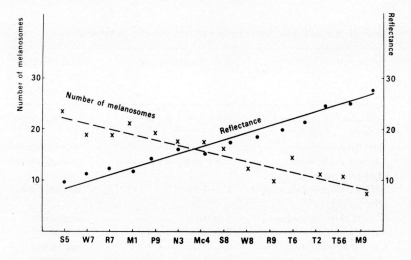

Fig. 2. Relationship between the number of melanosomes per malpighian cell and skin reflectance values in 14 Negroid subjects.

The Distribution Pattern of Melanosomes

Another determining factor in the skin color is the distribution pattern of melanosomes. Melanosomes that are aggregated in groups of two or more (e. g. as in Caucasoid skin) will contribute less to the scattering of the impinging wavelengths of light than when they are singly dispersed. In addition, the singly dispersed melanosomes in keratinocytes, because of effective absorption of the impinging light, will render more color to the skin than they do when they are dispersed in groups of two or more. This concept can be elaborated by the following two studies involving the distribution pattern of melanosomes in the keratinocytes and the reflection spectrophotometry of the skin treated with 4,5',8-trimethyl psoralen (TMP) and long-wave ultraviolet light.

After application of TMP (50 μg/3 cm^2) and a single exposure to long-wave ultraviolet-light radiation (320–400 nm, 19.4 × 10^7 ergs/cm^2) the distribution pattern of melanosomes was examined in the skin of one Caucasoid subject. Biopsies were obtained from the irradiated and non-irradiated sites at 4 and 7 days and at intervals of 1, 3, 6, and 9 months. In contrast to the control (non-irradiated) sites where the melanosomes were always aggregated within the keratinocytes, the melanosomes in the hyperpigmented, irradiated sites were non-aggregated. There was also an increase in the average size of the melanosomes (table II). The switch in the distri-

Table II. Size and distribution pattern of melanosomes in keratinocytes

Specimens	Size of melanosome	Distribution of melanosomes in keratinocytes
Follicular keratinocytes from all races	1.1–1.3 × 0.5–0.7 μ	non-aggregated
Dark Negroid skin	1.0–1.3 × 0.5–0.6 μ	non-aggregated
Psoriasis lesion from Mongoloid skin	1.0–1.2 × 0.5–0.6 μ	non-aggregated
TMP+UV-treated Caucasoid skin	1.1 × 0.5 μ	non-aggregated
Control skin (unirradiated)	0.6 × 0.3 μ	aggregated
Stripped Mongoloid skin	0.6–0.7 μ × 0.3–0.4 μ	aggregated
Caucasoid skin (unexposed)	0.6–0.7 μ × 0.3–0.4 μ	aggregated
Mongoloid skin (unexposed)	0.7 × 0.3 μ	aggregated
Negroid skin (unexposed)	1.0–1.3 × 0.5–0.6 μ	non-aggregated

bution pattern of melanosomes from aggregated to non-aggregated explains the long duration of hyperpigmentation induced by treatment with TMP and UVL. One might suggest that the differences in the turnover rates of keratinocytes of the non-irradiated and TMP-plus-UVL-irradiated sites might be responsible for the non-aggregated and aggregated patterns of melanosomal distribution. This possibility appears to be ruled out by the fact that stripping of the horny cells by the adhesive tape did not result in the alteration of the melanosomal pattern. It is well-known that stripping causes rapid turnover of keratinocytes and also results in hyperpigmentation of the skin. The pattern of melanosomal distribution in the stripped skin of Caucasoids and Mongoloids, however, did not change, and the melanosomes remained aggregated (table II).

Reflection Spectrophotometry and Skin Color

BUCKLEY and GRUM [1] analyzed variations of human skin color by reflection spectrophotometry. We also measured changes in skin color by reflection spectrophotometry after application of TMP and ultraviolet-light irradiation. A recording spectrophotometer with a diffuse reflectance sphere was used to measure skin reflection in the visible spectrum of light. Three hours after exposure, a marked change was observed in the skin reflection. This initial change correlated with the immediate pigment-darkening reaction (an oxidation reaction in melanin) that is well-known to occur when skin is exposed to ultraviolet and visible radiation. Within 24 h, when the initial pigment darkening reaction usually subsides, the reflection value in the long-wave visible spectrum (600-700 nm) did not change and was similar to the control (unirradiated skin) value. This indicated no new melanin formation in the skin within 24 h. The reflection value of the skin in the short-wave visible spectrum (400–600 nm), however, was markedly low (a decrease of about 8 %) indicating the onset of the erythema reaction that usually occurs in skin that is photosensitized by ultraviolet light plus TMP. Four days after the exposure (fig. 3), the percentage difference in the skin reflection in the range of 400–600 nm was the highest, indicating a persistence of marked vasodilatation of blood vessels due to the erythema reaction. One week after initial exposure, the reflection value in the short-wave visible-light spectrum (400–600 nm) tended to return to the normal levels of the control sites, indicating subsidence of the erythema reaction. On the other hand, four days after exposure, the reflection value in the spectrum ranging from 600–700 nm was significantly lower than the control value. This indicated new melanin formation in the skin. The skin reflection value in the region of 600–700 nm continued to remain markedly lower than the control value for three weeks after exposure. Hence, the degree of skin color change due to the formation of new melanin could be evaluated. In order to differentiate the erythema

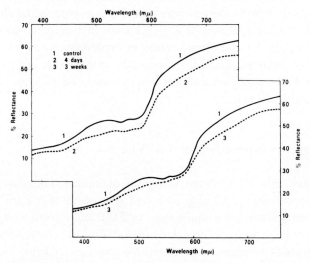

Fig. 3. Changes in skin reflectance of a Caucasoid subject after topical application of 4,5′,8-trimethylpsoralen and irradiation with long-wave ultraviolet light. The skin reflection curves were obtained from the control (unirradiated) and the irradiated sites on 4th and 20th day after exposure.

Table III. Skin color variation after TMP and UV[1]

	X	Y	Z	S	H′	V′	C′
Cont.	0.2999	0.2787	0.2038	0.7824	−2.6	58.8	30.9
3 h	0.2854	0.2667	0.1966	0.7487	−2.8	57.7	29.2
1 day	0.2797	0.2555	0.1865	0.7216	−1.0	56.6	32.2
3 days	0.2668	0.2406	0.1832	0.6906	0.77	55.2	31.8
4 days	0.2652	0.2406	0.1740	0.6798	0.55	55.2	32.6
1 week	0.2595	0.2369	0.1832	0.6796	0	54.8	29.5
3 weeks	0.2837	0.2635	0.1891	0.7362	2.77	57.4	30.8

1 X, Y, and Z, respectively, represent trichromatic co-ordinates of red, green and blue values obtained by measuring skin reflection on a recording spectrophotometer. S represents the sum of X, Y, and Z values. For an absolute white (100 % reflectance) standard X = 0.98, Y = 1.0, and Z = 1.18. H′, LV′, and C′ values, respectively, represent visual color notation of skin in terms of three attributes- hue, value, and chroma.

Skin Color: Its Ultrastructure and Its Determining Mechanism

component against the pigment component, however, these skin reflection changes were additionally subjected to evaluation by the determination of the color trichromatic co-efficients after computing the x, y, and z values of chromaticity co-ordinates. The data are presented in table III and figures 4 and

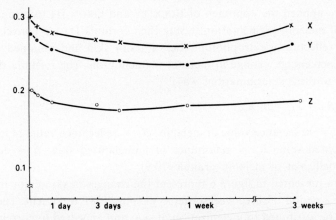

Fig. 4. Showing variation in skin color of a Caucasoid subject after topical application of TMP and UV irradiation. X, Y, and Z, respectively, represent trichromatic co-ordinates of red, green, and blue values obtained by measuring skin reflection on a recording spectrophotometer.

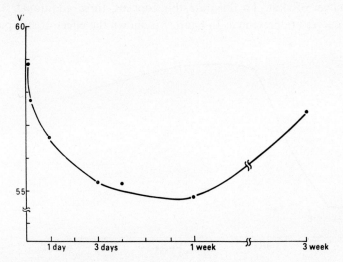

Fig. 5. This figure shows a plot of v′, the value notation that indicates the degree of darkness of visual color in relation to a neutral gray scale. The value symbol 0 is used for absolute black, and 100 for absolute white.

5. Despite careful mathematical treatment of trichromatic values, we were, however, unable to differentiate between the skin color changes due to melanin pigmentation and those due to erythema reaction, both of which usually occur together when skin is exposed to sunburn-producing radiation (290–320 nm) or to ultraviolet light (320–400 nm) in the presence of TMP. We therefore applied the approach of BUCKLEY and GRUM [1] to estimate the degree of melanin pigmentation. Using their formula [1] we converted the 650-nm reflection spectrophotometric values of skin and obtained the measure of percentage change in melanin pigmentation. The formula that was applied to these determinations was:

$$R_p = bR_m + (1 - b) R_y$$

in which: R_m = reflectance value of melanin; R_p = reflectance value of normally pigmented skin; R_y = reflectance of unmelanized skin; b = concentration coefficient of melanin granules (0.9).

The data presented in figure 6 represent the changes in skin color that occurred for a period of three weeks after UVL-plus-TMP treatment. These changes can be attributed to melanization of skin and appeared to correlate satisfactorily with the visual skin color grade and ultrastructural findings that were presented earlier. Hence, skin reflection data obtained by reflection spectrophotometry can be effectively used in estimating the degree of color change of skin. To illustrate this concept, three additional skin reflectance curves are presented. In figure 7 is shown the effect of stripping

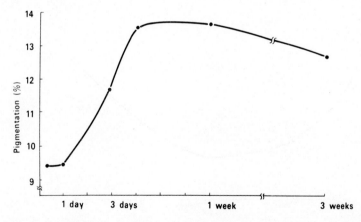

Fig. 6. Shows percent change in skin color for a period of three weeks after topical application of TMP and irradiation with ultraviolet light. This plot of percent change in melanin pigmentation was obtained using the formula of BUCKLEY and GRUM [1].

of the horny-cell layer by adhesive tape. At 24 h after stripping, no difference in skin reflection value at 650 nm was observed, although there was a noticeable change at 400 nm and also at 500–575 nm. These changes were related to the vasodilatation and erythema reaction. There was no increase in melanin

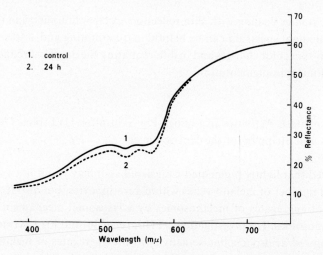

Fig. 7. Shows changes in skin color at 24 h after stripping of the horny-cell layers by adhesivet ape. Changes in melanin pigmentation are minimum.

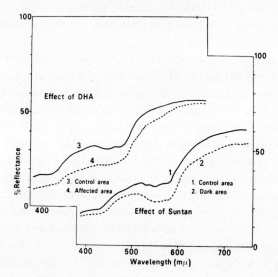

Fig. 8. Showing pigmentary changes and erythema reaction at 120 h in skin after sun exposure and after repeated application of dihydroxyacetone solution.

pigmentation of the skin. Contrary to this, the degree of pigmentation and erythema reaction in skin after exposure at 96–120 h (fig. 8), or after repeated application of dihydroxyacetone (DHA) that also causes darkening of the skin, can be assessed only when the data are analyzed by the just-mentioned formula.

Therefore, studies dealing with skin reflection spectrophotometry in the long-wave visible-light spectrum can be helpful in quantitating and assessing the variations in skin color changes and in differentiating the degree of redness (erythema) and melanin pigmentation.

Acid-Phosphatase Activities in Guinea-Pig Skin after UVL-plus-TMP Treatment and Stripping of the Horny Layer

It is believed that lightly pigmented Caucasoid skin may result not only from a reduced number of melanosomes within keratinocytes, but also from the formation of aggregates of melanosomes by a lysosomal mechanism in the form of phagosomes. The investigations of HORI et al. [6] and FITZPATRICK et al. [5, 11] showed acid-phosphatase activity in the aggregates of melanosomes, or 'melanosome complexes'. The degradation of melanosomes as a basis for the differences in and the control of melanin pigmentation was examined by the determination of acid-phosphatase activity in pigmented guinea-pig skin after treatment with TMP plus UVL and also after stripping of the horny layer.

The biochemical assay of acid-phosphatase activity was carried out as described by VALENTINE and BECK [13]. Homogenates of guinea-pig dorsal skin (5 g) were prepared identically on an equal wet-weight basis: the specimens were taken a) after TMP application and UV irradiation; b) after stripping of the horny-cell layers, and c) from non-irradiated controls. Skin homogenates were incubated with sodium β-glycerophosphate solution at pH 5.0. The hydrolytic reaction was terminated after 60 min of incubation. Phosphorus was determined by the method of FISKE and SUBBAROW [2].

The acid-phosphatase activity in the homogenates of guinea-pig skin after TMP application and UVL irradiation was found to undergo a marked change (fig. 9). After the initial drop of activity within the first 6–8 h, the acid-phosphatase activity increased, reaching the highest levels on the second day after irradiation. By 72 h, the acid-phosphatase activity started declining, and, on the seventh day after exposure, an additional rise in activity was observed. The acid-phosphatase activity in the homogenates of skin obtained after removal of the horny layer did not, however, show marked change. Increment in activity was observed only at 20–24 h after stripping.

The changes in acid-phosphatase activity are probably related to two processes: a) the erythema reaction and b) the degradation reaction of melanosomes by hydrolytic enzymes of the lysosomes.

A progressive increase in the fragility of lysosomes, manifested by the release of hydrolytic enzymes such as acid phosphatase, has been observed by JOHNSON et al. [7] in skin after the sunburn reaction was induced.

The investigations of HORI et al. [6] and TODA and FITZPATRICK [11] showed acid-phosphatase activity in the aggregates of melanosomes (melanosome complexes) and it is therefore believed that melanosome complexes are similar to lysosomes. This similarity is suggested by a) the presence of acid-phosphatase activity and b) observations that melanosomes within the melanosome complexes undergo degradation.

It is perhaps an oversimplification to regard degradation of melanosomes as a basis for the variation in pigmentation between Negroids and Australian Aborigines and people of other races. Our observations, however, indicate that the melanosome in the keratinocyte, whether disposed as a single melanosome or as a melanosome complex, shows acid-phosphatase activity and has autolytic or heterolytic function. Variations in the acid-phosphatase activity in melanosome complexes after irradiation or stripping appear to have some significance in regard to the fate of melanosomes after their discharge from melanocytes.

The Variation in the Degree of Melanization of Melanosomes

TODA and FITZPATRICK [11] outlined four stages in the development of the melanosomes. Stage I was characterized as a spherical membrane-limited vesicle that contained tyrosinase and, at times, some filaments, but no melanin. In stage II, the organelle is oval and shows numerous membranous filaments, with or without cross linkings, having distinct periodicity. In stage III, the internal structure of the stage II melanosome becomes partly obscured by the deposition of melanin. In stage IV, the oval organelle is electron-opaque, without discernible internal structure.

The variation in the degree of melanization of melanosomes also appears to be a determining factor in the gradation of the visual color of skin. Simple morphologic examination of melanosomes in various stages of development reveals three distinct grades of melanization (fig. 10). In grade I, the melanosome appears as it does in stage II and III of its development, but is not heavily laden with melanin. In grade II, the melanosome is as it is in stage IV of the development, but has no discernible translucent bodies. In grade III, the melanosome is also as it is in stage IV of development, but has translucent bodies that can be distinctly recognized.

The concept that the variation in the degree of melanization of melanosomes is also responsible for the determination of color intensity of skin, can be exemplified when the skin is studied. In our experiment, the unexposed skin of two normal Caucasoids was carefully examined by electron microscopy and serial sectioning. The number of melanosomes in the epidermis

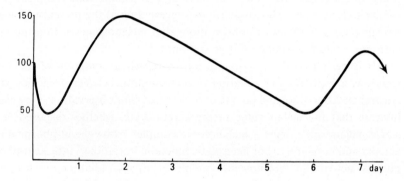

Fig. 9. Acid-phosphatase activity in guinea-pig skin after UV irradiation.

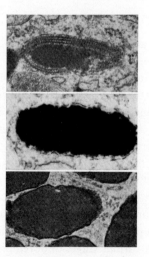

Fig. 10. Grade of melanization.

(in the basal cells, as well as in the malpighian cells) of these two individuals was almost identical, and the size of the melanosomes, which were aggregated in each specimen, was also similar in each individual, but the skin reflectance values were 26 and 35%. Electron micrographs of the skin biopsies of these two individuals, however, revealed differences in the morphologic appearance of the melanosomes. Melanosomes in the melanocytes and in the keratinocytes of the darker subject (skin reflectance 26%) had several translucent bodies (fig. 10), whereas the lighter subject (skin reflectance 35%) did not reveal any translucent bodies.

The melanosome is the carrier of melanin from the melanocytes to the keratinocytes. It is therefore apparent that the variation in the melanin content of the melanosomes can be a determining factor in the visual gradation of skin color.

Conclusions

Melanin pigmentation of the skin in *man* results from the close interaction between the epidermal melanocytes that synthesize melanosomes and the keratinocytes that acquire the melanosomes and serve in their transport towards the horny cells of stratum corneum [3]. The epidermal melanocyte–keratinocyte pool, known as the 'epidermal melanin unit', serves as a structural and functional unit for the understanding of the variations in skin color [3]. Each epidermal melanin unit consists of an epidermal melanocyte and the constellation of keratinocytes with which it maintains functional contact. The number of active epidermal melanin units can vary markedly between different regional sites in the same individual (due to regional variation in melanocyte population), but the ratio of keratinocytes to melanocytes within epidermal melanin units, however, remains constant [4].

Visible pigmentation of the skin in this epidermal melanin unit is governed by four biological processes: a) production or synthesis of melanosomes in melanocytes; b) melanization of melanosomes in melanocytes and possibly, within keratinocytes; c) transfer of melanosomes within lysosome-like organelles (melanosome complexes) in keratinocytes; d) degradation of melanosomes within lysosome-like organelles in keratinocytes.

Thus, variations in color are related to the number and functional activity of 'epidermal melanin units' in the skin. The number and functional activity of these epidermal melanin units are primarily under genetic control. Data presented in this study, however, also reveal that non-genetic factors, i.e., environmental factors such as ultraviolet irradiation, can also influence the functional activity of the epidermal melanin unit.

The functional activity of this epidermal melanin unit and thus the intensity of skin color can be influenced in the following four ways: a) the rate of melanogenesis within the melanocytes or the degree of melanization of the melanosomes in the melanocytes; b) the total number of melanosomes present within the keratinocytes and particularly in the basal-cell layer; c) the size of the melanosome synthesized; d) the distribution pattern of melanosomes in the epidermis.

In this brief presentation, the emphasis is placed on the non-genetic, i.e., environmental factors that can influence skin color. This variation in skin color does not occur at the structural level, but principally at the functional level of the epidermal melanin unit and is related to: a) the variation in the population density of secretory melanocytes; b) differences in the number of melanosomes in the epidermis, especially in the basal-cell layer; c) differences in the degree of melanization of the melanosomes; d) differences in the localization of melanosomes in the basal cells, malpighian cells, and horny cells; e) differences in the size of melanosomes in the epidermis (either in the aggregated or in the non-aggregated form); f) differences in the degradation of melanosomes due to variations in hydrolytic activity (e.g., acid phosphatase).

Skin reflection spectrophotometry can be helpful in quantitating color variation and differentiating between the erythema component and the melanin component of skin.

References

1 Buckley, W.R. and Grum, F.: Reflection spectrophotometry. Arch. Derm. *83:* 249–261 (1961).
2 Fiske, C.H. and SubbaRow, Y.: The colorimetric determination of phosphorus. J. Biol. Chem. *66:* 375 (1925); *81:* 629 (1929).
3 Fitzpatrick, T.B. und Breathnach, A.S.: Das epidermale Melanin-Einheit-System. Derm. Wschr. *147:* 481–489 (1963).
4 Fitzpatrick, T.B.; Quevedo, W.C., Jr.; Szabó, G. and Seiji, M.: Biology of the melanin pigmentary system; in T.B. Fitzpatrick *et al.* Dermatology in general medicine, pp. 369–401 (McGraw-Hill, New York 1971).
5 Fitzpatrick, T.B.; Hori, Y.; Toda, K.; Kinebuchi, S. and Szabó, G.: The mechanism of normal human melanin pigmentation and of some pigmentary disorders; in T. Kawamura, T.B. Fitzpatrick and M. Seiji, Biology of normal and abnormal melanocytes, pp. 369–401 (University of Tokyo Press, Tokyo 1971).

6 HORI, Y.; TODA, K.; PATHAK, M.A.; CLARK, W.H., Jr. and FITZPATRICK, T.B.: A fine structure study of the human epidermal melanosome complex and its acid phosphatase activity. J. Ultrastruct. Res. *25:* 109–120 (1968).
7 JOHNSON, B.E.; DANIELS, R. and MAGNUS, I.A.: Response of human skin to ultraviolet light; in Photophysiology, Vol. IV, pp. 139–202 (Academic Press, New York 1968).
8 SZABÓ, G.: Quantitative histological investigations on the melanocyte system of the human epidermis; in GORDON Pigment cell biology, pp. 99–125 (Academic Press, New York 1959).
9 SZABÓ, G.; GERALD, A.B.; PATHAK, M.A. and FITZPATRICK, T.B.: Racial differences in the fate of melanosomes in human epidermis. Nature, Lond. *222:* 1081–1082 (1969).
10 TODA, K.; PATHAK, M.A.; PARRISH, J.A. and FITZPATRICK, T.B.: Induction of pigmentation by psoralens and ultraviolet radiation. J. Invest. Derm. *56:* 255 (1971).
11 TODA, K. and FITZPATRICK, T.B.: The origin of melanosomes; in T. KAWAMURA, T.B. FITZPATRICK and M. SEIJI Biology of normal and abnormal melanocytes, pp. 265–278 (University of Tokyo Press, Tokyo 1971).
12 TODA, K.; PATHAK, M.A.; PARRISH, J.A.; FITZPATRICK, T.B. and QUEVEDO, W.C.: Alteration of racial differences in melanosome distribution in human epidermis after exposure to ultraviolet light. Nature, Lond. *236:* 143–145 (1972).
13 VALENTINE, W.N. and BECK, W.S.: Biochemical studies on leucocytes. J. Lab. clin. Med. *38:* 39 (1951).

Author's address: Dr. KIYOSHI TODA, Department of Dermatology, Tokyo Teishin Hospital, 2-16-1, Fujimi, Chiyoda, *Tokyo* (Japan)

Biochemical Characterization of Tyrosinase in Vertebrates[1]

Yu Min Chen and Anne Huo

Detroit Institute of Cancer Research, Division of the Michigan Cancer Foundation, Detroit, Mich.

Introduction

The comparative biochemical aspects of integumental and tumor tyrosinase activity in vertebrate melanogenesis has been studied [6–12]. In order to understand further the biochemical characterization of tyrosinase in vertebrates and mammalian tumors, it is necessary to demonstrate a more detailed subcellular location of tyrosinase and its isozyme patterns in the soluble and solubilized particulate fractions of these species. The resolution of tyrosinase isozymes in tumors has been initiated in recent years [1, 2, 15]. Utilizing these recently improved techniques, we have been able to obtain a more detailed picture of tyrosinase isozymes in tumors]13].

Amphiuma means, B-16 mouse melanoma and human melanoma have been chosen for this investigation. The skin of *Amphiuma means* had the highest tyrosinase activity of all vertebrates studied [6, 9]. Generally, melanoma displayed high tyrosinase activity.

Materials and Methods

A female *Amphiuma means,* weighing 545 g and obtained commercially, was sacrificed by decapitation. The ventral skin was removed, trimmed, weighed, and frozen ($-20\,°C$). The frozen skin was sliced on dry ice and then homogenized (2 min) with deionized water in an ice-chilled micro-Waring blendor. The homogenization was repeated 3 to

1   This investigation was supported in part by U.S. Public Health Service Research Grant No. CA 12731-01 from the National Cancer Institute and in part by an institutional grant to the Detroit Institute of Cancer Research, Division of the Michigan Cancer Foundation, from the United Foundation of Greater Detroit.

5 times to assure the complete disintegration of the skin. The mixture was then transferred to a pre-chilled tissue grinder and ground by hand. The crude homogenate was decanted and refrigerated at 0–4 °C. The residual skin contained only negligible tyrosinase activity.

B-16 mouse melanomas were obtained 18 days after transplantation into the back leg of virgin C57BL female mice (8–10 weeks old). The tumors were removed, trimmed and frozen as soon as the animal was sacrificed by cervical dislocation. The tumor was homogenized (2 min) with deionized water after thawing, as described above. The homogenate was refrigerated at 0–4 °C.

Melanotic human melanomas with high tyrosinase activity were obtained from oncology centers in Detroit. They were excised surgically from the patients, frozen on dry ice until they reached the laboratory. They were then stored in the freezer (−20 °C) until use. The homogenate of human melanoma was prepared in the same way as described for B-16 melanoma.

The homogenates from skin and melanomas were then centrifuged at 600 × g for 10 min (0–4 °C) to obtain clear supernatants which were then subjected to differential centrifugation, yielding particulate and soluble fractions at 20,000 × g (½ h), 105,000 × g (1 h) and 144,000 × g (40 min). The soluble fraction at lower centrifugation was used to obtain the particulate and soluble fractions at the next higher centrifugation. All the particulate fractions were suspended in deionized water and recentrifuged at the corresponding speed at which they were obtained. This washing process was repeated at least twice in order to remove the contamination of soluble activity in the particulate fraction.

The particulate tyrosinase was solubilized by lipase digestion [13] and the solubilized tyrosinase was obtained by the centrifugation of the lipase digested mixture at 144,000 × g for 40 min.

Acrylamide disc gel electrophoresis [1] modified by this laboratory [13] was utilized for resolution of tyrosinase isozymes in the soluble and solubilized tyrosinase preparations from different sources. The tyrosinase activity in the soluble and solubilized fractions was further concentrated by centrifugation of the proteins precipitated by 60% saturation of ammonium sulfate and then dissolving the pellet in a minimal amount of deionized water containing 5% sucrose. This solution was then applied to the gel for resolution by electrophoresis.

The determination of tyrosinase activity in all enzyme preparations obtained was performed by incubation of the enzyme preparation with ^{14}C-L-tyrosine, as described previously [4–6]. The development of melanin band on the acrylamide gel after electrophoresis was accomplished by incubation of the gel with DL-dopa [13]. The bands were then scanned by a densitometer.

Results and Discussion

The subcellular location of tyrosinase activity in the skin of *Amphiuma means*, B-16 mouse melanoma and human melanoma are shown in tables I, II, III and IV, respectively. All species utilized had high total tyrosinase

Table I. Subcellular location of tyrosinase activity in skin of *Amphiuma means*

Fractions[1] from differential centrifugation	Fresh tissue (cpm/g)	% of activity (S 600 × g)[2]	% of total activity[3]
S 144,000 × g	758,000	97.61	51.65
P 20,000 × g	300	0.04	0.02
P 105,000 × g	13,250	1.71	0.90
P 144,000 × g	5,015	0.65	0.34
S 600 × g	776,565		52.92
P 600 × g	691,000		47.08

1 S 600 × g and P 600 × g were the supernatant and pellet obtained at 600 × g by centrifugation of the crude homogenate. Centrifugation of S 600 × g at 20,000 × g, S 20,000 × g at 105,000 × g and S 105,000 × g at 144,000 × g gave fractions P 20,000 × g, P 105,000 × g, P 144,000 × g and S 144,000 × g.
2 Activity in S 600 × g was the sum of activities in S 144,000 × g, P 20,000 × g, P 105,000 × g and P 144,000 × g.
3 Total activity was the sum of activities in S 600 × g and P 600 × g.

Table II. Subcellular location of tyrosinase activity in B-16 mouse melanoma

Fractions[1] from differential centrifugation	Fresh tissue (cpm/g)	% of activity (S 600 × g)[2]	% of total activity[3]
S 144,000 × g	157,000	44.10	12.92
P 20,000 × g	15,000	4.21	1.23
P 105,000 × g	169,000	47.47	13.91
P 144,000 × g	15,000	4.21	1.23
S 600 × g	356,000		29.30
P 600 × g	859,000		70.70

1 See footnote 1 in table I.
2 See footnote 2 in table I.
3 See footnote 3 in table I.

activity (sum of activities in S 600 × g and P 600 × g). The total tyrosinase activity in cpm/g fresh tissue was 1.47×10^6 in the skin of *Amphiuma means*, 1.22×10^6 in the B-16 mouse melanoma, 3.00×10^6 in the metastatic Negro melanoma and 0.32×10^6 in the metastatic Caucasian melanoma. It is interesting that the Negro melanoma had tyrosinase activity almost ten fold that in the Caucasian melanoma. The total tyrosinase activity in the ventral skin of *Amphiuma means* [9] could be estimated as roughly 300-fold that in Cauca-

Table III. Subcellular location of tyrosinase activity in human melanoma (metastatic melanotic Negro melanoma)

Fractions[1] from differential centrifugation	Fresh tissue (cpm/g)	% of activity (S 600 × g)[2]	% of total activity[3]
S 144,000 × g	552,000	18.66	18.40
P 20,000 × g	270,000	9.13	9.00
P 105,000 × g	2,116,000	71.53	70.53
P 144,000 × g	20,000	0.68	0.67
S 600 × g	2,958,000		98.60
P 600 × g	42,000		1.40

1 See footnote 1 in table I.
2 See footnote 2 in table I.
3 See footnote 3 in table I.

Table IV. Subcellular location of tyrosinase activity in human melanoma (metastatic melanotic Caucasian melanoma)

Fractions[1] from differential centrifugation	Fresh tissue (cpm/g)	% of activity (S 600 × g)[2]	% of total activity[3]
S 144,000 × g	11,700	3.74	3.64
P 20,000 × g	36,300	11.60	11.28
P 105,000 × g	259,500	82.91	80.69
P 144,000 × g	5,500	1.76	1.71
S 600 × g	313,000		97.33
P 600 × g	8,600		2.67

1 See footnote 1 in table I.
2 See footnote 2 in table I.
3 See footnote 3 in table I.

sian human skin [6]. Thus, the extremely high tyrosinase activities contained in the skin and melanomas studied here were visualized.

The per cent of soluble (144,000 × g supernatant) enzymic activity in the 600 × g supernatant varied from different tissues; approximately 98% for skin of *Amphiuma means,* 40% for B-16 mouse melanoma, 20% for metastatic Negro melanoma and 4% for metastatic Caucasian melanoma. The per cent of particulate tyrosinase activity in the 600 × g supernatant was negligible in the skin of *Amphiuma means* but quite high in the melanomas. It is interesting to note that the activity in the P 105,000 × g (microsome-like particles) fraction contained the major part of the particulate activity in 600 × g supernatant (water solution) for both skin and melanomas studied.

SEIJI et al. [16] used an isotonic sucrose solution for the extraction of tyrosinase from B-16 mouse melanomas which yielded a nuclear pellet at 700 × g for 10 min, a large granule pellet (melanosomes) at 15,000 × g for 10 min, and a small granule pellet and soluble fraction at 105,000 × g for 60 min. The tyrosinase activity in the different fractions was determined by the oxygen consumption (manometric methods) which may not be specific for tyrosinase activity in crude enzyme preparations. These results showed that the distribution of tyrosinase activity was 28% in the nuclear fraction, 42.4% in the large granule fraction, 18.6% in the small granule fraction and 11.0% in the soluble fraction. The per cent activity of soluble fraction agrees closely to that (12.92%) found in this investigation. However, the percentages of activity distributed in the particulate fractions are different. It is probable that part of the pellet obtained by centrifugation at 600 × g in water remains in the supernatant when spun in isotonic sucrose and is spun down at the high g forces (15,000).

The tyrosinase isozyme patterns in the soluble and solubilized particulate fractions of skin and melanomas are shown in figures 1–3. In the skin of *Amphiuma means,* figure 1a, many isozymes appear at positions close to origin of sample application, indicating the presence of large molecular weight tyrosinases. Judging by the isozyme patterns of mushroom tyrosinase [14], these molecules may be polymers containing more than five monomer units. It would be interesting to see whether these high polymer tyrosinases appear in the skin of higher vertebrates. The washed P 600 × g fraction after lipase digestion showed two isozymes (fig. 1b). One had migrated a considerable distance from the origin; this peak may represent monomer tyrosinase. Another was close to origin and may be one of the higher polymers in figure 1a. The well-washed unground part of the skin showed one isozyme after lipase digestion which was close to origin (fig. 1c) and must be one of the high polymer tyrosinases in figure 1a.

In the B-16 mouse melanoma, three isozymes are present in the soluble fraction (fig. 2a); all migrated a considerable distance from the origin of application. The middle and dominant form may be similar to the tetramer found in the mushroom tyrosinases [14]. The smallest band of activity at the farthest position from origin may be the monomer tyrosinase which has been often missed because of its thermolability. It is interesting to note that the dominant form and the smaller peak on the side closer to origin

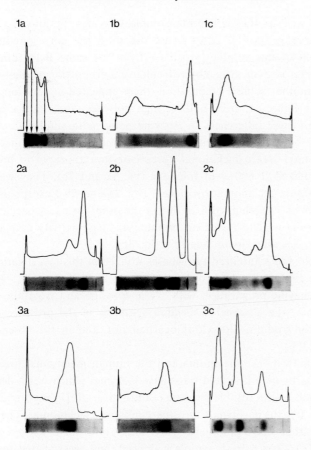

Fig. 1a–c. Tyrosinase isozyme patterns in the skin of *Amphiuma means*. – a) Soluble tyrosinases from 144,000 × g supernatant. – b) Solubilized tyrosinases from P 600 × g fraction. – c) Solubilized tyrosinase from unground skin part.

Fig. 2a–c. Tyrosinase isozyme patterns in B-16 mouse melanoma. – a) Soluble tyrosinases from 144,000 × g supernatant (appeared in most cases). – b) Solubilized tyrosinases from P 600 × g fraction. – c) Soluble tyrosinases from 144,000 × g supernatant.

Fig. 3a–c. Tyrosinase isozyme patterns in human melanoma. – a) Soluble tyrosinases from 144,000 × g supernatant of a metastatic Negro melanoma. – b) Solubilized tyrosinases from P 600 × g fraction of the melanoma in 3a. – c) Solubilized tyrosinases from P 600 × g of a partially pigmented human melanoma containing no activity in 144,000 × g supernatant.

The enzyme migrated from left to right. Each figure had scanning picture above and the corresponding picture of acrylamide gel electrophoresis below.

correspond very well to the melanoma soluble enzymes T_1 and T_2 as named by BURNETT et al. [1]. After lipase digestion the corresponding P 600 × g fraction of this soluble tyrosinase from the same B-16 tumor showed three similar isozymes (fig. 2b) with enhanced proportion of the two small forms. In another soluble preparation, there appeared in addition to the three forms in figure 2a, another three high polymer tyrosinases close to origin (fig. 2c). These differences in isozyme patterns in similar tumor preparations may be due to the degree of maturation of tumors utilized.

In the metastatic Negro melanoma, three isozymes appeared in both soluble and solubilized (P 600 × g) fractions (fig. 3a and 3b). They were similar to B-16 mouse melanoma tyrosinases (fig. 2a and 2b), except that the dominant form was overlapped with one of the other forms. A partially pigmented melanoma which showed no soluble tyrosinase activity gave six solubilized tyrosinases after lipase digestion (fig. 3c).

All the soluble and solubilized tyrosinases utilized in this investigation were specifically characterized according to the following criteria; incubation of the enzyme preparation with ^{14}C-L-tyrosine and isolation of radioactive melanin, the absence of melanin synthesis in the presence of tyrosinase inhibitor (sodium diethyldithiocarbamate), and in the absence of dopa catalyst [3, 6].

In both the B-16 mouse melanoma and the human melanoma, lipase digestion of the particulate liberated tyrosinase isozymes which were identical to those found in the soluble fraction. BURNETT et al. [1] used many chemical agents to free the particulate tyrosinase in mouse melanomas. These workers showed that the solubilized particulate tyrosinase existed in multiple forms (T_3) when a fresh sample was used, and reassociated to a complex unit if the sample was allowed to stand under refrigeration. Since these T_3 tyrosinase isozymes were not identical to those found in the soluble fraction, it must be considered that the chemical treatment may have brought about microsomal breakdown rather than extraction of the enzyme.

References

1 BURNETT, J. B.; SEILER, H. and BROWN, I. V.: Separation and characterization of multiple forms of tyrosinase from mouse melanoma. Cancer Res. 27: 880–889 (1967).
2 BURNETT, J. B. and SEILER, H.: Multiple forms of tyrosinase from human melanoma. J. Invest. Derm. 52: 199–203 (1969).
3 CHAVIN, W.: Fundamental aspects of morphological melanin color changes in vertebrate skin. Amer. Zool. 9: 505–520 (1969).

4 CHEN, Y. M. and CHAVIN, W.: Radiometric assay of tyrosinase and theoretical considerations of melanin formation. Anal. Biochem. *13:* 234–258 (1965).
5 CHEN, Y. M. and CHAVIN, W.: Incorporation of tyrosine carboxyl groups and utilization of D-tyrosine in melanogenesis. Anal. Biochem. *27:* 463–472 (1969).
6 CHEN, Y. M. and CHAVIN, W.: Comparative biochemical aspects of integumental and tumor tyrosinase activity in vertebrate melanogenesis; in W. MONTAGNA and F. HU Advances in biology of the skin, pp. 253–268 (Pergamon Press, Oxford 1967).
7 CHEN, Y. M. and CHAVIN, W.: Integumental tyrosinase activity in reptiles. Experientia *23:* 917–919 (1967).
8 CHEN, Y. M. and CHAVIN, W.: Utilization of D-tyrosine by vertebrate skin tyrosinase. Experientia *23:* 997–999 (1967).
9 CHEN, Y. M. and CHAVIN, W.: Integumental tyrosinase activity in amphibians. Experientia *24:* 31–33 (1968).
10 CHEN, Y. M. and CHAVIN, W.: Melanogenesis in frog skin. Experientia *24:* 332–334 (1968).
11 CHEN, Y. M. and CHAVIN, W.: Ontogenetic alterations in tyrosinase activity. J. Invest. Derm. *50:* 289–296 (1968).
12 CHEN, Y. M. and CHAVIN, W.: Distribution of integumental tyrosinase activity in red-eared turtle, *Pseudomys scripta elegans*. Experientia *26:* 717–718 (1970).
13 CHEN, Y. M. and HUO, A.: Preparation and resolution of tyrosinase isozymes from skin and melanoma tissues. (In preparation.)
14 JOLLEY, R. L., Jr.: The tyrosinase isozymes; in W. MONTAGNA and F. HU Advances in biology of the skin, vol. 8, pp. 269–281 (Pergamon Press, Oxford 1967).
15 POMERANTZ, S. H.: Separation, purification and properties of tyrosinases from hamster melanoma. J. Biol. Chem. *238:* 2351–2357 (1963).
16 SEIJI, M.; SHIMAO, K.; BIRBECK, M. S. C. and FITZPATRICK, T. B.: Subcellular localization of melanin biosynthesis. in RILEY and FORTNER, The Pigment cell, Ann. N. Y. Acad. Sci. *100:* 497–533 (1963).

Author's address: Dr. YU MIN CHEN, Detroit Institute of Cancer Research, Division of the Michigan Cancer Foundation, *Detroit, MI 48201* (USA)

Metabolism of Tyrosine and Its Control in Cultured Melanoma Cells[1]

A. OIKAWA, M. NAKAYASU and M. NOHARA

Biochemistry Division, National Cancer Center Research Institute, Tokyo

Introduction

Since MOORE et al. [4] established cultured cell lines of Syrian hamster melanoma, a number of cultured melanoma lines of hamster and mouse became available for studying melanogenesis in simplified systems. An assay method for tyrosinase activity of cell-free extract as well as that in living cells in culture [7] allowed us to follow the process of melanogenesis in melanoma cells in culture [5].

Results concerning the metabolism of tyrosine by tyrosinases and their application to studies on control of tyrosinase in melanoma cells will be discussed below.

Metabolism of Tyrosine by Tyrosinases

Assay of Tyrosinase Activity with L-tyrosine-3,5-^{3}H

Tyrosinase catalyzes the hydroxylation of L-tyrosine to L-dopa (L-3,4-dihydroxyphenylalanine) coupling with oxidation of dopa to dopa quinone. Polymerization of the latter to melanin proceeds without enzyme. Therefore, the enzyme activity is best expressed by the rate of tyrosine hydroxylation, as has been proposed by POMERANTZ [8]. He demonstrated that the release of tritium as water from L-tyrosine-3,5-^{3}H was stoichio-

---

[1] This investigation was supported in part by research grants Nos. 92019 and 92112 from the Ministry of Education, Japan

metric when it was hydroxylated to dopa [8, 9]. But in prolonged incubation of the tritiated tyrosine with crude extracts or living cells of melanoma, dopa and dopa quinone are further oxidized and polymerized to melanin, releasing a considerable part of the second tritium of L-tyrosine-3,5-^{3}H as water, which obscures the stoichiometric release of the first tritium atom.

Radioactive reaction products from L-tyrosine-3,5-^{3}H incubated with cell-free extract of melanotic melanoma cells in the presence of a catalytic amount of dopa, were extensively surveyed and only water, dopa and melanin were found in appreciable amounts with a satisfiable recovery of radioactivity, while cell-free extracts of amelanotic cells did not produce radioactive water, dopa or melanin [7].

When melanotic cells were cultured in the medium containing L-tyrosine-3,5-^{3}H, the radioactive products in culture were water, melanin and protein. Amelanotic cells again did not produce radioactive water or melanin but radioactive protein, confirming that radioactive water was produced only by hydroxylation of the labeled tyrosine as it was in cell-free extracts [7]. The quantitative relationships between these products are shown in figure 1.

$\triangle Ty/W$, the ratio of the radioactivity of tyrosine hydroxylated to that of water produced, was fairly constant for a variety of enzyme sources. Figure 2 shows the time courses of formation of reaction products. Mushroom tyrosinase (fig. 2A) showed the well-known reaction inactivation. Tyrosinase solubilized from Harding-Passey melanoma [3] (fig. 2B) and cell-free extracts of human malignant melanoma (fig. 2C) were not inactivated for a longer period than those of cultured mouse melanoma cells [7]. Despite these different time courses of reaction, $\triangle Ty/W$ was constant

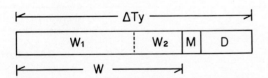

Fig. 1. Quantitative relationships between products of tyrosinase reaction on L-tyrosine-3,5-^{3}H, expressed by radioactivity. ΔTy = tyrosine hydroxylated; W = $W^1 + W^2$, water; $W_1 = \Delta Ty/2$, which is formed by tyrosine hydroxylation; W_2, water formed by autoxidative polymerization to melanin; M, melanin; D, dopa. $\Delta Ty/W$, an empirical constant, is used for estimating ΔTy from W, the easiest value to determine. $2(W_2 + M) = \Delta Ty - 2D$ is the amount of tyrosine converted to melanin. This is equal to ΔTy for cells in culture because D = 0 for living cells (fig. 2D).

during the reaction and almost the same for these entirely different sources of the enzyme. △Ty/W for living cells in culture was also constant (fig. 2D), although lower than those for cell-free extracts. These results, together with those previously reported [7], are summarized in table I.

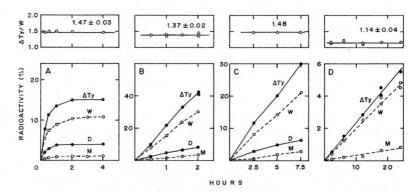

Fig. 2. Metabolism of L-tyrosine-3,5-^{3}H by various preparations of tyrosinase. Reaction mixtures contained 250 nmoles of L-tyrosine-3,5-^{3}H (5 μCi), 25 nmoles of L-dopa, 12.5 μg of chloramphenicol and (A) purified mushroom tyrosinase (2.5 μg as protein; Sigma, grade III), (B) solubilized tyrosinase of Harding-Passey mouse melanoma (212 μg as protein; kindly prepared [3] and provided by Dr. SEIJI), or (C) cell-free extract of human malignant melanoma (164 μg as protein) in 250 μl of 80 mM sodium phosphate buffer (pH 6.8). Incubations were carried out at 37.5°. (D) Cells (C_2M) of B16 melanoma were cultured in Eagle's minimum essential medium supplemented with 10% calf serum and 2μCi/ml of L-tyrosine-3,5-^{3}H at 37.5°. Radioactivities in reaction products were assayed as reported previously [7] and expressed as percent of radioactivity added as tyrosine. For symbols, see legend to figure 1.

Table I. △Ty/W and $W_2/(W_2 + M)$ values of tyrosinases from various sources

Enzyme	△Ty/W	$W_2/(W_2+ M)$
Living cells		
Cultured cells B16[1]	1.09 ± 0.04	0.84 ± 0.05
Cell-free extract		
Cultured cells B16	1.38 ± 0.05	0.69 ± 0.08
Harding-Passey melanoma[1]	1.43	0.66
Human melanoma	1.48	0.75 – 0.66
Soluble enzyme		
Mushroom, purified	1.43 ± 0.05	0.76 ± 0.04
Harding-Passey melanoma, solubilized	1.38 ± 0.11	0.76 ± 0.09

1 Data in ref. [7].

The lower $\triangle$Ty/W in living cells in culture suggests the higher participation of position 7 of indole type intermediates or position 5 of tyrosine derivatives in polymerization to melanin. This becomes clearer by comparing another parameter of the reaction, $W_2/(W_2+M)$, the fraction of polymerization units which lost their second tritium in newly synthesized melanin. As seen in table I, this value for cell-free extract of cultured cells of B16 melanoma was 69%, while that for living cells of the same strain in culture was 84%. The latter is in good agreement with results for tumors in host animals reported by HEMPEL [1], in which L-dopa-5-^{3}H was administered to melanoma-bearing mice and radioactivity in tumor was determined. This may suggest that the melanin synthesis in living cells proceeds more orderly than in cell-free systems.

The tyrosinase activity as the rate of tyrosine hydroxylation ($\triangle Ty$) can be estimated from the radioactivity of tritiated water (W) formed from L-tyrosine-3,5-^{3}H, by taking advantage of the constancy of $\triangle$Ty/W value. The tritiated water is easily separated from other radioactivity by adding trichloroacetic acid and Norit A to the sample medium and filtering it through Millipore filter. It should be also emphasized that the method is applicable to living cells in culture without interrupting culture because the incorporation of tyrosine in cellular protein does not interfere with the assay and only the used media are required for the assay [7].

Evolution of Carbon Dioxide

When cell-free extracts of melanoma cells were incubated with L-tyrosine-1-^{14}C, the main radioactive products were CO_2 and dopa, and a minor radioactivity was found in melanin. Recovery of radioactivity in these three products and unreacted tyrosine was nearly 100% for 24 h with 8% fluctuation. Decarboxylation of tyrosine by melanin synthesis was calculated from radioactivities of melanin and CO_2 (fig. 3). The degree of decarboxylation was about 80% in the first 4 h and almost 100% in the last 8 h, suggesting that melanin synthesized in the early phase contained more carboxyl groups or that the decarboxylation continued after oxidation products of tyrosine polymerized to 'melanin'. Experiments with L-tyrosine-U-^{14}C gave the similar results that the decarboxylation proceeded still linearly after the rate of melanin synthesis was largely slowed down, and CO_2 equivalent to more than 90% of α-carboxyl was released. A large discrepancy between the present results and those reported by HEMPEL [1],

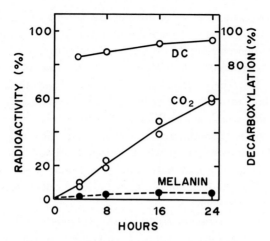

Fig. 3. Decarboxylation of tyrosine by cell-free extract of cultured melanoma cells C_2M. L-tyrosine-1-^{14}C was incubated with the cell-free extract (69 μg protein per tube). Composition of reaction mixtures was as in figure 2 except the label of tyrosine. Radioactivities are plotted as in figure 2. The degree of decarboxylation (DC) is expressed as percentage of radioactivity of CO_2 in sum of those of CO_2 and melanin.

which claimed 71 % retention of carboxyl group of dopa in melanin, may be ascribed to the difference between reactions *in vitro* and *in vivo* or to the re-utilization of bicarbonate formed by decarboxylation in tumor.

Control of Tyrosinase Activity in situ

Tyrosinase Activities of Cultured Cells Assayed with Living Cells in Cultures (TyC) and with Cell-Free Extracts (TyH)

These activities were determined by the methods described above. When cell-free extracts were made by freezing and thawing and homogenization in distilled water, fully activated TyH's were obtained as MENON reported [2]. As seen in figure 4, TyH/cell showed a maximum in the mid-exponential phase of growth, while TyC/cell had its maximum in the very early phase of growth. The rapid decrease in TyC/cell in exponentially growing cells in which TyH/cell was rather increasing, suggests some mechanism suppressing tyrosinase *in situ*.

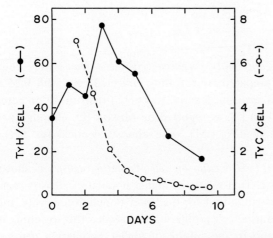

Fig. 4. Tyrosinase activities of cultured cells. Melanoma cells (C_2M) were cultured in Leighton tubes (flat area, 5 cm^2) with inocula of 1×10^5 cells per tube. After a temporary decrease in viable cell number, cells proliferated exponentially for 6 days with the doubling time of 19 h. Maximum population density was 2×10^6 cells per tube. Cultures with medium containing L-tyrosine-3,5-^{3}H (2 μCi/ml) were used for TyC assay and the other group of cultures was used for TyH assay (see legend to figure 2) by harvesting cells at the indicated times. TyH and TyC, expressed as nmoles tyrosine hydroxylated per 10^5 cells per day, were calculated from the radioactivity recovered as water and the values of ΔTy/W, 1.4 and 1.1 for TyH and TyC, respectively (table I).

Table II. TyH values assayed with cell-free extracts prepared with isotonic sucrose and distilled water[1]

Experiment number	Population density (10^5 cells/cm^2)	TyH-s	TyH-w	TyH-s/TyH-w
		(nmoles TYR/mg protein/h)		
1.	0.20	107	174	0.62
	3.2	15	43	0.35
2.	0.084	25.3	64.5	0.39
	0.72	13.5	66.0	0.22

[1] Freshly harvested cells from cultures of different population densities were divided into two portions. One was homogenized in 0.25 M sucrose, the other in distilled water and TyH's were assayed. In experiment 1, cells (C_2M) were harvested after 1- and 5-day cultivations; in experiment 2, cells (a subline of C_2M) were inoculated differently in size, and harvested after 3-day cultivation. Suffixes s and w mean 0.25 M sucrose and distilled water, respectively, as homogenizing medium.

Change in the States of Tyrosinase
in situ during Culture

Values of TyH with differently prepared cell-free extracts were compared. Cell-free extracts prepared from freshly collected cells with 0.25 M sucrose show much lower tyrosinase activity (TyH-s) than those prepared with distilled water (TyH-w) [7], as reported by MENON [2], although two parameters, $\triangle$Ty/W and $W_2/(W_2+M)$, were almost identical for TyH-s and TyH-w [7].

Dependence of TyH-s/TyH-w ratio on cell population density is shown in table II. In experiment 1, TyH-w for the higher population density (near fully grown) was much less than that for the lower population density, but the depression in TyH-s during cell proliferation exceeded this, resulting in a lower TyH-s/TyH-w ratio in the culture of higher population density. In experiment 2, cells were harvested from exponentially growing cultures with different population densities, but a lower value of TyH-s/TyH-w was obtained again for the culture with a higher population density. These results may be explained by a partial retention after isotonic homogenization of the suppressed state of tyrosinase *in situ,* which was seen as a lower TyC/TyH ratio in a more densely populated culture (fig. 4).

Control of Melanin Synthesis in Culture

Combining these results with the other fact that the TyC decreases reversibly by lowering the cultivation temperature [6], a suppressing mechanism of tyrosinase activity *in situ* is strongly suggested. This hypothetical mechanism should be characterized by its function to regulate the rate of melanin synthesis reversibly responding to the environmental factors such as the population density and cultivation temperature, and by its sensitivity to hypotonic conditions.

Inactivation by melanin deposit on the enzyme protein or melanosomes may explain the decrease in TyC/TyH or TyH-s/TyH-w in cultures of higher population density with some assumptions, but the reversible change of TyC by temperature shifts is hardly explained by this. It may be possible to assume a substance which reversibly inactivates tyrosinase by binding it, because a preliminary experiment showed that puromycin prevented the TyC depression caused by temperature shift-down [unpublished data].

References

1 HEMPEL, K.: Investigation on the structure of melanin in malignant melanoma with ^{3}H- and ^{14}C-dopa labelled at different positions; in G. D. PORTA and O. MUHLBOCK, Structure and control of the melanocyte, pp. 162–175 (Springer, Berlin 1966).
2 MENON, I. A. and HABERMAN, H. F.: Activation of tyrosinase in microsomes and melanosomes from B 16 and Harding-Passey melanomas. Arch. Biochem. Biophys. *137:* 231–242 (1970).
3 MIYAZAKI, K. and SEIJI, M.: Tyrosinase isolated from mouse melanoma melanosome. J. Invest. Derm. *57:* 81–86 (1971).
4 MOORE, G. E.; MOUNT, D.; TARA, G. and SCHWARTZ, N.: Culture of malignant tumors of the Syrian hamster. J. nat. Cancer Inst. *31:* 1217–1226 (1963).
5 OIKAWA, A.; NAKAYASU, M. and NOHARA, M.: Regulation of melanogenesis in cultured melanoma cells. Jap. J. develop. Biol. *24:* 116–117 (abstract) (1970).
6 OIKAWA, A.; NOHARA, M. and NAKAYASU, M.: Melanin synthesis in cultured melanoma cells; in T. KAWAMURA, T. B. FITZPATRICK and M. SEIJI 'Biology of normal and abnormal melanocytes', pp. 209–220 (Univ. of Tokyo Press, Tokyo 1971).
7 OIKAWA, A.; NAKAYASU, M.; NOHARA, M. and TCHEN, T. T.: Fate of L-tyrosine-3,5-^{3}H in cell-free extracts and tissue cultures of melanoma cells: a new assay method for tyrosinase in living cells. Arch. Biochem. Biophys. *148:* 548–557 (1972).
8 POMERANTZ, S. H.: Tyrosine hydroxylation catalyzed by mammalian tyrosinase: an improved method of assay. Biochem. biophys. Res. Comm. *16:* 188–194 (1964).
9 POMERANTZ, S. H.: Tyrosine hydroxylase activity of mammalian tyrosinase. J. biol. Chem. *241:* 161–168 (1966).

Author's address: Dr. ATSUSHI OIKAWA, Biochemistry Division, National Cancer Center Research Institute, *Tokyo 104* (Japan)

Subcellular Localization of Peroxidase-Mediated Oxidation of Tyrosine to Melanin[1]

M. R. OKUN, R. P. PATEL, BARBARA DONNELLAN and L. M. EDELSTEIN

Departments of Dermatology, Tufts University School of Medicine and Boston City Hospital, Boston, Mass., and Departments of Dermatology, University of Massachusetts School of Medicine and St. Vincent Hospital, Worcester, Mass.

Introduction

Histochemical [11–14] and biochemical [18] studies in our laboratory have demonstrated that peroxidase can mediate the synthesis of melanin from tyrosine, and that dopa is the first product of this synthesis. Dopa [11–14, 18] or dihydroxyfumarate [16, 18] were effective as co-factors for the hydroxylation of tyrosine by peroxidase.

Since dopa itself is oxidized to melanin by peroxidase, the use of dopa as co-factor in this reaction requires a control in which dopa alone is present [12, 14]. With optimum tyrosine-dopa ratio, differential pigment derived from tyrosine is visually evident by light microscopy. However, the tyrosine-dopa method is not suited for electron microscopy since differential pigment formation may be difficult to judge on an ultrastructural level. Previous light-microscopic histochemical studies in our laboratory [16] have shown that dihydroxyfumarate can be used as a co-factor in the direct demonstration of peroxidase-mediated oxidation of tyrosine to melanin.

In this study, ultrastructural localization of peroxidatic oxidation of tyrosine to melanin was investigated in granulocytes and melanoma cells, using dihydroxyfumarate as co-factor.

---

[1] Supported by USPHS Grant T1 AM 5220 and by the St. Vincent Hospital Research Foundation.

Materials and Methods

Light Microscopy

Neutrophils and eosinophils were studied in human peripheral blood. Smears were prepared in the standard way and were air-dried for 24 h, followed by prefixation in 10% buffered formalin (pH 7.4) for 15 min. 6 μ cryostat sections of a lightly melanized Harding-Passey melanoma (from Arthur D. Little, Inc., Cambridge, Mass.) were also prefixed in 10% buffered formalin.

Preparations were incubated in substrate solution containing 20 mg% L-tyrosine, 32 mg% dihydroxyfumaric acid and 10^{-4} or 10^{-5} M hydrogen peroxide in 0.1 M phosphate buffer (pH 7.4) at 37°C for 3 h [16]. Controls included a) exposure to 100°C wet heat for 15 min before incubation in substrate solution; b) omission of tyrosine, dihydroxyfumaric acid or hydrogen peroxide; c) preincubation of specimens in 10^{-2} M sodium diethyldithiocarbamate (DDC) for 2 h at 37°C before incubation in the substrate solution; d) incubation in substrate solution containing 0.1 mg/ml beef liver catalase (Sigma No. C-100); e) incubation of specimens in buffer alone.

For comparison, specimens were subjected to Laidlaw's dopa reaction [8], tyrosine-dopa reaction [15] and benzidine reaction of De Robertis and Grasso [19].

Electron Microscopy

For studies of neutrophils and eosinophils, heparinized blood was centrifuged at 400 g for 15 min. Plasma was withdrawn and replaced with Karnovsky's fixative. After 1 h the buffy coat was lifted so that the fixative would be in direct contact with its undersurface and fixation was allowed to proceed for an additional hour. The buffy coat was then washed with multiple changes of 0.1 M phosphate buffer, pH 7.4, and placed in this buffer overnight at 4°C. 40 μ cryostat sections of buffy coat were then subjected to the tyrosine-dihydroxy fumarate-hydrogen peroxide reaction, with controls as described above. Substrates were removed by repeated washing with buffer. Specimens were then post-osmicated, dehydrated and embedded in araldite in the usual manner.

Specimens of Harding-Passey melanoma were prefixed in Karnovsky's fixative and 40 μ cryostat sections were treated as described above.

1 μ sections of specimens were observed with phase microscopy and thin sections were obtained from appropriate areas. Thin sections stained with lead citrate for 2 min were examined with an RCA EMU-3G electron microscope. The reduced staining was to facilitate visualization of reaction product.

For comparison, the 40 μ sections of melanoma and buffy coat were subjected to the EM-dopa reaction [9] and EM-diaminobenzidine (DAB) reaction [7]. A 10-min incubation time was used with the DAB reaction.

Results (Table I)

Light Microscopy

After incubation in the solution of tyrosine, dihydroxyfumaric acid and hydrogen peroxide, neutrophils and eosinophils in smears had a

Table I. Oxidation of tyrosine to melanin in the presence of dihydroxyfumarate and H_2O_2

	Light microscopic distribution of reaction product	Electron microscopic distribution of reaction product	Correlation with benzidine reaction and DAB reaction	Effect of omission of exogenous H_2O_2	Effect of catalase	Effect of pre-incubation with DDC
Neutrophils	All cells positive	Azurophil granules and occasional specific granules	Same distribution as benzidine reaction and DAB reaction	Decreased reaction	Suppressed reaction	No effect
Eosinophils	All cells positive	Matrix of eosinophil granules	Same distribution as benzidine reaction and DAB reaction	Decreased reaction	Suppressed reaction	No effect
Harding-Passey melanoma cells	Groups of positive cells. Reacting cells mostly smaller tumor cells and cells near edges of section or near tissue tears	Melanosomes and GERL area	Same distribution as benzidine reaction. DAB reaction in melanosomes only	Partial suppression of reaction	Suppressed reaction	Suppression of reaction in GERL area and possibly in some melanosomes

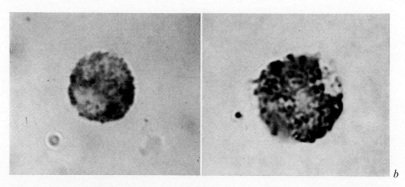

Fig. 1a and b. Neutrophil (a) and eosinophil (b) show grayish-black reaction product with granular distribution in their cytoplasm after incubation with tyrosine, dihydroxyfumarate and hydrogen peroxide. (No counterstain.) × 1740.

granular deposition of grayish-black induced melanin in their cytoplasm (fig. 1). The reaction was more intense in eosinophils than in neutrophils; use of 10^{-4} M hydrogen peroxide resulted in a more intense reaction than use of 10^{-5} M hydrogen peroxide. Boiled preparations showed suppressed reaction product. Preparations in which tyrosine or dihydroxyfumarate was omitted showed no reaction product. In some preparations reaction product was present when exogenous hydrogen peroxide was omitted, but this was suppressed by the presence of catalase. The dopa reaction, benzidine reaction and tyrosine-dopa reaction were also positive in these cells.

Cryostat sections of Harding-Passey melanoma showed groups of tumor cells with induced melanin (fig. 2); positive cells were most numerous near tissue tears and edges of sections. Grayish-black induced melanin was easily differentiated from golden-brown preformed melanin. Adjacent sections subjected to the benzidine reaction and tyrosine-dopa reaction showed a similar distribution of reacting cells. Adjacent sections subjected to the dopa reaction showed reacting cells corresponding to those in the benzidine reaction and tyrosine-dihydroxyfumarate-hydrogen peroxide reaction, but in addition showed positive cells not corresponding to positive cells in these reactions. The latter were usually larger tumor cells containing varying amounts of preformed melanin.

The positive tyrosine-dihydroxyfumarate-hydrogen peroxide reaction in cryostat sections of melanoma was eliminated by boiling or by omission of tyrosine or dihydroxyfumaric acid. Omission of hydrogen peroxide resulted in a decrease in the reaction, but did not eliminate it. However, the presence of catalase completely eliminated it. Preincubation with DDC

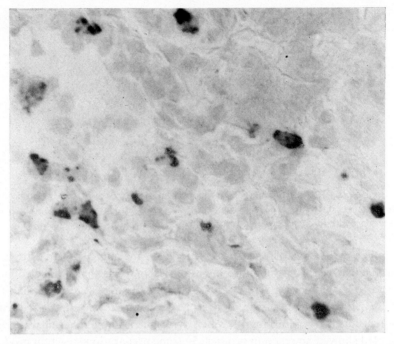

Fig. 2. Cryostat section of Harding-Passey melanoma shows tumor cells with grayish-black induced melanin formed after incubation with tyrosine, dihydroxyfumarate and hydrogen peroxide. Induced melanin was easily differentiated from preformed melanin which had a yellowish-brown color (H-E counterstain.) × 1570.

resulted in a partial supression of the tyrosine-dihydroxyfumarate-hydrogen peroxide reaction in some melanoma cells.

Electron Microscopy

After incubation in the solution of tyrosine, dihydroxyfumaric acid and 10^{-4} M hydrogen peroxide, electron-dense material representing melanin formed by the oxidation of tyrosine was observed in azurophil granules and occasional specific granules of neutrophils (fig. 3) as well as in granules of eosinophils (fig. 5a). Reaction product spared the crystalline core of the eosinophil granules (fig. 5a) so that this area was less electron-dense than the matrix, a reversal of the pattern in unreacting granules (fig. 5b). Cells varied in their degree of reaction, possibly because of variations in substrate penetration in the 40 μ sections.

The results of the control studies in neutrophils and eosinophils parallelled those observed by light microscopy. Reaction product of the

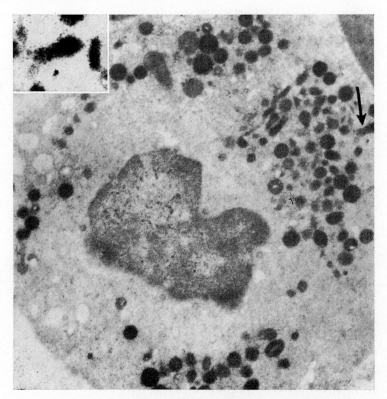

Fig. 3. Electron micrograph illustrates distribution of electron-dense melanin formed from tyrosine by peroxidase in a neutrophil. Reaction is mostly in azurophil granules, but some specific granules (arrow) also show reaction product. Inset shows higher magnification of specific granule indicated by arrow. (2-min lead citrate stain.) × 18,500. Inset: × 38,500

DAB reaction (fig. 4b and 5c) and melanin formed by the oxidation of dopa in these cells had the ultrastructural localization of melanin formed by the oxidation of tyrosine.

After the melanoma tissue was incubated in the tyrosine-dihydroxy-fumarate-hydrogen peroxide solution, reaction product was noted in groups of tumor cells. It was present in both melanosomes and in the Golgi-associated cisternae of smooth-surfaced endoplasmic reticulum (GERL) (fig. 6).

This reaction product was suppressed by boiling and by omission of either tyrosine or dihydroxyfumarate. It was partially suppressed by omission of hydrogen peroxide and totally suppressed by the presence of catalase. Preincubation with DDC resulted in almost complete elimination of

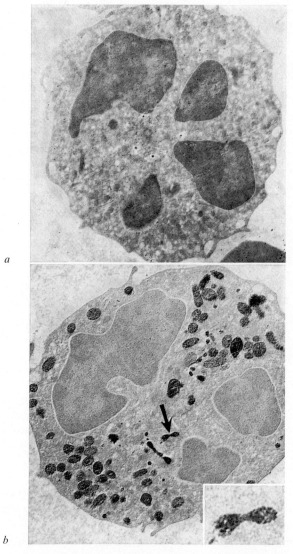

Fig. 4a and b. Electron micrographs illustrate appearance of neutrophil (a) incubated in buffer alone and (b) subjected to the DAB reaction for peroxidase. Without reaction product, cytoplasmic granules are barely discernible under the staining conditions cited. Reaction product of the DAB reaction parallels that of the peroxidatic tyrosine-melanin reaction in its distribution. As above, reaction product is mostly in azurophil granules, but occasional specific granules are also positive (arrow). Inset shows higher magnification of specific granule indicated by arrow. (2-min lead citrate stain.) × 11,500. Inset: × 38,500.

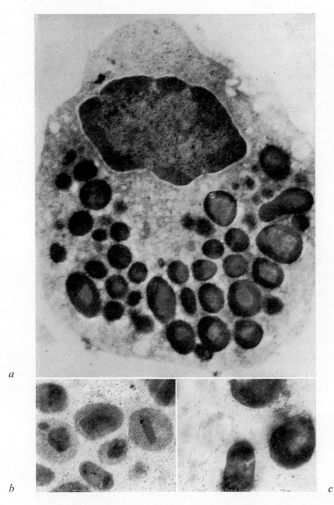

Fig. 5 a–c. Electron micrograph illustrates distribution of electron-dense melanin formed from tyrosine by peroxidase in an eosinophil (a). Reaction product spares the crystalline core so that this area is less electron-dense than the matrix, a reversal of the pattern in unreacting granules (b). Reaction product after exposure to DAB (c) shows a similar distribution to that of the tyrosine melanin. (2-min lead citrate stain.) a) × 12,000; b, c) × 34,000.

reaction product in the GERL but had no discernible effect on the reaction product in melanosomes.

The DAB reaction showed reaction product in melanosomes, but not in the GERL. The EM-dopa reaction showed reaction product both in the

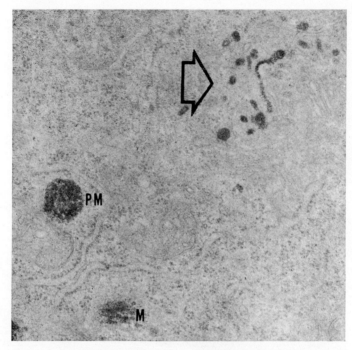

Fig. 6. Electron micrograph illustrates distribution of electron-dense melanin formed in Harding-Passey melanoma cell after incubation with tyrosine, dihydroxyfumarate and hydrogen peroxide. This reaction product is present in melanosomes (PM) and in the GERL (arrow). M = unreacting melanosome. Inhibitor studies indicated that the reaction product in the melanosomes was formed directly by peroxidase and that the reaction product in the GERL was formed by aerobic dopa oxidase from dopa synthesized by peroxidase. (2-min lead citrate stain.) × 31,000.

GERL of tumor cells and in melanosomes. Reaction product in the GERL was catalase-resistant and DDC-labile. Some reaction product in melanosomes was catalase-resistant and DDC-labile and other reaction product in melanosomes was catalase-labile and DDC-resistant. The results of the DAB and EM-dopa reactions in Harding-Passey melanoma parallelled those of our previous studies [14].

Discussion

Since dihydroxyfumarate is not melanogenic, melanin formed from tyrosine in its presence could be directly visualized. Melanin formed from tyrosine in the presence of dihydroxyfumarate was clearly the result of peroxidatic enzymatic oxidation in the cells studied. The reaction was heat-labile and entirely peroxide-dependent. Further confirmation of the peroxidatic nature of this reaction was the correlation of the subcellular distribution of its product with that of the DAB reaction for peroxidase in neutrophils and eosinophils. In the GERL of Harding-Passey melanoma cells and possibly in some melanosomes, there was melanin formed from tyrosine in the presence of dihydroxyfumarate which was both peroxide-dependent and DDC-labile. That this represented partial suppression of peroxidase by the chelating action of DDC is excluded by the absence of reaction product in the GERL after incubation with DAB; that this represented enzymatic action of 'tyrosinase' alone is excluded by its complete peroxide dependency. An explanation which is compatible with the data is that melanin formed in the GERL after incubation with tyrosine and dihydroxyfumarate was due to the action of 'tyrosinase' (dopa oxidase) on dopa formed by peroxidase catalysis in melanosomes. As noted in our results, melanin formed in the GERL after incubation with exogenous dopa was catalase-resistant and DDC-labile.

Previous histochemical studies in our laboratory [11-14] have suggested that mammalian 'tyrosinase' is actually a dopa oxidase in accordance with BLOCH's [1] hypothesis. This has been confirmed by biochemical studies in progress [17] in our laboratory which have shown that 'tyrosinase' in melanoma extracts is an aerobic dopa oxidase with no ability to oxidize tyrosine even in the presence of dopa co-factor.

The many studies of previous investigators purporting to demonstrate the existence of a mammalian tyrosinase in crude or partially purified melanoma extracts did not have a control for the presence of peroxidase. BURNETT [2] has recently described a purified preparation of 'tyrosinase' from Harding-Passey melanoma, but data pertaining to the ability of this preparation to oxidize tyrosine were omitted, although a method for assaying tyrosine oxidation was cited in her experimental procedure.

The frequency of reaction product in melanoma cells in areas of tissue damage suggests that demonstration of peroxidase activity in these cells exhibits latency (dependency of histochemical reaction on structural damage to effect adequate substrate-enzyme interaction).

Partial resistance of the formation of melanin from tyrosine with dihydroxyfumarate co-factor in melanoma cells to omission of exogenous hydrogen peroxide was due to the presence of endogenous hydrogen peroxide, since the reaction was totally suppressed by catalase.

On the basis of our studies, it appears that the ability to oxidize tyrosine to melanin *via* dopa synthesis is a generic property of heme protein peroxidases. In neutrophils and eosinophils it is likely that the peroxidatic oxidation of tyrosine to melanin was carried out by peroxidases in these cells which have been isolated and characterized. The peroxidase or peroxidase-like enzyme responsible for the oxidation of tyrosine to melanin in melanocytes remains to be isolated and characterized. It is not known whether other heme proteins, such as catalase, which under some circumstances can act as a peroxidase [5], can also oxidize tyrosine to melanin. There are no reliable histochemical methods for differentiating the peroxidatic action of catalase from that of peroxidase [3]. However, there is evidence that peroxidatic oxidation of tyrosine to melanin in melanoma cells is not mediated by catalase: electrophoretic studies [4, 11] in our laboratory have shown that peroxidase bands, responsible for the conversion of tyrosine to melanin, were distinct from the catalase band.

Hydrogen peroxide is both generated and consumed by peroxidatic oxidation of tyrosine to melanin: rapid auto-oxidation of dihydroxyindole formed in the terminal stages of melanin synthesis generates hydrogen peroxide [21] which is utilized in the enzymatic first two steps of the peroxidase-mediated oxidation of tyrosine to melanin (hydroxylation of tyrosine and oxidation of dopa to dopa quinone). Peroxidase-mediated oxidation of tyrosine to melanin can, therefore, be self-sustaining from the standpoint of peroxide-requirement. However, there must be enough hydrogen peroxide to initiate the reaction. When dopa is used as co-factor for the reaction, the hydrogen peroxide generated by its auto-oxidation [6] is enough to initiate the reaction [11–14]. Although our previous experiments [16] suggested that hydrogen peroxide had to be added when dihydroxyfumarate was used as co-factor, results of this study indicated that reaction product may be seen in the peroxidase-mediated oxidation of tyrosine to melanin using dihydroxyfumarate as co-factor, without added hydrogen peroxide. This indicated that hydrogen peroxide generated by auto-oxidation of dihydroxyfumarate and/or endogenous hydrogen peroxide was enough to initiate the reaction. In all instances, the peroxidatic nature of the reaction was confirmed by its suppression by catalase.

The ability of peroxidase to oxidize tyrosine to melanin *via* dopa synthesis and its presence at sites of melanin [11–14] and catecholamine [10] synthesis suggests that it may have roles in their biosynthesis *in vivo*. The importance of the peroxidase-dependent tyrosine to melanin pathway in granulocytes remains to be determined. Melanin is not commonly present in mammalian leukocytes, although it has been described [22]. Melanin is readily demonstrable in leukocytes in the peripheral blood of amphibians [22]. Regulation of melanin synthesis in granulocytes may be related to the availability of hydrogen peroxide, the presence of inhibitors such as ascorbate and thiol compounds, or the degradation of dopa in alternate pathways. Studies in our laboratory and elsewhere [20] have shown that thiol compounds in the incubation medium will inhibit peroxidase. Whether or not granulocytes contain other enzymes required for catecholamine synthesis is unknown.

References

1 BLOCH, B.: Das Problem der Pigmentbildung in der Haut. Arch. Derm. *124:* 129–207 (1917).
2 BURNETT, J.: The tyrosinases of mouse melanoma. J. biol. Chem. *246:* 3079–3091 (1971).
3 COTRAN, R. and LITT, M.: Ultrastructural localization of horseradish peroxidase and endogenous peroxidase activity in guinea pig peritoneal macrophages. J. Immunol. *105:* 1536–1546 (1970).
4 EDELSTEIN, L.: Unpubl. data.
5 FAHIMI, D.: Cytochemical localization of peroxidatic activity of catalase in rat hepatic microbodies (peroxisomes). J. Cell Biol. *43:* 275–288 (1969).
6 GILLETTE, J., WATLAND, D. and KALNITSKY, G.: The catalysis of the oxidation of some dihydroxybenzine derivatives by various metallic ions. Biochim. biophys. Acta. *15:* 526–532 (1954).
7 GRAHAM, R., Jr. and KARNOVSKY, M. J.: Glomerular permeability: ultrastructural cytochemical studies using peroxidases as protein tracers. J. exp. Med. *124:* 1123–1133 (1966).
8 GRIDLEY, M. (ed.): Manual of histologic and special staining technics, Armed Forces Inst. Path.; p. 113 (Blakiston Div., McGraw-Hill, New York 1960).
9 NOVIKOFF, A., ALBALA, A. and BIEMPICA, L.: Ultrastructural and cytochemical observations on B16 and Harding-Passey mouse melanomas; the origin of melanosomes and compound melanosomes. J. Histochem. Cytochem. *16:* 299–319 (1968).
10 OKUN, M.; DONNELLAN, B.; LEVER, W.; EDELSTEIN, L. and OR, N.: Peroxidase-dependent oxidation of tyrosine or dopa to melanin in neurons. Histochemie *25:* 289–296 (1971a).
11 OKUN, M.; EDELSTEIN, L.; HAMADA, G.; OR, N.; BLUMENTAL, G.; DONNELLAN, B. and BURNETT, J.: Oxidation of tyrosine to melanin mediated by mammalian peroxidase; the possible role of peroxidase in melanin synthesis and catecholamine synthesis *in vivo;* in V. RILEY Pigmentation, its genesis and biologic control, pp. 571–592 (Appleton-Century-Crofts, New York 1972).
12 OKUN, M.; EDELSTEIN, L.; OR, N.; HAMADA, G. and DONNELLAN, B.: Histochemical studies of conversion of tyrosine and dopa to melanin mediated by mammalian peroxidase. II. Life Sci. *9:* 491–505 (1970a).

13 OKUN, M.; EDELSTEIN, L.; OR, N.; HAMADA, G. and DONNELLAN, B.: The role of peroxidase vs. the role of tyrosinase in enzymatic conversion of tyrosine and dopa to melanin in melanocytes, mast cells and eosinophils; an autoradiographic-histochemical study. J. Invest. Derm. 55: 1–12 (1970b).

14 OKUN, M.; EDELSTEIN, L.; OR, N.; HAMADA, G.; DONNELLAN, B. and LEVER, W.: Histochemical differentiation of peroxidase-mediated from tyrosinase-mediated melanin formation in mammalian tissues; the biologic significance of peroxidase-mediated oxidation of tyrosine to melanin. Histochemie 23: 295–309 (1970c).

15 OKUN, M.; EDELSTEIN, L.; NIEBAUER, G. and HAMADA, G.: The histochemical tyrosine-dopa reaction and its use in localizing tyrosinase activity in mast cells. J. Invest. Derm. 53: 39–45 (1969).

16 OKUN, M.; PATEL, R.; DONNELLAN, B.; LEVER, W.; EDELSTEIN, L. and EPSTEIN, D.: Dopa compared with dihydroxyfumarate as co-factor in the peroxidase-mediated oxidation of tyrosine to melanin. Histochemie 27: 331–334 (1971b).

17 PATEL, R. and OKUN, M.: Unpubl. data.

18 PATEL, R.; OKUN, M.; EDELSTEIN, L. and EPSTEIN, D.: Biochemical studies of peroxidase-mediated oxidation of tyrosine to melanin: demonstration of hydroxylation of tyrosine by plant and human peroxidase. Biochem. J. 124: 439–441 (1971).

19 PEARSE, A.: Histochemistry, theoretical and applied; p. 903 (Little, Brown, Boston 1960).

20 RANDALL, L.: Reaction of thiol compounds with peroxidase and hydrogen peroxide. J. biol. Chem. 164: 521–527 (1946).

21 SWAN, G. and WRIGHT, D.: A study of the evolution of carbon dioxide during melanin formation, including the use of 2(3:4-dihydroxyphenyl) (1-^{14}C)- and 2-(3:4-dihydroxyphenyl) (2-^{14}C) ethylamine. J. chem. Soc., pp. 381–384 (1954).

22 WASSERMANN, H.: Extension of the concept 'vertebrate epidermal melanin unit' to embrace visceral pigmentation and leukocytic melanin transport. Nature, Lond. 213: 282–283 (1967).

Author's address: Dr. MILTON R. OKUN, Dermatology Laboratory, Boston City Hospital, 818 Harrison Ave., *Boston, MA 02118* (USA)

Tyrosinase Inhibition Studies in Human Malignant Melanoma Grown *in vitro*

Marvin M. Romsdahl and Peggy A. O'Neill

M. D. Anderson Hospital and Tumor Institute, Houston, Tex.

Introduction

Inhibition of tyrosinase activity was postulated as an important factor in melanogenesis as early as 1946 [6]. While chemicals of many types have been implicated as inhibiting the conversion, by enzyme, of tyrosine to dopa and dopa to melanin, little work has been done to identify the naturally occurring inhibitor in pigment producing cells. Riley and his coworkers, in 1953 [4], observed a natural inhibitor separable by column chromatography. More recently, Wilde and Flawn [7] described a similar substance which they isolated from guinea-pig skin and partially characterized as being a heat stable, alkali labile protein of approximately 6000 molecular weight. This inhibitor was able to prevent tyrosine conversion to melanin. Chian [2] also found a naturally occurring soluble protein in Cloudman S91 mouse melanoma capable of inhibiting tyrosinase activity. It was characterized as having a molecular weight of less than 10,000, dialyzable, and heat stable.

While not isolating nor extensively characterizing such an inhibitor, evidence is submitted for the existence of such a factor in human pigment-producing cells. We have demonstrated that melanin production is not strictly related to tyrosinase activity and that there is a natural inhibitor in human malignant melanoma cells grown *in vitro* which is presumed to play a role in pigment production. While the quantity of enzyme may be similar in a low-pigment producing strain and a highly pigmented strain, the resultant activity is not identical due to interaction with an inhibitor. We have demonstrated this by mixing experiments with melanoma cell homogenates, and utilizing gel electrophoresis, column chromatography, and dialysis. These experiments support the claim that a natural inhibitor performs a substantial role in pigment regulation.

Materials and Methods

Tissue Culture Cells

In 1967, we established a permanent line of human malignant melanoma cells in tissue culture [5]. This line, designated Line LeCa, was later cloned to yield a number of strains varying in morphology, pigment production, and tyrosinase activity. Strain 19-4, a well-differentiated, heavily pigmented strain having the highest tyrosinase activity, was used in the mixing experiments as the source of tyrosinase. Other strains, ranging from low-pigmented Strain 26-4 and Strain 39-4 to moderately pigmented Strain 9-4, served as sources of inhibitor. A non pigment-producing human amniotic cell line, Line Fl, served as a control. In all experiments homogenates of cells were prepared at 4 °C utilizing a ground glass homogenizer and an initial concentration of cells in distilled water of 1:2 (w/v). Dilutions in distilled water ranging downward to 1:32 were made for testing purposes.

Tyrosinase Assay

Tyrosinase activity in these cell strains was determined by means of a fluorimetric method [1]. The assay utilizes tyrosine as substrate, with a catalytic amount of dopa, and measures production of 5,6-dihydroxyindole after a carefully controlled time period.

Melanin Determination

Melanin content of these strains and the Line Fl was compared by hydrolysis of the cells in an alkaline solution. Cells were placed in 0.1N sodium hydroxide at a concentration of 30 mg of cells/1 ml of sodium hydroxide and incubated, with occasional agitation, for 5 h at 37 °C. Optical density of the resultant solution was read at 420 mμ versus a sodium hydroxide blank. Comparative readings were recorded as representative of melanin content.

Cell Homogenate Mixing Experiments

To determine if tyrosinase activity was related to amount of enzyme present or to a natural inhibitor, tissue culture cell homogenate mixing experiments were performed. Using a constant concentration of homogenate of Strain 19-4 as enzyme source, dilutions of homogenates from each of the other cell strains were added individually, incubated, and assayed for tyrosinase activity. Dilutions of 'inhibitor' strains and Line Fl ranged from 1:4 to 1:16.

Electrophoresis

Analysis of tyrosinase activity of the different strains was carried out, using acrylamide gel disc electrophoresis. Cell homogenates were prepared as described and subjected to electrophoresis in the standard disc acrylamide gel system for protein separation, utilizing a 7% gel in tris-glycine buffer, pH 9.0 [3]. Following protein separation, the gels were placed in a solution of 0.3% dopa in 0.1M phosphate buffer at pH 6.8 for 1 h. At the end of this time, a dark brown band could be seen at a distance of approximately 1 to 1.5 cm from the origin, presumptively representing tyrosinase. Semi-quantitation of the band was achieved by scanning spectrophotometry at 420 mμ and planimetry.

Column Chromatography

Homogenates of human malignant melanoma tumors were subjected to column chromatography on Sephadex G-50 with 0.1M phosphate buffer. Following elution, serial fractions were analyzed for protein and tyrosinase activity, the latter utilizing the fluorimetric technique previously described.

Results

The strains from the original Line LeCa have been shown to have different tyrosinase activities, as determined fluorimetrically [1], ranging from production of 35 mμM of dopa from tyrosine per hour by Strain 19-4 to 2 mμM per hour by Strain 39-4. Figure 1 shows enzyme activity at various cell homogenate concentrations for six melanoma cell strains. The control cells, Line Fl, demonstrated no tyrosinase activity by fluorimetric assay.

The enzyme activity of the strains, however, did not always correspond to the degree of pigmentation, as indicated by the melanin content as shown in table I. While Strain 9-4 demonstrated intermediate melanin production, enzyme activity was low. Very minimally pigmented strains, Strains 39-4 and 26-4, on the other hand, were shown to have intermediate values of tyrosinase activity.

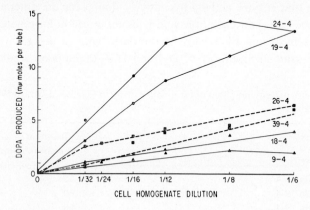

Fig. 1. Tyrosinase activity in cloned strains of a human malignant melanoma cell line, LeCa, grown *in vitro*.

Table I. Melanin content of normal and human malignant melanoma cells grown in vitro

Cell line	O.D. at 420 mµ	
	1	2
Line Fl	0.245	0.238
Line LeCa, Strain 26-4	0.348	0.352
Line LeCa, Strain 39-4	0.360	0.348
Line LeCa, Strain 9-4	0.468	0.471
Line LeCa, Strain 19-4	0.542	0.550

In mixing experiments, using cell homogenates of moderate and low-pigmented strains as the inhibitor, one of the low-pigmented cell strains, Strain 26-4, gave a maximum inhibition of tyrosinase activity of 26%. The other low-pigmented strain, Strain 39-4, inhibited tyrosinase activity by 60%. Both the moderately pigmented strain, Strain 9-4, and the control, Line Fl, gave a maximum inhibition of enzyme activity of 66% (table II). In all experiments, the inhibitory effect was linear with increasing concentrations of inhibitor cell homogenate. Figure 2 compares observed values of enzyme activity for various mixtures of Strain 19-4 homogenate with Strain 9-4 as inhibitor with expected values of the same mixtures. Inhibition of Strain 19-4 by Strain 9-4 is clearly evident. Figure 3 shows percent inhibition for each mixture of Strains 19-4 and 9-4, illustrating linearity with concentration of inhibitor cell homogenate.

Analysis of tyrosinase activity, utilizing acrylamide gel disc electrophoresis, indicates that enzyme activity is relatively similar for the different strains. Utilizing the same concentrations and volumes of cell homogenates, enzyme activity as measured by density of the dark band in the gel was approximately the same in Strains 26-4, 39-4 and 19-4. Preliminary studies

Table II. Maximum percent inhibition of tyrosinase activity in Line LeCa, Strain 19-4 by other cell strains

Cell homogenates	Maximum percent inhibition
Line LeCa, Strain 26-4	26
Line LeCa, Strain 39-4	60
Line LeCa, Strain 9-4	66
Line Fl	66

indicate that the natural inhibitor migrates at a slightly slower rate than albumin on acrylamide gel electrophoresis.

Chromatographic data utilizing Sephadex G-50, has indicated that the soluble inhibitor has a molecular weight of slightly more than 10,000. As similarly observed by RILEY et al. [4], melanoma tumor homogenates showed increased tyrosinase activity after being subjected to Sephadex filtration. Substantiating the size and solubility of the inhibitor is the observation that human melanoma tumor homogenates showed increased activity following dialysis.

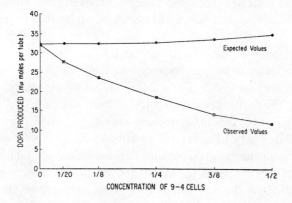

Fig. 2. Expected and observed values of tyrosinase activity for mixtures of Line LeCa, Strain 19-4 with varying concentration of Line LeCa, Strain 9-4.

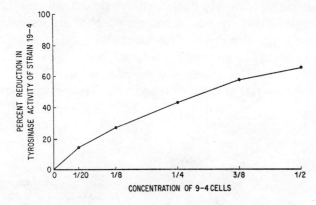

Fig. 3. Inhibition of tyrosinase activity in Line LeCa, Strain 19-4 by homogenates of Line LeCa, Strain 9-4.

Discussion

Following concentration by centrifugation of a suspension of cells from each of the various strains isolated from malignant melanoma Line LeCa, one can observe differences in pigment content between the individual cell strains. Strain 19-4 produces a large quantity of melanin, Strain 9-4 an intermediate amount, and Strains 26-4 and 39-4 a very small amount of pigment. The Line Fl produces no melanin at all, having a very white pellet in contrast to the gradation from gray to black observed in strains obtained from Line LeCa. Attempts to correlate melanin production with tyrosinase activity in these strains indicates that these two properties are not always proportional. Consequently, pigmentation was not strictly related to degree of enzyme activity.

The mixing experiments indicate that there is an inhibitor present in the cells; secondly, there appears to be some type of interaction or equilibrium between inhibitor and enzyme which results in regulation of the quantity of melanin produced. The fact that homogenates of human Line Fl cells, a non melanin-producing line, inhibits tyrosinase activity to a high degree suggests that the inhibitor may be a natural constituent of a wide variety of cells. In pigment producing cells, the relationship between the inhibitor and enzyme may be closely regulated by certain genetic factors. It may be possible that the inhibitor is the actual regulator of pigment production, rather than the enzyme, tyrosinase.

Electrophoresis data indicate that there is as much enzyme present in the low-pigmented strains as in the highly-pigmented one. However, it is assumed that it is being suppressed by inhibitor within the cell. By utilization of the acrylamide gel column, tyrosinase and the inhibitor are separated, allowing the enzyme to function maximally, and permitting a determination of the true quantity of enzyme present. The solubility and small molecular weight may suggest a complex substrate competitor of some type as a regulator product, but no real evidence to support this has been advanced.

References

1 ADACHI, K. and HELPRIN, K. M.: A sensitive fluorimetric assay method for mammalian tyrosinase. Biochem. biophys. Res. Comm. *26:* 241–246 (1967).
2 CHIAN, L. T. Y. and WILGRAM, G. F.: Tyrosinase inhibition: its role in suntanning and in albinism. Science *155:* 198–200 (1967).

3 DAVIS, B. J.: Disc electrophoresis. II. Method of application to human serum proteins. Ann. N. Y. Acad. Sci. *121:* 404–427 (1964).
4 RILEY, V.; HOBBY, G. and BURK, D.: Oxidizing enzymes of mouse melanomas: their influence, enhancement, and chromatographic separation. Pigment cell growth. Proc. 3rd Int. Pigment Cell Conf. (Academic Press, New York 1953).
5 ROMSDAHL, M. M.: Characteristics of cloned strains from human malignant melanoma cell lines grown *in vitro.* Fed. Proc. *27:* 720 (1968).
6 ROTHMAN, S.; KRYSA, H. F. and SMILJANIC, A. M.: Inhibitory action of human epidermis on melanin formation. Proc. Soc. exp. Biol. Med. *62:* 208–209 (1946).
7 WILDE, P. F. and FLAWN, P. C.: Partial characterization of a natural dopa autoxidation inhibitor and its mode of action. Abstr. 7th Int. Pigment Cell Conf., p. 27 (1969).

Author's address: Dr. MARVIN M. ROMSDAHL, M. D. Anderson Hospital and Tumor Institute, *Houston, TX 77025* (USA)

Melanogenic Inhibition by Protein Derivatives

M. L. COOPER

Departments of Dermatology, Wayne State University School of Medicine, Detroit, Mich., and Veterans Administration Hospital, Allen Park, Mich.

For more than twenty-five years now, investigators of the biology and biochemistry of pigmentation have described the presence of naturally occurring melanogenic inhibitors. These various inhibitors have been described as heat-stable, dialyzable, sulfhydryl compounds [7, 12] which bind copper, heat-stable, non-sulfhydryl substances [1, 11, 13], whose mode of action is unclear, and substances which bind the tyrosinase substrates, tyrosine and dopa, making them unavailable for melanin synthesis [2, 5, 14]. Some authors have concentrated their investigations on the naturally occurring inhibitors of dopa autoxidation, thus eliminating interaction with tyrosinase as a cause of inhibition, and have suggested that their described inhibitors function through substrate binding [5] or the binding of heavy metal impurities, which have been shown to stimulate the non-enzymic oxidation of dopa [8]. It is now clear that there are several naturally occurring melanogenic inhibitors, and at least one of them has been found in every tissue examined [6].

In 1968, utilizing the tyrosinase radio-assay [10, 2] method, I found evidence for a substrate binding inhibitor in a commercial preparation of partially purified mushroom tyrosinase. Incubations containing a large amount of the tyrosinase formed less melanin than more dilute preparations at the time of substrate exhaustion. The addition of additional tyrosinase and post incubation resulted in no increase in melanin, while the addition of tyrosine followed by post incubation did cause increased melanin production. Thus, although active tyrosinase remained, the substrate had been completely exhausted in the more concentrated preparations, even though they had produced less melanin than the more dilute preparations. This suggested that this substrate binding inhibitor might be intimately associated with the tyrosinase molecule. KARKHANIS has reported an inhibitor which remains with mushroom tyrosinase until the final steps of purification, but he believed this inhibitor to have a higher affinity for tyrosinase than for tyrosine. When this inhibitor was removed from his tyrosinase preparation, no lag was seen in the oxidation of tyrosine. Addition of the inhibitor induced a lag in the oxidation of tyrosine but not in the oxidation of dopa [9]. Since the inhibitor which MISHIMA and I had found in melanomas [2] functioned by binding substrate, and since the commercial mushroom tyrosinase also contained a substrate binding inhibitor, I speculated that this inhibitor might be a portion of the tyrosinase molecule, possibly separated substrate binding sites, which could bind substrate, but not oxidize it [3]. Although attempts to produce an inhibitor by the fragmentation of tyrosinase molecules with x-irradiation were unsuccessful, treatment of tyrosinase with γ-chymotrypsin produced, as shown in table I, an inhibitor which functions by making tyrosine unavailable for enzymic oxidation. In the untreated mushroom tyrosinase preparation, the 1:1 (0.4 mg tyrosinase) incubate produced somewhat more C^{14}-melanin than a 1:10 dilution. The addition of 1 mg γ-chymotrypsin to the 1:1 preparation with its subsequent dilution for the 1:10 incubate, resulted in a distinct inhibition in melanin production in the more concentrated preparation, with little change in the diluted preparation. When these tyrosinase-chymotrypsin mixtures were incubated for an additional 4 hours, only a slight increase in melanin was seen in the 1:1 incubate. This remained unchanged even when additional tyrosinase was supplied after the initial 16 hours. The addition of more tyrosine, however, was able to increase the amount of melanin produced. Inhibition was thus due to substrate binding. This was further substantiated by paper chromatography of the 1:1 and 1:10 incubates containing chymotrypsin after 16 hours incubation, without the addition of trichloro-acetic acid (TCA), as shown in figure 1. The C^{14}-melanin remains at the origin, while the inhibitor-tyrosine complex moves near the solvent front. Control chromatography of tyrosine in this system shows the tyrosine peak at 12 cm. The presence of dopa-binding inhibitors in some commercial mushroom tyrosinase preparations which showed no evidence of tyrosine binding indicated the presence of at least two distinct inhibitors [4], one binding tyrosine and the other binding dopa (fig. 2). In this working hypothesis, the substrate binding sites of the tyrosinase molecule which are split off from the molecule by protease, inhibit melanogenesis by binding the substrate, thus preventing its oxidation.

The heat stability of this tyrosinase-chymotrypsin-produced inhibitor was investigated by incubating 8 mg mushroom tyrosinase with 5 mg

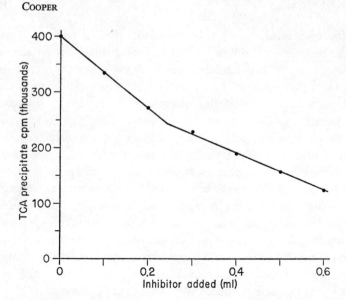

Fig. 1. Paper chromatography in 30% dimethyl sulfoxide of 1:1 (0.4 mg) and 1:10 dilutions of mushroom tyrosinase incubated with chymotrypsin (1 mg for 1:1 incubate) for 16 h with 2.6×10^{-5} M DL-tyrosine-C^{14} as substrate. No TCA was added prior to chromatography. Tyrosine itself migrates to the 12 cm position in this system.

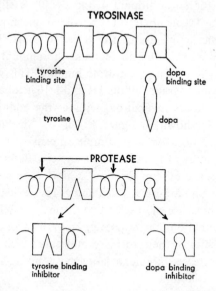

Fig. 2. A working hypothesis for the genesis of melanogenic substrate-binding inhibitors from chymotrypsin-treated mushroom tyrosinase.

Table I. Radioassay of total melanin produced during incubation with 2.6×10^{-5} M DL-tyrosine-C^{14} as substrate

DPM of the TCA precipitate after 16 h incubation

1:1	1:10	1:1 CH	1:10 CH
260×10^3	221×10^3	85×10^3	213×10^3
243	238	87	237
271	229	84	248

DPM of the TCA precipitate after 20 h incubation

1:1 CH	1:10 CH	1:1 CH+E	1:10 CH+E	1:1 CH+S	1:10 CH+S
99×10^3	242×10^3	88×10^3	272×10^3	158×10^3	269×10^3
97	240	90	256	163	272
93	259	93	254	160	268

1 mg chymotrypsin (for 1:1), 0.4 mg additional mushroom tyrosinase or an additional original amount of substrate were added as indicated. 1:1 equal 0.4 mg tyrosinase. When additional tyrosinase or additional substrate were added, this was done after the completion of 16 h incubation, followed by 4 h post-incubation. CH = chymotrypsin added; E = tyrosinase added; S = C^{14}-tyrosine added.

chymotrypsin in 8 ml water at 37°C for 15 min, followed by boiling for 15 min. When this is centrifuged to remove coagulated protein, and various amounts of supernatant are incubated with C^{14}-tyrosine (2.5×10^{-5} M), the dose response seen in figure 3 is obtained.

Further work, however, has shown that chymotrypsin is able to produce melanogenic inhibitors not only from tyrosinase, but from such proteins as albumin and fibrinogen (table II), although albumin and boiled chymotrypsin themselves possess no inhibitory properties.

While the inhibitor produced by the action of chymotrypsin on albumin is heat stable, it is rapidly destroyed during storage, as seen in table III. Storage for 20 h at room temperature reduces the amount of inhibition by greater than 60%. Furthermore, this inhibitor is able to induce a lag period in the oxidation of dopa (table III, fig. 4) which is parallel to the inhibition of tyrosinase activity. This induction of a lag in the oxidation of dopa suggests that this inhibitor contained in the boiled supernatant of chymotrypsin-treated albumin may be functioning as a reducing agent as well as inhibiting by some other mechanism. It seems clear, how-

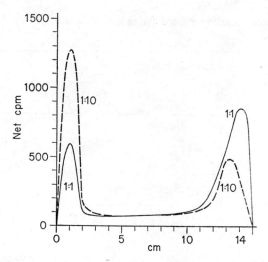

Fig. 3. Inhibition of melanin production by the boiled supernatant of mushroom tyrosinase (20 mg) incubated with chymotrypsin (8 mg) for 15 min in 8 ml water. Total assay volume 1.2 ml.

Table II. Radioassay of total melanin produced during 16 h incubation with 2.6×10^{-5} molar tyrosine-C^{14} as substrate

Control	60,000
	63,000
BAC	39,000
	33,000
BFC	30,000
	30,000
Albumin	67,000
	71,000

BAC = boiled supernatant of 20 mg albumin incubated 15 min with 8 mg chymotrypsin in 8 ml water, then boiled for 15 min. 0.3 ml was added for a total incubation assay volume of 0.6 ml; BFC = same as above except that fibrinogen was used in place of albumin. Assay with native albumin contained 0.6 mg albumin.

ever, that either the mechanism of melanogenic inhibition by chymotrypsin-treated tyrosinase, which binds substrate, is different from that of albumin or fibrinogen treated with chymotrypsin, or that the chymotrypsin digests of these nonmelanogenic-specific proteins are in some way able to bind tyrosine and dopa.

Since the submission of this manuscript, the work of ROSTON [J. Biol. Chem. *235:* 1002–1004 (1960)] showing the reaction of sulfhydryl groups with an oxidation product of dopa has come to my attention. While I can

Table III. Spectrophotometric tyrosinase assay measuring dopachrome formation at 475 mμ with 4×10^{-4} M L-dopa as substrate

	Percent activity	Lag (min)
Control	100	0
BAC (fresh)	29	5
BAC (frozen 20 h)	43	4
BAC (22°, 20 h)	71	1
Albumin (8 mg/ml)	100	0

BAC (boiled albumin-chymotrypsin) was produced by incubating 2 g albumin with 40 mg chymotrypsin in 40 ml water at 37° for 1 h, then 100° for 15 min followed by centrifugation and filtration through glass wool. 1.0 ml was added for a total incubation volume of 3.0 ml.

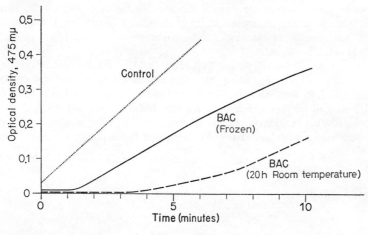

Fig. 4. Spectrophotometric tyrosinase assay with 4×10^{-4} M L-dopa as substrate. BAC = Boiled albumin-chymotrypsin supernatant prepared as in table III.

not rule out a role of sulfhydryl binding in the inhibition reported here, iodoacetamide, at a final concentration of 1.8 mg/cc, had no effect on the inhibitory activity of BAC measured by the radioassay procedure.

References

1 CHIAN, L. T. J. and WILGRAM, G. F.: Tyrosinase inhibition; its role in suntanning and in albinism. Science *155:* 198–200 (1967).
2 COOPER, M. and MISHIMA, Y.: Substrate limiting melanogenic inhibitor in malignant melanomas. Nature, Lond. *216:* 189–190 (1967).
3 COOPER, M.: Inhibitory activity of enzyme fragments. J. Cell Biol. *47:* 40a (1970).
4 COOPER, M. and MISHIMA, Y.: Substrate limiting melanogenic inhibitor. Further characterizations. J. Cell Biol. *43:* 26a (1970).
5 FLAWN, P. C. and WILDE, P. F.: A study of the mechanism of action and role of a natural inhibitor of dopa autoxidation isolated from guinea pig skin. J. Invest. Derm. *55:* 159–163 (1970).
6 FLAWN, P. C. and WILDE, P. F.: Isolation and partial characterization of the natural inhibitor of dopa autoxidation in skin. J. Invest. Derm. *55:* 153–158 (1970).
7 FLESCH, P.: Inhibitory action of extracts of mammalian skin on pigment formation. Proc. Soc. exp. Biol. Med. *70:* 136–140 (1949).
8 HIRSCH, H. M.: Inhibition of melanogenesis by tissues and the control of intracellular autoxidations; in M. GORDON Pigment cell biology, pp. 327–358 (Academic Press, New York 1959).
9 KARKHANIS, Y. D.: The inhibition of mushroom tyrosinase by its apoenzyme, a protein inhibitor, and certain reagents; Ph. D. thesis, Florida State Univ. (1961).
10 KIM, K. and TCHEN, T. T.: Tyrosinase of the goldfish *Carassius autatus* L. I. Radioassay and properties of the enzyme. Biochim. biophys. Acta *59:* 569–576 (1962).
11 QUEVEDO, W. C. and ISHERWOOD, J. E.: Mammalian pigmentation. I. Inhibitors of dopa autoxidation in extracts of human and mouse skin. J. Invest. Derm. *34:* 309–315 (1960).
12 ROTHMAN, S.; DRYSA, H. F. and SMILJANIC, A. M.: Inhibitory action of human epidermis on melanin formation. Proc. Soc. exp. Biol. Med. *62:* 208–209 (1946).
13 SATOH, G. J. Z. and MISHIMA, Y.: Tyrosinase inhibitor in Fortner's amelanotic and melanotic malignant melanoma. J. Invest. Derm. *48:* 301–303 (1967).
14 SIDMAN, R. L.; PEARLSTEIN, R. and WAYMOUTH, C.: Pink-eyed dilution (p.) gene in rodents: increased pigmentation in tissue culture. Develop. Biol. *12:* 93–116 (1965).

Author's address: MICHAEL L. COOPER, Veterans Administration Hospital, *Allen Park, MI 48101* (USA)

Some Properties of the Outer Melanosomal Membrane[1]

M. H. Van Woert

Departments of Internal Medicine and Pharmacology, Yale University School of Medicine, New Haven, Conn.

Introduction

Most of the tyrosinase activity in mammalian melanocytes is localized in cytoplasmic organelles, known as melanosomes, where melanin is synthesized and stored [11]. Tyrosinase in these melanosomes has been demonstrated to be in a latent or partially inactive state [7, 12, 14]. Various physical and chemical agents which labilize biological membranes (e.g., phenothiazines, bile acids, neutral steroids, proteolytic enzymes, sonication and detergents) increase melanosomal tyrosinase activity [7, 12, 14], whereas chloroquine, which has a membrane stabilizing action, inhibits this increase in tyrosinase activity [14]. From this experimental evidence, we have postulated that the permeability of the outer unit membrane which surrounds the melanosome determines the quantity of the substrate L-tyrosine which is transported into the melanosome to react with the tyrosinase molecule. Alteration of melanosomal membrane permeability may be one mechanism controlling the rate of melanin synthesis.

This paper presents additional experimental evidence in support of this hypothesis. X-irradiation, aromatic amino acids, and metabolic inhibitors were investigated for their effect on the latent tyrosinase activity in B-16 mouse melanoma.

Materials and Methods

Preparation of B-16 Melanoma Large Granule Fraction Containing Melanosomes.
B-16 mouse melanomas in C57Bl/6J mice were excised 11–13 days after transplantation and a melanosome-rich fraction was prepared fresh daily for each experiment.

---

[1] This work was supported by USPHS grant NS 07542 and USAEC contract AT (30-1) 3960.

The melanomas were weighed and homogenized gently in a Potter-Elvehjem homogenizer (teflon pestle) with 10 volumes of 0.25 M sucrose containing 0.08 M sodium phosphate buffer, pH 6.8 at 4°C. The nuclear fraction was removed by centrifuging the homogenate for 5 min at 700 g. A large granule fraction containing most of the melanosomes was prepared by centrifuging the supernatant at 11,000 g for 10 min. The pellet was resuspended in 0.25 M sucrose with 0.08 M sodium phosphate buffer, pH 6.8 (¼ of the original volume). This large granule fraction also contained mitochondria and lysosomes, which do not have tyrosinase activity.

In vitro X-Irradiation

The x-ray machine used was a Siemens Stabilipan 250. Melanosome-rich suspensions were placed in glass tubes close to the exit window and x-irradiated at 250 kV 15 ma, without filtration. The total exposure varied from 25 to 375 krads. The dose rate measured by ferrous sulfate dosimetry was 4.6 to 5.4 krads per minute for volumes from 7 ml to 4 ml respectively. Suspensions were kept in an ice bath during x-irradiation and were stirred slowly with an overhead glass stirrer.

Tyrosinase Assay

Tyrosinase was measured by the method of POMERANTZ [9]. The tritiated water formed when L-tyrosine-3,5-^{3}H is hydroxylated by tyrosinase was separated with a Norit A-Celite column and counted in a liquid scintillation spectrometer with BRAY's solution [3]. Blanks containing boiled enzyme (100°C for 10 min) were run simultaneously. The incubation time was 1 h, temperature 37°C and each assay was carried out in duplicate.

Since tyrosine hydroxylase and peroxidase can also hydroxylate L-tyrosine under special conditions, the specificity of the Pomerantz method for measuring tyrosinase activity in the B-16 melanoma has been investigated and the details will be reported elsewhere [13]. The Pomerantz method does not detect tyrosine hydroxylase activity in the absence of the essential cofactor 2-amino-4-hydroxy-6,7-dimethyltetrahydropteridine (DMPH$_4$). This tyrosinase assay also could not detect the low levels of peroxidase activity found in the B-16 melanoma.

Results

In vitro Effect of X-Irradiation on B-16 Melanosomal Tyrosinase Activity

Suspensions of B-16 melanoma melanosomes were x-irradiated at doses ranging from 25 to 375 krads in an ice bath. Tyrosinase activity was assayed either immediately after irradiation or, in other samples, after incubation at 4°C for 20 h post-irradiation in air. Immediately after x-irradiation there was no change in tyrosinase activity even after doses as high as 325 krads (fig. 1). After incubation at 4°C for 20 h post-irradiation in air, tyrosinase activity increased proportional to the radiation dose, up to 125 krads. Between 125 and 375 krads, tyrosinase activity was more than 4-fold greater than in sham-irradiated control suspensions. There was no difference in

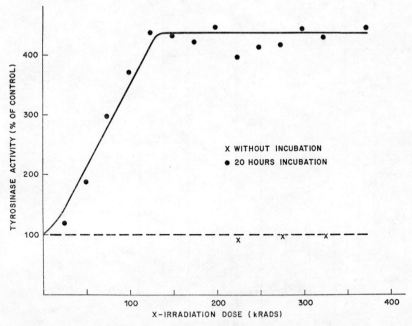

Fig. 1. Effect of *in vitro* x-irradiation on melanosomal tyrosinase activity. The solid dots represent the tyrosinase activity of melanosome suspensions incubated at 4°C for 20 h post-irradiation. The values are the mean of 1 to 3 experiments expressed as per cent of sham-irradiated control tyrosinase incubated under identical conditions. The crosses (X) are the tyrosinase activity of melanosome suspensions x-irradiated in the same manner but assayed immediately post-irradiation.

tyrosinase activity before and after 20 h of incubation in air at 4°C, in sham-irradiated control suspensions. The effect of 20 h incubation in nitrogen was compared with incubation in air under the same conditions (fig. 2). Melanosome suspensions were sham-irradiated, or x-irradiated at 75 krads and 100 krads. One half of each suspension was flushed with nitrogen, the remaining half flushed with air. The tubes were stoppered, incubated for 20 h at 4°C and the tyrosinase activity measured. Incubation in nitrogen significantly diminished the x-ray-induced increase in tyrosinase activity at 75 and 100 krads but had no effect on the sham-irradiated control suspensions (fig. 2).

As previously reported [14], incubation of melanosome suspensions with the non-ionic detergent nonylphenoxypolyethaneoxyl-ethanol (Igepal) also produces a 4-fold increase in tyrosinase activity. This increase in tyrosinase activity is attributed to solubilization of the melanosomal

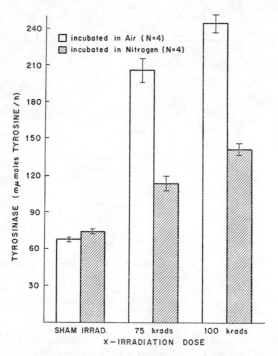

Fig. 2. Comparison of post-irradiation incubation (20 h, 4 °C) in room air (empty bars) and nitrogen (shaded bars) on melanosomal tyrosinase activity. The perpendicular lines are the SEM and N indicates the number of experiments.

membrane and release of some of the bound enzyme. In order to determine if there was any direct effect of x-irradiation on the tyrosinase molecule, the melanosome suspensions were incubated with 1% Igepal and x-irradiated at 50, 100, 150, 200, 250, and 300 krads. Post-irradiation, the suspensions were incubated for 20 h at 4 °C in air prior to assaying tyrosinase activity. There was no significant effect of these doses of x-irradiation on detergent-treated melanosomal tyrosinase activity. The tyrosinase molecule was radioresistant; it was not activated or denatured by the x-irradiation doses used in these experiments.

In vitro Effect of Chemical Agents on B-16 Melanosomal Tyrosinase Activity

Table I shows the effect of various compounds on tyrosinase activity in intact melanosomes and in melanosomes disrupted by the detergent

Table I. Effect of various compounds on untreated melanosome suspensions and detergent-treated (1% Igepal) melanosome suspensions[1]

Compound	Concentration	Tyrosinase activity (% of control)	
		Intact melanosomes	Detergent-treated melanosomes
L-phenylalanine	4×10^{-3}M	74	84
L-tryptophan	4×10^{-3}M	89	96
Na azide	10^{-3}M	55	37
Na iodoacetate	10^{-3}M	98	
2,4-dinitrophenol	10^{-3}M	103	
Ouabain	10^{-3}M	99	

1 Each value is the mean of 3 experiments.

Igepal. L-phenylalanine inhibited the tyrosinase activity in both intact and Igepal-treated melanosomes. L-tryptophan was a less potent inhibitor than phenylalanine and the inhibition was significant only in the intact melanosomes. The molar concentrations of both amino acids were 5-fold greater than that of L-tyrosine in the assay solution. Sodium azide inhibited the tyrosinase activity in both intact and detergent-treated melanosomes. Ouabain, iodoacetate, and 2,4-dinitrophenol had no effect on the tyrosinase activity in the intact melanosomes and were, therefore, not tested in the detergent-treated melanosomes.

As seen in table II, potassium cyanide had an unexpected action on melanosomal tyrosinase from B-16 melanoma. In a range of 10^{-5} to 10^{-4}M, KCN produced an approximate 50% increase in tyrosinase activity in intact melanosomes. The percentage increase in tyrosinase activity induced by KCN was less when the melanosomes were damaged by sonication or detergents. However, it is interesting that the absolute increase in tyrosinase activity (in mμ moles tyrosine hydroxylated/h) produced by KCN was similar in intact and disrupted melanosomes. At higher concentrations of KCN (10^{-3}M), melanosomal tyrosinase activity was inhibited. As reported by other investigators, all of these concentrations of KCN (10^{-3} to 10^{-5}M) inhibited mushroom tyrosinase.

Discussion

Under physiological conditions, tyrosinase may be in a latent state in the melanocyte. Several possible mechanisms may contribute to the partially

Table II. Effect of potassium cyanide on tyrosinase activity in melanosome suspensions and purified mushroom tyrosinase preparation[1]

Concentration of KCN	Melanosomal tyrosinase activity				Detergent-treated		Mushroom tyrosinase activity	
	Untreated		Sonicated					
	mμ moles tyrosine/h	% of control	mμ moles tyrosine/h	% of control	mμ moles tyrosine/h	% of control	mμ moles tyrosine/h	% of control
0	76.6 ± 6.1 (6)	–	232.8 ± 16.2 (4)	–	372.9 ± 16.9 (5)	–	63.8 ± 1.8 (6)	–
10^{-3}M	28.3 ± 2.4[3] (6)	37	54.9 ± 4.7[3] (2)	24	128.3 ± 4.7[3] (2)	34	0* (6)	0
10^{-4}M	114.9 ± 3.2[3] (6)	150	266.9 ± 9.9[4] (4)	115	406.2 ± 11.6[4] (5)	109	34.7 ± 3.3[3] (6)	54
5×10^{-5}M	115.0 ± 5.5[3] (7)	150	283.3 ± 19.1[4] (2)	122	409.2 ± 17.6[4] (3)	110	–	–
10^{-5}M	113.5 ± 7.0[2] (5)	148	–	–	–	–	55.3 ± 0.7[2] (5)	87
10^{-6}	82.7 (1)	108	–	–	–	–	63.6 ± 1.2[4] (6)	100

1 The values are the mean ±SE and the number of experiments are listed in parenthesis.
2 $p > 0.005$.
3 $p > 0.001$.
4 p not significant.

inactive nature of tyrosinase: a) the presence of endogenous inhibitors such as sulfhydryl compounds (e.g., glutathione) may inhibit tyrosinase activity [5]. b) Mammalian tyrosinase may exist as an inactive proenzyme state, as has been reported to be present in grasshopper eggs [1] and frog skin [6]. c) The outer melanosomal membrane could limit the influx of L-tyrosine into the melanosome, preventing substrate saturation of melanosomal tyrosinase. Our results suggest that this last mechanism could explain the increase in melanosomal tyrosinase induced by x-irradiation.

One of the major biological effects of x-irradiation is injury to extracellular and intracellular membranes [4, 15]. Phospholipids containing unsaturated fatty acids are known to be essential components of biological membranes. All subcellular particles which have been examined developed membrane damage from lipid peroxide formation when incubated *in vitro* for a sufficient period of time after x-irradiation [4, 16]. WILLS and WILKINSON [16] observed that acid phosphatase, cathepsin and β-glucuronidase are released from rat liver lysosomes when the membrane structure is disrupted by irradiation *in vitro*. Liberation of splenic and hepatic lysosomal enzymes by whole-body x-irradiation has also been observed [8, 10]. The x-ray dose and incubation time used to produce these radiation effects in lysosomes were similar to the conditions required to activate melanosomal tyrosinase in our experiments. When the melanosomal membrane structure is disrupted by the detergent Igepal, x-irradiation has no effect on the tyrosinase activity.

The reduction of the x-ray-induced increase in tyrosinase activity by nitrogen also suggests that membrane injury produced by lipid peroxidation is the mechanism responsible for the increase in enzyme activity during incubation in air.

Since other aromatic amino acids, such as L-phenylalanine and L-tryptophan, might compete with L-tyrosine for transport into the melanosome, the effect of these amino acids on tyrosinase activity was determined. L-phenylalanine has been reported to inhibit tyrosinase [2]. Our results indicate that L-phenylalanine produced slightly greater inhibition of the tyrosinase activity in intact melanosomes compared to the detergent-treated melanosomes. In addition to the direct inhibition of the tyrosinase molecule, L-phenylalanine may interfere with L-tyrosine transport across the melanosomal membrane. L-tryptophan had no significant effect on the disrupted melanosomal tyrosinase activity, but did decrease slightly intact melanosomal tyrosinase activity. Although the inhibition by tryptophan was minimal it also may indicate competition with L-tyrosine for membrane transport. The inhibition of tyrosinase by sodium azide is probably unrelated to the melanosome structure and due to binding of copper in the tyrosinase molecule. Three other metabolic inhibitors, sodium iodoacetate, 2,4-dinitrophenol and ouabain, had no effect on the tyrosinase activity in the intact melanosome. Of interest are the unexpected results obtained with potassium cyanide. Whereas cyanide is effective as an inhibitor of mushroom tyrosinase at levels of 10^{-5} to 10^{-4}M, these concentrations increase the tyrosinase activity of B-16 melanosomes. The mechanism by which cyanide activates B-16 melanoma tyrosinase activity is not apparent. It is not clear from our data whether cyanide caused any alterations in structural organization of the melanosome; however, this possibility might account for the greater percentage activation of tyrosinase in the intact melanosomes compared to detergent-treated melanosomes.

References

1 BODINE, J. H.; TAMISIAN, T. N. and HILL, D. L.: Effect of heat on protyrosinase, heat activation, inhibition, and injury of protyrosinase and tyrosinase. Arch. Biochem. *4:* 403–412 (1944).
2 BOYLEN, J. B. and QUASTEL, J. H.: Effects of L-phenylalanine and sodium phenylpyruvate on the formation of melanin from L-tyrosine in melanoma. Nature, Lond. *193:* 376–377 (1962).
3 BRAY, G. A.: A simple efficient liquid scintillator for counting aqueous solutions in a liquid scintillation counter. Anal. Biochem. *1:* 279–285 (1960).
4 DESAI, I. D.; SAWANT, P. L. and TAPPEL, A. L.: Peroxidation and radiation damage to isolated lysosomes. Biochim. biophys. Acta *86:* 277–285 (1964).
5 FLESCH, P. and ROTHMAN, S.: Role of sulfhydryl compounds in pigmentation. Science *108:* 505–506 (1948).
6 McGUIRE, J. S.: Activation of epidermal tyrosinase. Biochem. biophys. Res. Comm. *40:* 1084–1089 (1970).
7 MENON, I. A. and HABERMAN, H. F.: Activation of tyrosinase in microsomes and melanosomes from B-16 and Harding-Passey melanomas. Arch. Biochem. Biophys. *137:* 231–242 (1970).
8 PIERUCCI, O. and REGELSON, W.: Effects of whole-body irradiation on the activity of acid desoxyribonuclease of mouse tissues. Radiat. Res. *24:* 619–630 (1965).
9 POMERANTZ, S. H.: The tyrosine hydroxylase activity of mammalian tyrosinase. J. biol. Chem. *241:* 161–167 (1966).
10 ROTH, J. S. and HILTON, S.: The effect of whole-body x-irradiation on the distribution of acid desoxyribonuclease and β-galactosidase in subcellular fractions of rat spleen. Radiat. Res. *19:* 42–49 (1963)
11 SEIJI, M.; FITZPATRICK, T. B.; SIMPSON, R. T. and BIRBECK, M.S.C.: The melanosome: a distinctive subcellular particle of mammalian melanocytes and the site of melanogenesis. J. Invest. Derm. *36:* 243–252 (1961).

12 VAN WOERT, M. H.: Effect of phenothiazines on melanoma tyrosinase activity. J. Pharm. exp. Ther. *173:* 256–264 (1970).
13 VAN WOERT, M. H.: Manuscript under preparation.
14 VAN WOERT, M. H.; KORB, F. and PRASAD, K. N.: Regulation of tyrosinase activity in mouse melanoma and skin by changes in melanosomal membrane permeability. J. Invest. Derm. *56:* 343–348 (1971).
15 WILLS, E. D. and WILKINSON, A. E.: Effects of irradiation on sub-cellular components. II. Hydroxylation in the microsomal fraction. Int. J. Radiat. Biol. *17:* 229–236 (1970).
16 WILLS, E. D. and WILKINSON, A. E.: Release of enzyme from lysosomes by irradiation and the relation of lipid peroxide formation to enzyme release. Biochem. J. *99:* 657–666 (1966).

Author's address: Dr. MELVIN H. VAN WOERT, Departments of Internal Medicine and Pharmacology, Yale University School of Medicine, *New Haven, CT 06510* (USA)

Microsomal Cytochrome b_5 and Electron-Transfer System of Mouse Melanoma

M. HIRAGA, K. NAKAJIMA and F. K. ANAN

Division of Biochemistry, Institute for Cardio-Vascular Diseases, Tokyo Medical and Dental University, Tokyo

The functional role of mitochondria is exclusively oxidative phosphorylation to produce ATP at the expense of reducing equivalents and oxygen, except in cases such as adrenal cortical mitochondria, which have hydroxylating functions besides ATP formation related to their dual cytochrome systems and energy-dependent transhydrogenase action [6, 19]. In comparison with the extensive studies on the mitochondrial electron-transfer system, little work has been done on the redox system of microsomes. In the studies on microsomal electron-transfer systems, the most fruitful finding is that of the role of cytochrome P-450, which exerts mixed-function oxidation catalyzing steroid hydroxylation and drug metabolism [3, 12]. On the other hand, although described earlier, the precise role of cytochrome b_5 has not yet been established [18]. It has recently been suggested that stearyl-CoA desaturase activity is related to cytochrome b_5 [13] and that it has a role of second electron donor in sequential electron transfer in association with cytochrome P-450 and substrate [4]. Occurrence of cytochrome b_5 has been reported in the cells of liver and brain. Its subcellular distribution in liver cells is in membranes of microsomes, mitochondria, Golgi, and nuclei [1, 5, 8], as judged from its characteristic spectral properties. This paper reports the occurrence of cytochrome b_5 and NADH- and NADPH-linked electron-transfer systems in the microsomes of melanotic B-16 melanoma.

Preparation of Microsomes

The microsomal fraction of mouse melanoma was prepared by the procedures outlined in figure 1. During the preparation, 1.15% KCl was

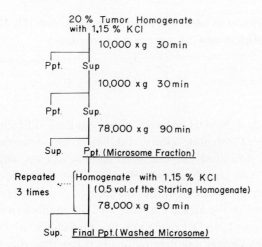

Fig. 1. Microsomal preparation of B-16 melanoma. All manipulations were performed at 2–4°.

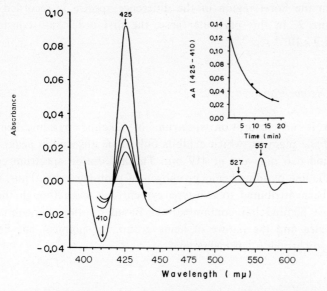

Fig. 2. Difference spectra of cytochrome b_5. The system consisted of 83 mM K-phosphate buffer, pH 7.5, 1.5 μM rotenone, 0.96%KCl, and 28.3 mg protein/ml. The base line, oxidiezed form vs. oxidized form, was run after sufficient aeration of the microsomes in the presence of rotenone. Reduction of the cytochrome b_5 was performed in the final concentration of 0.2 mM NADH. In the insert, difference of absorbance between 425 mμ and 410 mμ is plotted as a function of time. Temperature 27°. Light path, 10 mm.

used to minimize the adsorption of hemoglobin on the microsomes [11, 14]. The yield of microsomal protein was 2.2 mg per g of wet tumor.

Difference Spectra

As shown in figure 2, a NADH-reduced minus oxidized difference spectrum in the presence of rotenone clearly exhibits three absorption peaks, α-band at 557 mμ, β-band at 527 mμ, Soret band at 425 mμ, and a deep trough at 410 mμ. A wide trough around 450 mμ is likely due to flavin moieties of flavoprotein dehydrogenase(s) of the microsomes. A similar spectral pattern is obtained even in the absence of rotenone. In addition to the rotenone-insensitive pathway of NADH oxidation, this pigment is reduced by NADPH, although the extent of pigment reduction is less than that by NADH. The pigment is also autoxidizable. After exhaustion of the added NADH, the reduced pigment is gradually oxidized by sufficient aeration. The change in the Soret region of the difference spectra is recorded as shown in figure 2. In this particular case, the first order rate constant calculated is 8.9×10^{-4} sec^{-1}.

Pyridine Hemochromogen

Figure 3 is an absorption spectrum of alkaline pyridine hemochromogen of the pigment, which exhibits only three absorption peaks, at 557 mμ, around 526 mμ, and at 419 mμ. The absorption spectrum corresponds to alkaline pyridine hemochromogen of protoheme IX. Thus, the pigment has been attributed to a b-type cytochrome, even though there might be some hemoglobin contamination. Based on the characteristic difference spectra and the nature of heme group, the pigment has been found to be cytochrome b_5 of melanoma.

Reductase Activities

The microsomal preparation has reductase activities with artificial electron or hydrogen acceptors, using NADH and NADPH as the reducing agents. As given in table I, NADH-cytochrome c and -ferricyanide reductase activities are one order higher than those with NADPH, while NADH-DCIP

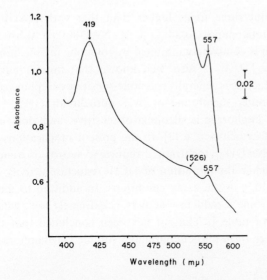

Fig. 3. Absorption spectrum of alkaline pyridine hemochromogen of the microsomal pigment(s). The system consisted of 0.05 N NaOH, 20% pyridine, 1% Na-deoxycholate, 0.81% KCl and 28.5 mg protein/ml. Reduction of the pigment(s) was performed by the addition of about 1 mg dithionite/ml. Temperature, 27°. Light path, 10 mm.

Table I. Reductases activities

Acceptors / Donors	NADH	NADPH
Cytochrome c	114 (6)	7.7 (6)
	109a (6)	- - - - -
Ferricyanide	598 (4)	49.0 (2)
DCIP	31.8 (3)	9.3 (3)
	29.3b (3)	9.0b (3)

100 mM K-phosphate buffer, pH 7.5, and 0.1 mM of either NADH or NADPH, were common constituents to the systems of all activity measurements. Other constituents were 0.05 mM horse heart cytochrome c, or 0.5 mM ferricyanide, or 0.05 mM 2,6-dichlorophenol-indophenol, respectively. Activities were measured spectrophotometrically as the change of absorbance. In the case of cytochrome c reductase activities, 0.33 mM cyanide in the final concentration was added to the reaction mixture immediately before starting the reaction. The average values of activities are expressed as mμmole acceptor reduced per min per mg protein at 27°. *Note:* in parentheses = number of preparations; a = 1 μM rotenone; b = 2 μM dicoumarol; succinate-cytychrome c reductase activity = 0.4 (4).

reductase activity is about three times higher than that with NADPH. As reported in a previous paper, oxidation of NAD-linked substrate through respiratory chain is completely inhibited by rotenone in amelanotic melanoma mitochondria [7]. It is also well known that mitochondrial NADH-cytochrome c reductase is generally rotenone-sensitive except in yeast mitochondria [10] and outer membrane of liver mitochondria [15, 17]. The soluble enzyme DT diaphorase is dicoumarol-sensitive, and does not catalyze the reduction of cytochrome c [2]. In the present preparations of melanoma microsomes, NADH-cytochrome c reductase activity is almost insensitive to rotenone, while the inhibition of DCIP reductase activity by dicoumarol is less than 10% in the assay conditions. In addition to these criteria, succinate-cytochrome c reductase activity is extremely low, which is shown in the legend of table I. Thus, it has been concluded that the microsomes are almost free from mitochondria and soluble DT diaphorase.

Vitamin K_3-Dependent Oxidation of NADPH

Microsomal oxidation of NADPH, as indicated in figure 4, is remarkably accelerated by the addition of catalytic amounts of vitamin K_3. Under

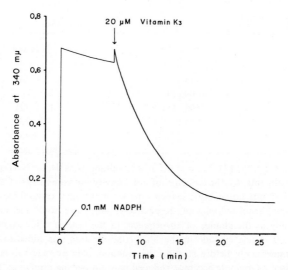

Fig. 4. Vitamin K_3-dependent NADPH oxidation. The system consisted of 250 mM K-phosphate buffer, pH 6.5, 0.19% KCl and 5.6 mg protein/ml. Temperature, 27°. Optical path, 10 mm.

anaerobic conditions, no oxidation of NADPH occurs. Moreover, a considerable increase in oxygen consumption is polarographically observed by the addition of vitamin K_3. At present, terminal oxidase such as cytochrome P-450 is unknown in the microsomes of melanoma. Nevertheless, like microsomes of liver [16] and brain [9], melanoma microsomes also contain naphthoquinone-dependent NADPH-oxidase system.

Microsomal Constituents

Some constituents of the microsomes are given in table II. The cytochrome b_5 content has been estimated by using an extinction coefficient of 163 mM^{-1} cm^{-1} for the absorption difference between 425 mμ and 410 mμ. The average content of cytochrome b_5 is 0.024 mμmole per mg protein, which is close to that of brain microsomes [9]. Although cytochrome b_5 content varied in diffrent preparations, similar values are obtained with cytochrome b_5 and RNA in the particular cases, A, B and C, on the basis of phospholipid phosphorus. This similarity implies that the cytochrome b_5 of the microsomes is closely related to membrane phospholipid of the endoplasmic reticulum.

Table II. Contents of microsomal constituents[1]

	per mg Protein	per mg Phospholipid-P		
		Prep. A	Prep. B	Prep. C
Cytochrome b_5	0.024 mμmole (7) 0.011~0.042[2]	3.68	3.23	3.64 mμmole
RNA	0.361 mg (3) 0.327~0.399	52.4	50.4	46.6 mg
Phospho-lipid-P	0.0072 mg (3) 0.0065~0.0077			

1 7, The first column shows average values per mg protein. The values between squiggles are the extremes obtained. The number of preparations is given in parentheses. In the particular preparations of A, B and C the contents are expressed per mg phospholipid-phosphorus.

2, NADH reduced. NADPH-red/NADH-red = 0.72 (3)
  0.63~0.85

The results thus obtained with the melanoma microsomes support the following tentative scheme,

$$NADH \longrightarrow f_{P_D} \longrightarrow cytochrome\ b_5 \dashrightarrow O_2$$
$$NADPH \longrightarrow f_{P_T} \longrightarrow$$

where f_{P_D} and f_{P_T} are denoted as flavoprotein dehydrogenases for NADH and NADPH, respectively.

Acknowledgements

The authors wish to thank Miss K. Morimoto for her technical assistance and Drs. F. Hu and K. Adachi for their supply of B-16 melanoma. This work is supported in part by the Scientific Research Fund of the Ministry of Education, Japan.

References

1 Berezney, R. and Crane, F. L.: Cytochromes in bovine liver nuclear membranes. Biochem. biophys. Res. Comm. *43:* 1017–1023 (1971).
2 Ernster, L.: DT diaphorase in R. W. Estabrook and M. E. Pullman, Methods in enzymology, vol. 10, pp. 309–317 (Academic Press, New York 1967).
3 Estabrook, R. W.; Cooper, D. Y. and Rosenthal, O.: The light reversible carbon monoxide inhibition of the steroid C21-hydroxylase system of the adrenal cortex. Biochem. Z. *338:* 741–755 (1963).
4 Estabrook, R. W.; Hildebrandt, A. G.; Baron, J.; Netter, K. J. and Leibman, K.: A new spectral intermediate associated with cytochrome P-450 function in liver microsomes. Biochem. biophys. Res. Comm. *42:* 132–139 (1971).
5 Fleischer, S.; Fleischer, B.; Azzi, A. and Chance, B.: Cytochrome b_5 and P-450 in liver cell fractions. Biochim. biophys. Acta *225:* 194–200 (1971).
6 Harding, B. W.; Wong, S. H. and Nelson, D. H.: Carbon monoxide-combining substances in rat adrenal. Biochim. biophys. Acta *92:* 415–417 (1964).
7 Hiraga, M. and Adachi, K.: Preparation and some properties of amelanotic melanoma mitochondria. Cancer Res. *30:* 1453–1458 (1970).
8 Ichikawa, Y. and Yamano, T.: Cytochrome b_5 and CO-binding cytochromes in the Golgi membranes of mammalian livers. Biochem. biophys. Res. Comm. *40:* 297–305 (1970).
9 Inouye, A. and Shinagawa, Y.: Cytochrome b_5 and related oxidative activities in mammalian brain microsomes. J. Neurochem. *12:* 803–813 (1965).
10 Ohnishi, T.; Kawaguchi, K. and Hagihara, B.: Preparation and some properties of yeast mitochondria. J. biol. Chem. *241:* 1797–1806 (1966).
11 Omura, T. and Sato, R.: The carbon monoxide-binding pigment of liver microsomes. J. biol. Chem. *239:* 2370–2378 (1964).

12 ORRENIUS, S.: On the mechanism of drug hydroxylation in rat liver microsomes. J. Cell Biol. *26:* 713–723 (1965).
13 OSHINO, N.; IMAI, Y. and SATO, R.: A function of cytochrome b_5 in fatty acid desaturation by rat liver microsomes. J. Biochem. *69:* 155–167 (1971).
14 PAIGEN, K.: Hemoglobin as the red pigment of microsomes. Biochim. biophys. Acta *19:* 297–299 (1956).
15 PARSON, D. F.; WILLIAMS, G. R.; THOMPSON, W.; WILSON, D. F. and CHANCE, B.: Improvements in the procedure for purification of mitochondrial outer and inner membrane: comparison of the outer membrane with smooth endoplasmic reticulum; in Mitochondrial structures and compartmentation, pp. 29–70 (Adriatica Ed., Bari 1967).
16 SATO, R.; NISHIBAYASHI, H. and OMURA, T.: Naphthoquinone-dependent oxidation of reduced triphosphopyridine nucleotide by liver microsomes. Biochim. biophys. Acta *63:* 550–552 (1962).
17 SOTTOCASA, G. L.; KUYLENSTIERNA, B.; ERNSTER, L. and BERGSTRAND, A.: Occurence of an NADH-cytochrome c reductase system in the outer membrane of rat liver mitochondria; in E. QUAGLIARIELLO, S. PAPA, E. C. SLATER and J. M. TAGER Mitochondrial structures and compartmentation, pp. 74–89 (Adriatica Ed., Bari 1967).
18 STRITTMATTER, C. F. and BALL, E. G.: A hemochromogen component of liver microsomes. Proc. nat. Acad. Sci., Wash. *38:* 19–25 (1952).
19 WILSON, L. D.; NELSON, D. H. and HARDING, D. W.: A mitochondrial electron carrier involved in steroid hydroxylations. Biochim. biophys. Acta *99:* 391–393 (1965).

Author's address: Dr. MASAZUMI HIRAGA, Institute for Cardio-Vascular Diseases, Tokyo Medical and Dental University, No. 5–45, 1-chome, Yushima, Bunkyo-ku, *Tokyo* (Japan)

Chemistry of Melanin

Marsupial Pigments

E. M. NICHOLLS and K. G. RIENITS

Schools of Human Genetics and Biochemistry, University of New South Wales, Kensington, N.S.W.

It is suggested that the marsupials arrived in Australia from South America via Antarctica some time prior to 43 million years ago when the three continents were believed to constitute one land mass [4]. The 170 species occupy all of the ecological niches which are occupied on other continents by placental mammals.

This paper is primarily concerned with the pigmentation of two species, the red kangaroo, *Megaleia rufa*, and the grey possum, *Trichosurus vulpecula*. These animals have perhaps not received the amount of attention which unusual features of their pigmentation seem to justify. In 1886 WEBER [11] reported on the red pigment of the red kangaroos which were available to him in the Amsterdam Zoo at that time. In the 1940's BOLLIGER [1] and BOLLIGER and HARDY [3] studied the brown pigment of the grey possum as part of more general studies of that animal. The histology of macropod skins was described by MYKYTOWYCZ and NAY [5]. Recently, NICHOLLS and RIENITS [7] reported on aspects of the chemistry and other features of the pigment in kangaroo and possum.

Most of the body hair of a male red kangaroo is orange-red in colour, while the female, often referred to as the 'blue flyer', has blue-grey hair. However, there may be great variation in both sexes, particularly in the relative amounts of orange-red and blue-grey hair. Both male and female grey possums are generally grey or grey-brown in colour, although there is also considerable variation over the wide geographic range of the species. The pigment of particular interest in each species is found external to the hairs and is most noticeable in males. It is most intense on the ventral neck and upper sternum in a diamond shaped, circular or irregular distribution. The pigment in this area of the kangaroo is bright red in colour, and in the possum it is brown or red-brown. The adult male kangaroo has orange coloured pigment adherent to hairs on dorsal and lateral aspects of the body. Some male possums have brown pigment staining the otherwise grey pelt over the

Fig. 1. Hairs from the neck of a male red kangaroo showing flake of free pigment.

back. The female red kangaroo has a small amount of free pigment dispersed over much of the body surface but does not have a clearly defined neck patch. The female grey possum generally has a small pigmented patch or some stained hairs on the neck. In both species the females have free pigment in the pouch.

The red and orange pigments of the red kangaroo are found on the root sheaths, which are stained intensely, and along the shafts of the hairs, where they are usually found in loose flakes or matting hairs together (fig. 1). They are absent from the distal parts of the hairs from which they are lost by attrition or presumably also when it rains, as they are soluble in water. In the possum the pigment is not found on the root sheath of the hair nor on the intra-epidermal part of the hair. It stains the free part of the shaft of hairs and often it is difficult to be sure whether it is on the hair, or within it. However, the appearances are consistent with it being attached to the outside of the hairs and it is possible to get large flakes of pigment from the outside of hairs. The belief that it is on the outside of hairs rather than in hairs is further supported by the ease with which solutions of the pigment may be obtained by immersing the hair in concentrations of HCl greater than 0.1 M, although not all of the pigment is extracted by this means. Possum hair root sheaths have fluorescent material on them and also the shafts of the hairs have a coat of fluorescent material on them which is attached to the outside of the hairs.

Extraction of Pigments (See [7] for full details)

Red Pigment of Kangaroo

The kangaroo hair was washed with 0.1 M $NaHCO_3$ at pH 8 to give a deeply pigmented solution. After acidification the pigments were taken into ethyl acetate and separated and purified by thin-layer chromatography (T.L.C.) on silicagel.

Brown Pigment of Possum

Hair was extracted by boiling in 0.1 M HCl for 10 min. This extract was concentrated by boiling and the pigments separated and purified by T.L.C. on silicagel.

Fluorescent Compounds of Possum Hair

Possum hair was extracted with boiling water. The extract was concentrated, adjusted to pH 6 and extracted with anaesthetic ether. Fluorescent materials were separated and purified by T.L.C.

Results and Discussion

Previous workers have drawn attention to unique histological features in skin of the ventral surface of the neck of possums and kangaroos [2, 5, 11]. The possum has a grossly enlarged sebaceous gland and sweat gland attached to each group of hairs. In the kangaroo the sebaceous glands are not noticeably different from expected size, whilst the sweat gland is similar morphologically and in size to that in the possum. It is a very large secretory gland. On other parts of each animal the glands are normal in size.

Our observations with frozen sections of glands of possum and kangaroo examined by fluorescence miscroscopy showed in the possum that the area surrounding the sweat glands fluoresced with a blue colour and blue fluorescent particles were seen in the hair follicles and scattered on the slide. In the kangaroo a dull blue fluorescence with no histological localization was noted.

A major component of the fluorescent material from possum hair was shown to be 3-hydroxyanthranilic acid (table I). On possum hair this material

Table I. Spectral properties of some substances extracted from hair

	Solvent	λ max. (nm)	Identification[1]
Red pigment from kangaroo	Ethyl acetate	255, 422, 445	Cinnabarinic acid
	Aqueous, pH 7.5	240, 445	
	5N HCl	236, 465, 496	
Yellow pigment from kangaroo	Ethyl acetate	255, 412, 434	Not known
Fluorescent compound from possum	Aqueous, pH 2.0	296	3-hydroxy-anthranilic acid
	5.0	320	
	7.0	312	
	11.0	332	

1 The identification rested on a comparison of spectral properties and thin-layer chromatographic behaviour in seven solvent mixtures with known materials.

Table II. Pigments and fluorescent substances adherent to the surface of hair roots and hairs of various mammals[1]

Name		Source of hair	Fluorescence[2]	Free pigment[3]
Eastern native cat	[4a] *Dasyurus viverrinus*	Back	+++	−
Tasmanian devil	[4a] *Sarcophilus harrisii*	Back	−	Brown and black++
Spotted cuscus	[4b] *Phalanger maculatus*	Back, neck	+	Red-brown+
Grey cuscus	[4b] *Phalanger orientalis*	Back	+	−
Grey possum	[4b] *Trichosurus vulpecula*	Back, neck	++++	Red-brown+++
Rainforest brushtail	[4b] possum *Trichosurus vulpecula johnstonii*	Back, neck, abdomen	+++	Red-brown+
Pigmy possum	[4b] *Cercartetus nanus*	Back	+	Red-brown+
Sugar glider	[4b] *Petaurus breviceps*	Back, neck	−	Red+
New Guinea glider	[4b] *Petaurus breviceps papuanis*	Back	−	Red and orange+
Ringtail possum	[4b] *Pseudocheirus peregrinus*	Back, neck	+	−
Common wombat	[4c] *Phascolomis ursinus*	Back, abdomen	−	Brown and black++
Hairy-nosed wombat	[4c] *Lasiorhinus latifrons*	Back	−	Brown and black+
Rufous pademelon	[4d] *Thylogale billardieri*	Back, neck, abdomen	+++	Red-brown++
	Thylogale billardieri	Margin of pouch	−	Red and brown+++
Red legged pademelon	[4d] *Thylogale stigmatica*	Back, abdomen, leg	+++	Orange, brown, black+
New Guinea pademelon	[4d] *Thylogale bruijni*	Back, neck	+++	−
	Thylogale bruijni	Margin of pouch	−	Red-brown++
Red necked wallaby	[4d] *Wallabia rufogrisea*	Back, neck	−	−
Parma wallaby	[4d] *Wallabia parma*	Back, neck	+++	−
Black glove wallaby	[4d] *Wallabia irma*	Back, abdomen	−	Brown and black+
	Wallabia irma	Pigment spot in front of ear	+	Red-brown++

Grey kangaroo[4d]	*Macropus major*	Back, neck	+	Red++
		Pouch	+	Orange++
Red kangaroo[4d]	*Megaleia rufa*	Back, neck	+	Red and orange++++
Goodfellow's tree[4d] kangaroo	*Dendrolagus goodfellowi*	Margin of pouch	–	Red-brown+
		Back, neck	++++	Brown+
Dorcopsis wallaby[4d]	*Dorcopsis veterum*	Back, abdomen	++	Red-brown+
Rufous rat kangaroo[4d]	*Aepyprymnus rufescens*	Back, head, abdomen,	–	Brown and black+
		Margin of pouch		
Northern hopping mouse[5]	*Notomys aquilo*	Back	–	–
Cat[6]	*Felis catus*	Back, neck	–	–
Dog[7]	*Canis familiaris*	Back	–	–
Man[8]	*Homo sapiens*	Scalp, eyebrow, arm	–	–

1 Classification after WALKER [10].
2 Refers to bright fluorescent coating of the hair root, bright fluorescent coating of the hair shafts, or free fluorescent particles in quantity in the mounting medium. The bright fluorescence is mainly blue, sometimes spots of yellow. The internal fluorescence of colourless hairs is not recorded.
3 Particles or layers of pigment staining the hair roots, attached to the outside of hair shafts or floating free in the mounting medium.
4 *Marsupialia*: a) *Dasyuridae*; b) *Phalangeridae*; c) *Phascolomidae*; d) *Macropodidae*.
5 *Rodentia*.
6 *Felidae*.
7 *Canidae*.
8 *Primates*.

is quite stable and remains unoxidized even in samples of hair kept in the laboratory over several years. However, 3-hydroxyanthranilic acid is relatively easily oxidized by a variety of oxidants to yield cinnabarinic acid (2-amino-3oxo-3H phenoxazine 1:9 dicarboxylic acid) and other pigments.

Fluorescence microscopy indicates that in the possum 3-hydroxyanthranilic acid is around the roots and on the shafts of hairs. In the red kangaroo these locations contain the red or orange pigment, both of which contain cinnabarinic acid (table I) as well as other pigments, as yet unidentified, but thought to be chemically related to cinnabarinic acid, i.e., phenoxazine derivatives.

Cinnabarinic acid has not been previously shown to be a hair pigment although it is present in cetain fungi [3]. There is present in the nuclear fraction of rat liver an enzyme which will convert 3-hydroxyanthranilic acid to cinnabarinic acid [9]. It is presumed that the red kangaroo is able to bring about the oxidation of 3-hydroxyanthranilic acid to cinnabarinic acid although we have no direct evidence to offer at this stage. The possum, on the other hand, does not do this and 3-hydroxyanthranilic acid remains on the hair. In the possum, after an injection of pilocarpine, droplets of fluorescent sweat were seen on a shaved area of the neck, suggesting that the fluorescent material may be secreted from the sweat glands.

The pigments of possum hair are not derived from 3-hydroxyanthranilic acid. They are identical with pigments which may be synthesised non-enzymatically by treatment of dopa and cysteine with ferric chloride [6] or which may be extracted from red feathers or human red hair with boiling HCl. The location in the possum of much of this pigment external to the hair shafts suggests that it is applied to the hair as water soluble precursor material and that pigment is deposited as the material dries. NICHOLLS [6] suggested a model of melanogenesis which involved a reaction between dopa, cysteine and ferric iron or similarly functioning material. The relative proportions of red, brown and black pigments produced depended upon the pH of the reaction. The copious secretion arising from both sweat and sebaceous glands in the possum had been noted previously [2]. It may be that these glands secrete between them the necessary components which then form pigment externally on the hair.

A clear distinction is being made here, i.e., that the brown pigment of the possum is a tyrosine derived pigment and that the red pigment of the kangaroo is a tryptophan derived pigment. The data presented in table II indicate that Australian marsupial hair is not infrequently possessed of brightly fluorescent material external to the hair *or* flakes of reddish pigment

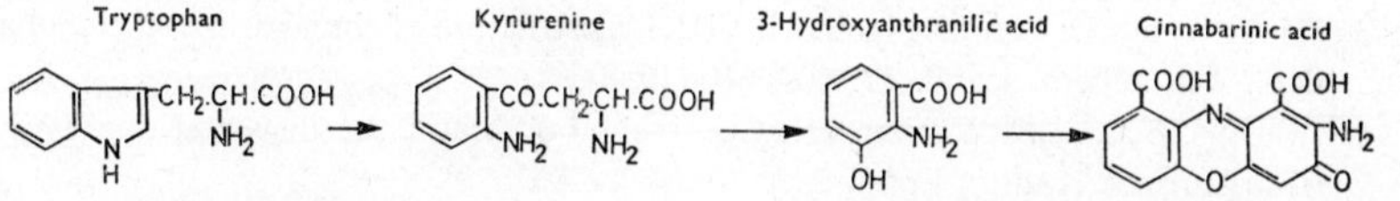

Fig. 2. Scheme linking cinnabarinic acid to tryptophan through metabolites which have been identified in marsupial hair.

external to the hair. Rarely are both seen to a marked degree at the same time. In addition it has been shown [7] that both kynurenine and 3-hydroxyanthranilic acid are components of the fluorescent material of the tree kangaroo *(dendrolagus goodfellowi)*. Albino rat hair contains kynurenine and other tryptophan metabolites [8] but fluorescence microscopy indicates these to be *within* the hair shaft.

Figure 2 shows a scheme, involving known metabolic reactions, which relates the observations recorded so far. It assumes that the red kangaroo possesses an active mechanism for oxidizing tryptophan through to cinnabarinic acid on the skin whereas other species show evidence of partial or complete blocks of the pathway.

Conclusions

The Australian marsupials provide a unique opportunity to study some aspects of the production of epidermal pigments. Tyrosine and tryptophan derived pigments are found external to the hairs in a number of animals. The amount of free red or orange pigments produced tend to bear an inverse relationship to the amount of fluorescent compounds found external to hairs, and also to the amount of free brown or red-brown pigments which may be found. Sexual dimorphism in pigment production in those animals which have been most studied may suggest relevance of the pigments and fluorescent compounds to reproduction.

References

1 BOLLIGER, A.: On the fluorescence of the skin and the hairs of *Trichosurus vulpecula*. Austr. J. Sci. *7:* 35 (1944).
2 BOLLIGER, A. and HARDY, M. H.: The sternal integument of *Trichosurus vulpecula*. J. Proc. roy. Soc. N.S.W. *78:* 122–133 (1944).

3 Gripenberg, J.: Fungus pigments. VIII. The structure of cinnabarin and cinnabarinic acid. Acta. chem. scand. *12:* 603–610 (1958).
4 Jardine, N. and McKenzie, D.: Continental drift and the dispersal and evolution of organisms. Nature, Lond. *235:* 20–24 (1972).
5 Mykytowycz, R. and Nay, T.: Studies of the cutaneous glands and hair follicles of some species of *Macropodidae*. C.S.I.R.O. Wildl. Res. *9:* 200–217 (1964).
6 Nicholls. E. M.: Dopa and the red, brown and black pigments of hair and feathers. J. Invest. Derm. *53:* 302–309 (1969).
7 Nicholls, E. M. and Rienits, K. G.: Tryptophan derivatives and pigment in the hair of some Australian marsupials. Int. J. Biochem. *2:* 593–603 (1971).
8 Rebell, G.; Lamb, J. H.; Mahvi, A. and Lee, H. R.: The identification of L-kynurenine as the cause of the fluorescence of the hair of the laboratory rat. J. Invest. Derm. *29:* 471–477 (1957).
9 Subba Rao, P. V.; Jegannathan, N. S. and Vaidyanathan, C.S.: Enzymic conversion of 3-hydroxyanthranilic acid into cinnabarinic acid by the nuclear fraction of rat liver. Biochem. J. *95:* 628–632 (1965).
10 Walker, E. P.: Mammals of the world; 2nd ed. (Johns Hopkins Press, Baltimore, Md. 1968).
11 Weber. M.: Über neue Hautsecrete bei Säugetieren. Arch. micr. Anat. EntwMech. *31:* 499–507 (1886).

Author's address: Dr. E. M. Nicholls, Schools of Human Genetics and Biochemistry, University of New South Wales, *Kensington, NSW 2033* (Australia)

Current Knowledge of Melanin Structure

G. A. SWAN

Department of Organic Chemistry, The University of Newcastle upon Tyne, Newcastle upon Tyne

In this paper I shall confine my attention entirely to the so-called eumelanins, i.e., black pigments containing nitrogen, derived from tyrosine, dopa, dopamine, etc., and which occur in skin, hair, and melanoma.

Before the structure of a natural product can be elucidated, it is usually necessary to isolate the compound in the pure state. In the case of a melanin, one does not know how to achieve such isolation, and even if it were achieved, we have no real criterion for the purity of a melanin. In the natural state the black pigments are usually attached to protein, and if the latter is split off by hydrolysis, one has no guarantee that the structure of the melanin pigment has not been altered during the process.

The most extensive research on the structure of a natural melanin is that on sepiomelanin by NICOLAUS and his collaborators in Naples [12]. However, their samples were 'purified' by treatment with concentrated hydrochloric acid for long periods. It is true that significant differences were observed between the degradation products of natural and synthetic melanins, but this does not prove that the natural melanins were unchanged by treatment with acid. NICOLAUS' conclusions have also been criticised because the yields of pyrrolecarboxylic acids formed on oxidation were low [11, 13].

HEMPEL [6] sought to obviate the need for isolation of melanin by injecting mice bearing melanoma simultaneously with specifically tritiated dopa and [α-^{14}C]-dopa, and measuring the ^{3}H:^{14}C ratio in a homogenate of the melanoma. Unfortunately the interpretation of the results is uncertain, because the homogenate may have contained dopa which was not part of the melanin polymer [13].

We and others have studied melanogenesis *in vitro*, but even then the melanin may have protein attached, derived from the enzyme. KIRBY and

OGUNKOYA [9] gave a preliminary account of experiments, which are commendable in that by using double labelling, removal of the enzyme was unnecessary. Their results were in good agreement with some of ours [8], but this work has apparently not been pursued.

If the RAPER [16] scheme were correct, it would be expected that the same melanin would be obtained from tyrosine, dopa, dopamine, and 5,6-dihydroxyindole. However, we found that these melanins were not identical [18]. Most of our tracer experiments have been carried out on dopa or dopamine. We prepared melanins *in vitro* a) enzymatically at pH 6.8, using a highly purified polyphenoloxidase isolated from mushrooms [1], and b) by autoxidation at pH 8. The enzymic product had attached protein, and in the case of tracer experiments this protein had to be removed by long boiling with dilute hydrochloric acid before isotopic measurement.

We disproved BU'LOCK and HARLEY-MASON's [4] regular structure (fig. 1, II) in the case of synthetic dopa-melanin by demonstrating retention of deuterium in the melanin formed from [β-^{2}H]-dopa (fig. 1, i) [17].

From experiments using [$^{14}CO_2H$]-dopa (fig. 1, III), we [5, 19] showed, contrary to the Raper scheme, that approximately one out of every five polymer units contained a carboxy-group derived from the original amino-

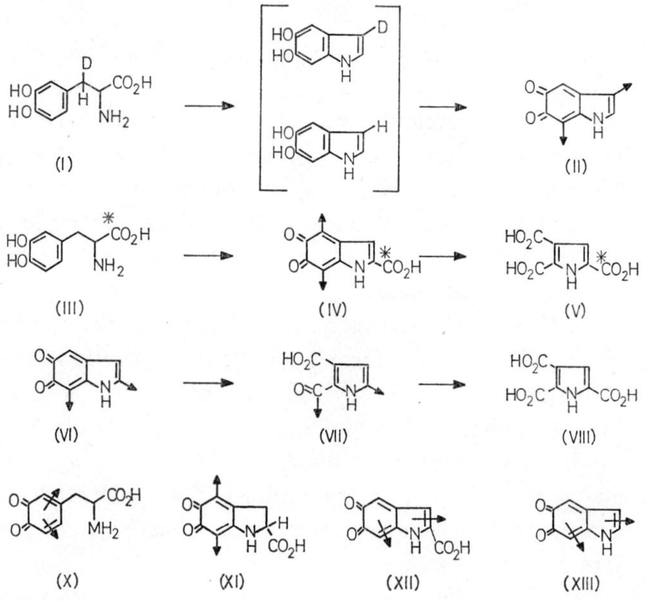

Fig. 1. (See text.)

acid (e.g. fig. 1, IV), and that part of the carbon dioxide evolved during melanogenesis was derived from one or more of the other carbon atoms of the dopa molecule. We found that dopamine also evolved carbon dioxide during melanogenesis.

During the autoxidation of an o-diphenol to an o-quinone, hydrogen peroxide is formed, and this can attack the 6-membered ring of an indole-5,6-quinone unit (e.g.fig. 1 VI) in the polymer, resulting in the elimination of carbon atoms 5- and 6- (and, to some extent 4-) as carbon dioxide, leaving a pyrrolecarboxylic acid unit (fig. 1, VII) in the polymer. We [20] demonstrated this by using samples of dopamine (fig. 2, IX), specifically labelled with ^{14}C at each of the α- and β-positions of the side-chain, and the 3-, 4-, and 5-positions of the benzene ring (separately). Nearly two thirds of the evolved carbon dioxide originated from carbon atoms 3- and 4- of dopamine (i.e., 5- and 6- of indole-5,6-quinone) (fig. 2).

Melanin prepared from, [$^{14}CO_2H$]-dopa (fig. 1, III) was oxidised, and pyrrole-2,3,5-tricarboxylic acid (fig. 1, V) was isolated from the products [19]. From the specific activity of the latter it was calculated that for every molecule which arises by oxidation of units (fig. 1, IV) to compound (fig. 1, V), 1.3 molecules must arise by oxidative fission of the 6-membered ring of indole units which are linked through the 2-position to other polymeric units, i.e., by oxidation of units (VI) through (VII) to (VIII) (fig. 1).

From experiments [19] on the decarboxylation of melanin derived from [$^{14}CO_2H$]-dopa (fig. 1, III), it was concluded that the melanin might contain not only indole-5,6-quinone (fig. 1, XIII), indole-5,6-quinone-2-carboxylic acid (fig. 1, XII), and indoline-5,6-quinone-2-carboxylic acid (fig. 1, XI) units, but also units (fig. 1, X) which had not undergone cyclisation. These experiments also gave a measure of the pyrrolecarboxylic acid units (fig. 1, VII) in the melanin.

Treatment of dopa-melanin with methanolic hydrogen chloride, or with diazomethane resulted in esterification of the carboxy-group, or methylation of all hydroxy-groups, respectively. From the methoxy-content of the

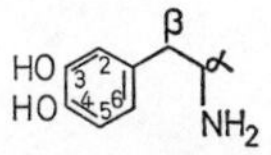

(IX)

Fig. 2. Evolution of carbon dioxide during the autoxidation of specifically ^{14}C-labelled dopamine.

Position in dopamine	α	β	3	4	5
% of total evolved CO_2	1.9	2.3	29.8	27.2	3.5

products, we concluded that only approximately one half of the melanin units were in the quinonoid form, the remainder being in the dihydroxy form [19].

We then deduced a combination of units (fig. 3) in terms of which our various results on autoxidative dopa-melanin could be explained. We believe our melanins to be irregular polymers, containing a number of different types of units linked in various ways. The fully methylated melanin was treated with [^{14}COCl]-benzoyl chloride, and the specific activity of the product agreed with that expected from the above if only the primary amino group (fig. 3, XIV) reacted [19].

To determine the proportion of indoline-type units in the melanin, and the relative numbers of linkages at different positions of the polymeric units, we converted samples of dopa, specifically deuteriated at each of the α- and β-positions of the side-chain, and the 2-, 5-, and 6-positions of the benzene ring (separately) into melanins, and compared the enrichment of deuterium in the precursor and melanin [8]. As an isotope effect could obscure the interpretation of the results, we carried out two series of experiments; in one series the precursor contained only a tracer concentration of deuterium, while in the other the relevant position of the molecule was deuteriated to as nearly 100 % as possible. We further carried out each series both by autoxidation and enzymically.

Melanins prepared from α- and β-deuteriated dopa were boiled with 2N-HCl for various periods, after which their deuterium contents were measured. The results suggested that two distinct exchange processes occurred, which could be explained on the basis of the above structure and the values checked with those expected. The deuterium retention in enzymic melanins after long boiling with acid were not greatly different from the corresponding values on autoxidative melanins, which had been boiled with acid for similar periods. There was nothing in our results to suggest a

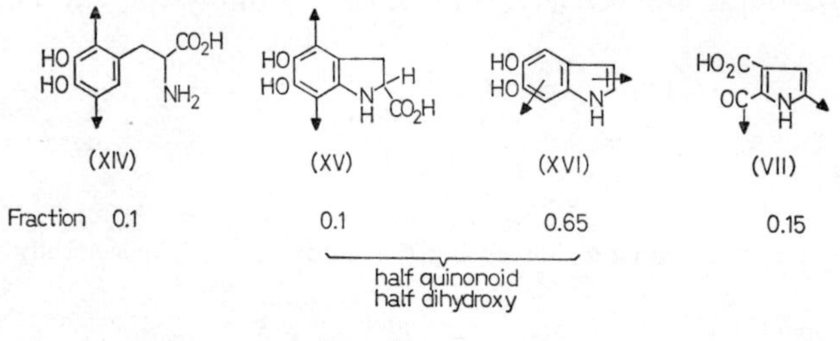

Fig. 3. Autoxidative dopa-melanin.

fundamental difference in structure between autoxidative and enzymic dopa-melanins.

There was little difference in the R-values (ratio of deuterium enrichments in melanin and precursor) between the tracer and 100% deuterium series, which suggests that an isotope effect cannot have any great influence.

If it is assumed that units (XIV), (XV) and (VII) (fig. 3) are linked as shown, it can be calculated that the following fractions of the main units (XVI) are linked at the position indicated.

Position	2	3	4	7
Fraction	0.36	0.37	0.34	0.28

Thus the linkages at these positions appear to be fairly evenly shared. If one adds up the number of polymeric linkages to each average polymer unit, one obtains a result of 2.05, which implies that the majority of units are linked to two other units.

In dopamine-melanin the proportion of uncyclised units (approximately one in three) appears to be greater than in dopa melanin [2]. In the case of melanins derived from specifically deuteriated dopamine, the R-values were in almost all cases higher than for samples of dopa-melanin prepared and treated under identical conditions. We concluded that in dopamine-melanin there must be a considerable number of polymeric linkages at positions other than those thought to be involved in dopa-melanin. These extra linkages could be through oxygen or nitrogen atoms and we produced some experimental evidence of the feasibility of the latter. Our various results could be accommodated reasonably well in terms of a set of units shown in figure 4.

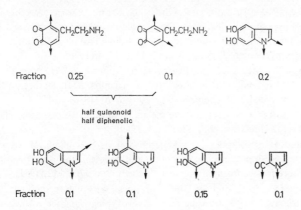

Fig. 4. Autoxidative dopamine-melanin.

From the preceding it may be concluded that no conclusive evidence as to the structure of natural melanin is at present available. Nevertheless evidence from different sources points to the probability that dopa-type melanins derived from natural sources, as well as synthetic dopa-melanins, are irregular polymers, built up from several types of units. BLOIS [3] has also concluded from his studies that melanin is a unique biopolymer, random or irregular in structure. The conclusions drawn by NICOLAUS and by HEMPEL were similar, although MASON [11, 13] has defended the homopolymer theory. It would seem rather likely that the polymerisation process involves radicals. WATERS [21] has recently suggested that biochemical oxidative coupling of phenols may proceed via either radicals or phenonium cations.

Finally, it should be noted that those of us who have attempted to study the structure of melanin formed enzymically *in vitro* have almost always used mushroom polyphenol-oxidase, which is much less specific in its activity than is tyrosinase of mammalian origin [7, 10, 15]. The recent work of OKUN *et al.* [14] has shown the ability of peroxidase to effect melanogenesis. Clearly it would be of importance to know whether these other enzymes give rise to the same melanin as does mushroom tyrosinase.

The research which I have outlined involved a great deal of experimental work, which was carried out by the research students and other collaborators, whose names appear in references 1, 2, 5, 8, 19, and 20, where full details of the methods and results are to be found.

References

1 BINNS, F.; CHAPMAN, R. F.; ROBSON, N. C.; SWAN, G. A. and WAGOTT, A.: Studies related to the chemistry of melanins. VIII. The pyrrole-carboxylic acids formed by oxidation or hydrolysis of melanins derived from 3,4-dihydroxyphenethylamine or ($\pm$)-3,4-dihydroxyphenylalanine. J. chem. Soc. C: 1128–1134 (1970).
2 BINNS, F.; KING, J. A. G.; MISHRA, S. N.; PERCIVAL, A.; ROBSON, N. C.; SWAN, G. A. and WAGOTT, A.: Studies related to the chemistry of melanins. XIII. Studies on the structure of dopamine-melanin. J. chem. Soc. C: 2063–2070 (1970).
3 BLOIS, M. S.: Biological free radicals and the melanins; in S. J. WYARD Solid state biophysics, pp. 243–262 (McGraw-Hill, New York 1969).
4 BU'LOCK, J. D. and HARLEY-MASON, J.: Melanin and its precursors. II. Model experiments in the reactions between quinones and indoles, and consideration of a possible structure for the melanin polymer. J. chem. Soc.: 703–712 (1951).
5 CLEMO, G. R.; DUXBURY, F. K. and SWAN, G. A.: Formation of tyrosine melanin. III. The use of carboxyl-labelled tyrosine and dihydroxyphenylalanine in melanin formation. J. chem. Soc.: 3464–3468 (1952).

6 HEMPEL, K.: Investigation on the structure of melanin in malignant melanoma with ^{3}H- and ^{14}C-dopa labelled at different positions. Proc. Symp. Structure and Control of the Melanocyte, pp. 162–175 (Springer, Berlin 1966).
7 KERTÉSZ, D.: The phenol-oxidizing enzyme of human melanomas; substrate specificity and relationship to copper. J. nat. Cancer Inst. *14:* 1081–1091 (1954).
8 KING, J. A. G.; PERCIVAL, A.; ROBSON, N. C. and SWAN, G. A.: Studies related to the chemistry of melanins. XI. The distribution of the polymeric linkages in dopa-melanin. J. chem. Soc. C: 1418–1422 (1970).
9 KIRBY, G. W. and OGUNKOYA, L.: Structure of melanin derived from (±)-3,4-dihydroxy-[^{14}C,^{3}H]phenylalanine by oxidation with tyrosinase. Chem. Comm.: 546–547 (1965).
10 LERNER, A. B.; FITZPATRICK, T. B.; CALKINS, E. and SUMMERSON, W. H.: Mammalian tyrosinase: action on substances structurally related to tyrosine. J. biol. Chem. *191:* 799–806 (1951).
11 MASON, H. S.: The structure of melanin; in Advances in biology of skin, Vol. VIII: The pigmentary system, pp. 293–312 (Pergamon Press, Oxford 1967).
12 NICOLAUS, R. A.: Melanins (Hermann, Paris 1968).
13 NICOLAUS, R. A.; HEMPEL, K. and MASON, H. S.: Comments on Howard S. Mason's paper 'The Structure of Melanin'; in Advances in biology of skin, Vol. VIII: The pigmentary system, pp. 313–317 (Pergamon Press, Oxford 1967).
14 PATEL, R. P.; OKUN, M. R.; EDELSTEIN, L. M. and EPSTEIN, D.: Biochemical studies of the peroxidase-mediated oxidation of tyrosine to melanin; demonstration of the hydroxylation of tyrosine by plant and human peroxidases. Biochem. J. *124:* 439–441 (1971).
15 POMERANTZ, S. H.: Separation, purification, and properties of two tyrosinases from hamster melanoma. J. biol. Chem. *238:* 2351–2357 (1963).
16 RAPER, H. S.: The aerobic oxidases. Physiol. Rev. *8:* 245–282 (1928).
17 SWAN, G. A.: Chemical structure of melanins. Ann. N.Y. Acad. Sci. *100:* 1005–1016 (1963).
18 SWAN, G. A.: Some studies on the formation and structure of melanins. C.R. Acad. Sci. fis. mat., Napoli (4) *31:* 1–20 (1964).
19 SWAN, G. A. and WAGOTT, A.: Studies related to the chemistry of melanins. X. Quantitative assessment of different types of units present in dopa-melanin. J. chem. Soc. C: 1409–1418 (1970).
20 SWAN, G. A. and WRIGHT, D.: A study of melanin formation by use of 2-(3:4-dihydroxy[3-^{14}C]phenyl-, 2-(3:4-dihydroxy[4-^{14}C]phenyl-, and 2-(3:4-dihydroxy[5-^{14}C]phenyl)-ethylamine. J. chem. Soc.: 1549–1557 (1956).
21 WATERS, W. A.: Comments on the mechanism of one-electron oxidation of phenols: a fresh interpretation of oxidative coupling reactions of plant phenols. J. chem. Soc. B: 2026–2029 (1971).

Author's address: Dr. G. A. SWAN, Department of Organic Chemistry, The University of Newcastle upon Tyne, *Newcastle upon Tyne, NE1 7RU* (England)

Structure of Melanins[1]

Y. T. THATHACHARI

Stanford University, Stanford, Calif. and Indian Institute of Science, Bangalore

Introduction

Melanins are complex polymers that are difficult to characterize. As no single technique gives unambiguous information, results from a number of physical studies are pooled together to gain an understanding of their molecular structure. Unlike many other biological molecules, it has not been possible to prepare melanins with any degree of crystallinity. The x-ray diffraction patterns of melanins, at least on a first look, show very few details and so, until recently, have not been examined with a view to obtain structural information. However, preliminary studies undertaken some time ago [7] indicated that some information could be derived from the x-ray diffraction patterns of a number of natural and synthetic melanins. These were characterized by diffuse halos with maxima corresponding to Bragg spacings 3 to 4 Å. The maxima were at about 3.4 Å for all animal and synthetic melanins examined. Since they were known to be chemically different and consisted of monomeric units of different sizes it was concluded that they had a common structural feature. It was thought probable that the planar monomeric groups, all of thickness 3.4 Å tended to be nearly parallel. In ustilago melanin the maximum was at a larger spacing of about 4.2 A.

It must be pointed out that the x-ray data are not influenced by the presence of impurities. The interpretation of the results does not require a knowledge of the chemical composition and structure of the polymers: all we need to know is that they are composed mostly of light atoms –

---

[1] This work was supported in part by N.I.H. grant CA-08064 and by the Advanced Research Project Agency through the Center for Materials Research at Stanford University.

C, N, O and H – and contain planar groups. The information that can be obtained exclusively by x-ray diffraction techniques relates to the nature and extent of order present in these polymers at the molecular level. This may be of more than casual significance in the understanding of their biological role.

To derive maximum information from the x-ray data it is not enough to consider only the positions of the diffraction peaks, since considerable scattering occurs at all angles. More refined techniques therefore become necessary. They have been described by a number of authors [e.g. 8]. This paper describes some of the results of sophisticated diffraction studies on melanins, supplemented by electron microscopic and mass-spectrometric investigations. A possible explanation for the reported high electron density of melanin granules has been suggested.

Experimental and Computational Techniques

Ten different samples of melanins were pressed into pellets about 6 mm diameter and ½ to 1 mm thick. The diffraction data were recorded using a computer-controlled four circle Hilger and Watts diffractometer employing the symmetric transmission technique. Radiation from a stablized fine focus Philips x-ray generator with a molybdenum target was reflected by a doubly bent lithium fluoride monochromator to provide the K α1 radiation. The data were recorded automatically from a Bragg angle of $\Theta = 1°$ to $60°$ in steps of .25°. The background due to air and slit scattering was determined by running blanks for the same scattering angles. For six of the samples the entire data including the background were redone and the reproducibility was within 2%. The attenuation factors were experimentally determined for all the pellets for various scattering angles from $\Theta = 0$ to $60°$ in steps of 5°. There was little variation of the linear absorption coefficient with the scattering angle and in each case it was found to be in agreement with the value derived from the assumed elemental composition and the mass absorption coefficients [3].

Squid melanin granules were imaged using a Philips EM 200 electron microscope. The sample was mounted on a carbon substrate or on a holey grid. Selected area electron diffraction was recorded from single granules. Estimation of trace elements was made using a spark source mass spectrometer.

The diffraction data were stored in a disk file and processed by IBM 360/50 computer and ACME system. The background were subtracted and the intensities were corrected for polarization, absorption and multiple scattering. The corrected intensities were placed on an absolute scale by matching them against the total independent scattering normalized to a single average atom. The scale factors were also derived by other techniques [8, 6, 1] and were found to be consistent. The incoherent scattering was then subtracted from the scaled intensities and the scattering curves were all placed on the same scale by dividing them by the scattering at $\Theta = 0°$. It is convenient to present the diffraction data as plots of the intensity of scattering thus derived against $2 \sin \Theta/\lambda$ (referred to as s). s is in A^{-1} units. Its reciprocal, referred to as d, is the associated Bragg spacing in Angstrom units. The plottings were done by a Calcomp X-Y plotter employing the ACME and IBM 1800

Data Acquisition systems, after checking out and composing the curves in a TV display. The radial distribution curves were derived from the scattering curves by Fourier techniques [8]. The scattering factors used were obtained by linear interpolation from tables [2] with entries at .05 A^{-1} intervals in $\sin \Theta/\lambda$. Similar tables for the incoherent scattering intensities were derived from data in literature [4] by polynomial interpolation. The elemental compositions for some samples were taken from literature [5]. Hydrogen contributions were ignored as the error involved was less than 1%. Wide variations by as much as 20% in the proportions of single elements in melanins have been reported [5]. However, since the shapes of the scattering factors for C, N and O are quite close, these variations hardly affect the average scattering factors. Diffraction curves were computed for various monomeric units. The effects due to changes in the mutual dispositions of adjacent units were studied by molecular simulations using the ACME system. To calculate the contrast observed in electron micrographs of melanins, electron scattering amplitudes were taken from the International Tables [3].

Results

Curves (a) and (b) in figure 1 present the scattering information for squid and ustilago melanins. The dotted curves known as 'independant scattering curves' represent the scattering that would result if the atoms in the polymers were distributed uniformly as in a gas. In a real case deviations

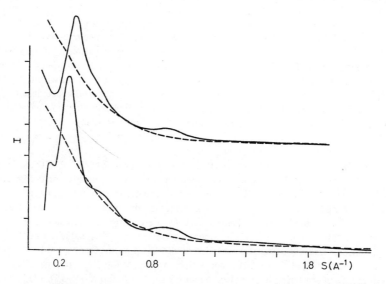

Fig. 1. Scattering curves for squid melanin(top) and ustilago melanin(lower). The broken curves are the independent scattering curves.

occur from this uniform distribution due to the finite size of the atoms, the shape and size of the monomeric units and varying degree of order present at different levels of organizations. The observed diffraction curve, therefore, deviates from the 'independent scattering curves'. These deviations tell us about the nature and extent of the order present. The 'radial distributions' derived from the observed scattering curves are associated with the variations in the density with the distance from any chosen reference atom.

Broadly, four regions can be identified in the observed scattering curves:

1. A peak at about $s = .9$ A^{-1}. This is seen in all melanins and the profiles are practically the same. The corresponding peak in the radial distribution is at 1.4 A. This is obviously the first neighbor distance with a value of about 1.22 A for $C = O$ and about 1.54 A for $C - C$, but mostly about 1.4 A in the aromatic units. This is what one would expect of a polymer composed mostly of catechol or related groups.

2. A peak (or rather a shoulder) at about $s = .5$ A^{-1}. This peak is also found in all melanins, although the profiles vary slightly toward larger Bragg spacings. The corresponding radial distribution peak is at about 2.5 A. This is the average distance between pairs of nonbonded atoms that are attached to a third atom by a covalent bond. Distances beyond say 2.7 A may also correspond to atoms in adjacent units that are not bonded.

3. A peak in the region $s = .2$ A^{-1} to $.4$ A^{-1}. This is the principal peak of melanins, seen even in simple Laue type diffraction photographs as a prominent halo. Table I lists the positions of this peak for the ten melanins studied,

Table I

Source of melanin	A^{-1}	d A
Catechol (auto-oxidized)	.295±.015	3.4±.15
L-dopa (auto-oxidized)	.295±.025	3.4±.30
Squid	.295±.015	3.4±.15
Mouse melanoma	.280±.025	3.6±.30
Phaeomelanin[1]	.300±.010	3.3±.10
Methyl hydroquinone[1]	.285±.015	3.5±.20
Luffa cylindrica seeds[1]	.265±.025	3.8±.35
Sunflower seeds[1]	.240±.040	4.1±.65
Ustilago myadis[1]	.250±.020	4.0±.30
Aspergillus niger	.235±.015	4.2±.30

The limits correspond to angles at which scattering drops to 95 % of the peak values.
1 Samples purified by Prof. Nicolaus by the method described in his monograph.

both in terms of s (A^{-1}) and the Bragg spacing d (A). All the animal and synthetic melanins studied have practically the same peak position at a Bragg spacing of 3.4 A. The plant and fungal melanins seem to have larger Bragg spacings. The profiles of the maximum for melanins from squid, catechol and L-dopa match very closely in all respects. The profiles of mouse melanoma and methylhydroquinone melanins also seem to be rather close to them with a shift of the maxima to slightly larger spacings. The greater sharpness of the phaeomelanin peak was confirmed. The corresponding region in the radial distribution has multiple peaks in the region 3.7 A to 5.0 A. As mentioned earlier, this region may contain information regarding the dispositions of adjacent monomers.

4. The region s less than about .2 A^{-1} (or Bragg spacings greater than about 5 A). All the melanins showed weak but detectable humps in the region s = .9 to 1.1 A^{-1} (d = 11 to 9 A). These are hardly noticeable in the plant melanins but are quite significant in phaeomelanins (at d = 14 A) and melanin from methyl hydroquinone (d = 10 A). This region may be associated with the size of the monomers and may also suggest order at longer ranges.

The peak at the Bragg spacing of 3.4 A in the animal and synthetic melanins studied was ascribed to the tendency of adjacent planar groups to aggregate in a near parallel stack – the planar separation being 3.4 A, the thickness of the units. However, this can not be proved unequivocally by the radial distribution curves, as the peaks also correspond to interatomic distances within a monomer. Scattering curves were, therefore, computed for various monomers, assuming that adjacent units can take all possible mutual orientations, as in a gas of the monomer. These curves were found to superpose with the experimental curves in regions (1) and (2) and they had no peaks in regions (3) and (4). In other words, the agreement between the two sets of curves is confined to the regions relating to a single monomer. It strengthens the belief that the peaks in regions (3) and (4) are due to certain preferred mutual orientations of the adjacent planar units. The observed scattering curves for these melanins were compared with the curves for carbon [1]. The agreement was very good except that the carbon peaks were sharper and taller. This is what one would expect, as carbons consist of a single atomic species; the planar groups involved may be much larger in size, and the parallel stacking may be much more regular. The tendency in these melanins for adjacent planar units to aggregate in a near parallel stack suggests a high density, as was indeed observed. Pellets of L-dopa melanins were found to have a density of over 1.5. Due to possible presence of voids, this would represent the lower limit of the density of the polymer. Selected area electron diffraction from single granules of squid melanins clearly showed diffraction halos centered at a Bragg spacing of 3.4 A, suggesting that any 'graphitic structure' present in these melanins is not induced by the process of pelletization.

The principal diffraction peaks of the plant and fungal melanins were at larger Bragg spacings (3.5 to 4.2 A) suggesting a more loose structure. The measured densitities of these melanins were found to be significantly low. Ustilago melanin, for instance, could not be pressed into pellets of density greater than 1. It has not yet been possible to link these diffraction maxima with any specific feature of the mutual orientations of adjacent monomers in a unique way. In a large number of lignins, diffraction halos centered at Bragg spacings of 4.2 A have been reported [9]. These authors have also reported for these materials diffraction peaks that would correspond to the regions (1) and (2) of melanins. However, they have not recognized the true significance of these peaks.

It has not been possible to assign uniquely any structural feature to the long spacings found in some of the melanins in region (4) with Bragg spacings 10–14 A.

The high electron density reported for melanin granules has been attributed in the past to the presence of heavy elements. For the ten samples studied by us, x-ray attenuation and fluorescence measurements ruled out the presence of elements heavier than oxygen in proportions greater than 2% by weight. Mass spectrometry has detected zinc and copper in proportions of about .5 and .05% respectively. Other trace elements, if at all present, were in much smaller amounts. Calculations readily show that if equal paths are assumed for both the melanin granules as well as the surrounding tissues, the amount of heavy metals estimated is not adequate to produce any significant contrast, unless the density of the granules is significantly higher than that of the surrounding tissues – as is indeed the case. As mentioned earlier, melanin pellets were found to have densities higher than 1.5 while the surrounding tissues may have a density less than 1.3. The density difference of this order is more than adequate to produce the contrast observed. These findings are consistent with the estimated variations in the contrasts of melanin granules of various sizes. At present these results are quite qualitative and more refined measurements are necessary to understand the nature of the observed contrast. It is, however, reasonable to conclude that the contrast is due to the high density of melanin granules and not to the presence of heavy elements like zinc and copper.

Conclusions

From the x-ray diffraction data it can be concluded that there is a short range order at the molecular level in all melanins. The separation between adjacent planar groups in the plant and fungal melanins seems to be much larger than in the case of the animal and synthetic melanins studied by us. The density measurements on the melanin pellets seem to support this conclusion. The adjacent planar groups in the synthetic and animal melanins are probably parallel. They also may have order extending to longer ranges. The electronic properties of these melanins, including their black color, may be related to the possible presence of 'graphitic structures'. The high density of these melanins may be responsible for the observed contrast in the electron micrographs of melanin-containing tissues.

Acknowledgements

I would like to thank Dr. M. S. BLOIS for his interest, Prof. R. A. NICOLAUS for providing melanin samples, Mrs. LINA TASKOVITCH and Rev. WILLIAM CARROLL S.J. for the preparation and purification of the samples, Dr. T. E. HOPKINS for help in collecting some of the data, and Mrs. MADHURI THATHACHARI for assistance in computation. The trace elements analysis by mass spectrometry was done in collaboration with Prof. V. S. VENKATASUBRAMANIAN.

References

1. ERGUN, S.: Chemistry and physics of carbon, vol. 3, Ed. P. L. WALKER (Marcel Dekker Inc., New York 1968).
2. HANSON, H. P.; HERMAN, F.; LEA, J. D., and KILMAN, S. S.: HFS scattering factors. Acta Cryst. *17:* 1040 (1964).
3. International tables for crystallography, vol. III, pp. 116–149 (Kynoch Press, Birmingham, England (1962).
4. MIRKIN, L. I.: Handbook of x-ray analysis of polycrystalline materials (Consultants Bureau, New York) 666–668 (1964).
5. NICOLAUS, R. A.: Melanins (Herman, Paris 1968).
6. NORMAN, N.: The Fourier transform method for normalizing intensities. Acta Cryst. *10:* 370 (1957).
7. THATHACHARI, Y. T. and BLOIS, M. S.: Physical studies on melanins – II Biophys. J. *9:* 77–89 (1969).
8. WARREN, B. E.: X-ray diffraction (Addison-Wesley, New York 1969).
9. ZAHN, H. and LAUTSCH, W.: X-ray studies of lignins. Kolloid Z. *110:* 82 (1948).

Author's address: Dr. Y. T. THATHACHARI, Department of Dermatology, University of California, *San Francisco, CA 94122* (USA)

Chemical Composition of Ten Kinds of Various Melanosomes

J. Duchǒn, J. Borovansky and P. Hach

Department of Biochemistry and Department of Embryology, Faculty of Medicine, Charles University, Prague

Melanosomes are specific melanin-forming and melanin-carrying particles of pigment cells. They have been studied by many authors whose work has brought much valuable information, especially from the morphologic and genetic point of view [1, 3, 5, 7]. It is, therefore, surprising that there are only very limited data available on the chemical composition and structure of these interesting organelles [6, 8, 10] especially their protein part.

Because of the insolubility of the melanin polymer itself in any suitable solvent and because of its very strong linkage to protein, it has not hitherto been possible to separate this protein as a whole. Therefore, as yet the only main way for the characterization of the protein component of melanosomes is the splitting-off of the protein in the form of its basic units, amino acids, using the method of total hydrolysis. It was decided to perform firstly this procedure to obtain initial information for the further study of melanosomal protein, or perhaps better, proteins.

Melanosomes were isolated from the large-granule fraction obtained by the differential centrifugation of Harding-Passey (HP), B 16, Cloudman S 91 and 'Stanford' (ST) [2] mouse melanoma homogenates, of human (HU) and horse (EQ) melanoma homogenates, of chicken embryos' retinal pigment epithelium (RP) [4] and ox chorioid (OX) homogenates, of the squid *Loligo pealii* ink sac (SQ) [9] and the *Sepia officinalis* ink sac (SE) homogenates, using the sucrose density-gradient ultracentrifugation method according to Seiji *et al.* [7]. The purity of the isolated melanosomes has been checked by electron microscopy and it has been found that reasonably pure preparations were obtained in all cases (fig. 1).

Isolated melanosomes (repeatedly washed by water to remove sucrose) were dried to a constant weight and hydrolyzed by 6 N hydrochloric acid

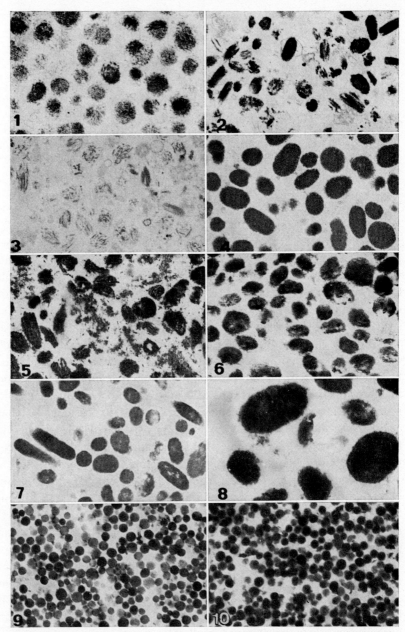

Fig. 1. Electron micrographs of melanosomes isolated from HP (1), B 16 (2), S 91 (3), ST (4), HU (5), EQ (6), RP (7), OX (8), SQ (9) and SE (10). Abbreviations used: see text. × 12,000.

in sealed tubes at 110°C for 24 h. Thereafter the insoluble melanin was quantitatively separated from the hydrolysate and dried to a constant weight again.

It has been found that melanin forms 18–72% of the total weight of the melanosomes, with respect to their source (table I). The amount of chemically recovered melanin[1] seems to be in agreement with the degree of melanization (electron density) shown in the electron micrographs. As far as the amount of recovered protein is concerned, it is approximately the same in HP, B 16, S 91, HU and OX melanosomes, while, it is much lower in ST, EQ and RP melanosomes and extremely low in SQ and SE melanosomes. The percentage of the non-melanin and non-protein residue is, however, in all cases nearly the same.

The hydrolysates were then analyzed using Beckman's amino acid analyzer, model 120 C. Since the usual kind of presentation of the amino acid composition of proteins, i.e., in μM/g of protein or initial material was not appropriate for the comparison of the results obtained in different

Table I. The amount of melanin and protein recovered from ten kinds of various melanosomes

Melanosomes[1]	% of melanin ± S.D.[2]	% of protein ± S.D.[3]	% of residue[4]
HP	30.3 ± 1.2	52.6 ± 1.4	17.1
B 16	32.0 ± 0.5	44.4 ± 1.6	23.6
S 91	17.9 ± 0.9	61.4 ± 1.8	20.7
ST	56.4 ± 0.5	22.5 ± 0.5	21.1
HU	30.1 ± 0.9	49.0 ± 0.3	20.9
EQ	57.4 ± 2.2	24.6 ± 1.2	18.0
RP	51.6 ± 1.8	19.3 ± 1.8	29.1
OX	28.9 ± 0.4	48.6 ± 0.4	22.5
SQ	72.3 ± 0.3	5.4 ± 0.1	22.3
SE	58.1 ± 1.5	7.6 ± 0.3	34.3

1 Abbreviations used: see text.
2 Determined by weighing of dry melanosomes and of recovered dry melanin.
3 Expressed as sum of all amino acids recovered by amino acid analyzer.
4 Determined by subtraction of the sum of melanin and protein from the total weight of the melanosomes.

---

1 There were no significant differences in the amount of recovered melanin after 24 and 72 h hydrolysis. For example: HP 24 h – 30.3%, 72 h – 30.1%; HU 24 h – 30.1%, 72 h – 30.8%; EQ 24 h – 57.4%, 72 h – 58.3%; SE 24 h – 58.1%, 72 h – 56.8%.

kinds of melanosomes (due to the different content of melanin as well as of protein in different melanosomes), the results were expressed in percentages of every individual amino acid from the total (100 %) of all amino acids (without ammonia) recovered by the amino acid analyzer (table II). The results can be summarized as follows:

1. The protein component of all ten kinds of melanosomes studied consists of all 18 (20) current amino acids present in proteins: Ala, Arg, Asp, Cys/2, Glu, Gly, His, Ile, Leu, Lys, Met, Phe, Pro, Ser, Thr, 'Trp', Tyr, Val (NH_3, $CySO_3H$)[2].

2. Traces of an unusual amino acid, 3,4-dihydroxyphenylalanine (dopa), approximately in 0.2–0.8 % concentration, have been found in the HP, ST, HU, RP, OX and SE hydrolysates.

3. Another unusual amino acid, 2-aminoethane-1-sulphonic acid (taurine), has been found in the SQ and SE hydrolysates only.

4. Generally, there are no considerable differences in the quantitative amino acid composition of melanosomes from HP, B 16, S 91 and HU melanomas, and also perhaps from the EQ and RP. However, the amino acid composition of ST and OX, and, especially of SQ and SE melanosomes differs considerably from that of the above melanosomes. For example, the high content of alanine and the very low content of lysine (and also of half-cystine) in SQ and SE melanosomes is striking. In the case of OX (and also ST) melanosomes, the extremely high content of glycine and proline and the low amount of leucine and valine is abnormal. The level of lysine seems to be indirectly proportional to the content of melanin. A more detailed analysis of the values of the other amino acids (e.g., the ones containing sulphur) could also be very interesting.

According to the authors' opinion, these results could be an expression of the genetic and species differences in the composition of melanosomal proteins in the course of phylo-, organo- and ontogenesis.

A more complete account of this work will be published elsewhere.

Acknowledgements

The work reported in the present paper has been initiated by Dr. DUCHŎN at the Department of Dermatology, Harvard Medical School, Massachusetts General Hospital,

2 A very high content of ammonia has been found in SQ and SE hydrolysates. In average it was 5 times higher than the amount of ammonia in all the other hydrolysates (SQ 5.1, SE 4.9 mM NH_3/g of protein; in all the other cases 0.7–1.5 mM NH_3/g of protein).

Table II. The amino acid composition of melanosomes isolated from ten various sources

Amino acid	Kind of melanosomes[1]									
	HP	B 16	S 91	ST	HU	EQ	RP	OX	SQ	SE
Ala	7.29	7.41	7.84	7.07	7.87	5.81	8.39	10.69	16.84	14.22
Arg	5.59	5.79	5.76	6.96	5.32	7.77	5.68	6.25	4.43	4.92
Asp	9.59	10.02	9.11	7.68	8.91	10.56	8.69	6.83	11.21	11.69
Cys[2]	2.07	2.04	1.43	4.99	3.07	4.83	1.75	0.71	0.89	0.77
Dopa	+	–	–	+	+	–	+	+	–	+
Glu	10.22	10.89	11.05	10.01	7.51	7.89	10.34	8.14	9.07	6.57
Gly	8.62	9.25	8.94	18.62	8.38	9.77	10.82	27.94	7.69	9.59
His	2.71	2.45	2.49	2.28	3.71	3.88	2.38	1.61	1.24	1.85
Ile	4.45	4.55	4.74	2.75	3.36	3.09	4.16	2.27	4.69	5.71
Leu	8.36	8.44	8.59	5.49	9.44	6.65	7.57	4.71	7.16	8.23
Lys	6.81	6.45	7.57	4.64	7.20	4.86	5.17	4.27	2.20	3.28
Met	1.64	0.81	1.67	1.11	2.84	0.62	1.24	1.18	0.73	1.09
Phe	4.20	3.96	3.93	2.67	4.33	4.15	4.03	2.01	3.28	4.75
Pro	6.20	6.02	5.42	8.96	5.00	6.92	6.48	9.70	4.23	4.08
Ser	6.02	6.23	6.16	5.37	6.52	6.51	6.68	4.87	4.96	6.10
Tau[3]	–	–	–	–	–	–	–	–	4.58	3.17
Thr	5.65	5.42	5.33	4.14	5.97	5.63	5.85	3.49	5.94	8.10
'Trp'[4]	+	+	+	+	+	+	+	–	+	+
Tyr	3.06	3.05	2.69	2.12	2.83	3.49	2.84	1.18	2.17	1.89
Val	6.93	6.68	6.79	4.55	7.72	6.86	6.67	4.13	5.04	6.90

The values are expressed as % of moles (M) of every individual amino acid from the sum (100 %) of all amino acids recovered after hydrolysis. (Ammonia was not included in the total.) The mean values from 5 (HP), 4 (B 16), 3 (S 91), 3 (ST), 2 (HU), 2 (EQ), 3 (RP), 2 (OX), 3 (SQ) and 3 (SE) independent determinations.

1 Abbreviations used: see text; 2 Cys = half-cystine + cysteic acid; 3 Tau = taurine; 4 'Trp' = peak considered to be a degradation product of tryptophan; + = present; – = absent.

Boston, Mass., U.S.A., during the tenure of an Eleanor Roosevelt Cancer Fellowship of the American Cancer Society awarded by the International Union Against Cancer.

The authors are indebted to Dr. G. SZABÓ for providing the ink sacs of the squid and for the electron microscopy of SQ melanosomes, to Dr. Y. HORI for the electron microscopy of HP, B 16, S 91, ST and RP melanosomes, to Dr. K. TODA for the preparation of RP melanosomes, and to Mr. H. SEILER (Boston) and Mr. J. ZBROZEK (Prague) for assistance in the determination of the amino acids.

Finally, one of the authors (J.D.) gratefully remembers the friendly reception extended to him by all members of the Department of Dermatology of Harvard Medical School, and especially by Dr. T. B. FITZPATRICK, as well as the numerous and valuable discussions on the fascinating problem of melanins, melanosomes, and melanogenesis.

References

1 BIRBECK, M. S. C.: Electron microscopy of melanocytes: the fine structure of hair-bulb premelanosomes. Ann. N. Y. Acad. Sci. *100:* 540–547 (1963).
2 BLOIS, M. S. and KALLMAN, R. F.: Incorporation of C^{14} from 3,4-dihydroxyphenylalanine-2-C^{14} into the melanin of mouse melanomas. Cancer Res. *24:* 863–868 (1964).
3 DROCHMANS, P.: Ultrastructure of melanin granules; in W. MONTAGNA and F. HU. Advances in biology of skin, vol. VIII, pp. 169–177 *(Pergamon Press, Oxford 1967)*.
4 MIYAMOTO, M. and FITZPATRICK, T. B.: On the nature of the pigment in retinal pigment epithelium. Science *126:* 449–450 (1957).
5 MOYER, F. H.: Genetic variations in the fine structure and ontogeny of mouse melanin granules. Amer. Zool. *6:* 43–66 (1966).
6 SEIJI, M.; FITZPATRICK, T. B.; SIMPSON, R. T. and BIRBECK, M. S. C.: Chemical composition and terminology of specialized organelles (melanosomes and melanin granules) in mammalian melanocytes. Nature, Lond. *197:* 1082–1084 (1963).
7 SEIJI, M.; SHIMAO, K.; BIRBECK, M. S. C. and FITZPATRICK, T. B.: Subcellular localization of melanin biosynthesis. Ann. N. Y. Acad. Sci. *100:* 497–533 (1963).
8 STEIN, W. D.: Chemical composition of the melanin granule and its relation to the mitochondrion. Nature, Lond. *175:* 256–257 (1955).
9 SZABÓ, G.; FITZPATRICK, T. B. and WILGRAM, G.: Studies on melanin biosynthesis in the ink sac of the squid *(Loligo pealii)*. IV. Biochemical studies of the ink gland and the ink. Biol. Bull. *125:* 394 (1963).
10 TAKAHASHI, H. and FITZPATRICK, T. B.: Large amounts of dihydroxyphenylalanine in the hydrolysate of melanosomes from Harding-Passey mouse melanoma. Nature, Lond. *209:* 888–890 (1966).

Author's address: Dr. J. DUCHŎN, Department of Biochemistry, Faculty of Medicine, Charles University, *Prague* (Czechoslovakia)

Fluorimetry of a Dopa Peptide and Dopa Thioethers

H. RORSMAN, A.-M. ROSENGREN and E. ROSENGREN

Department of Dermatology, Department of Biochemistry and Department of Pharmacology, University of Lund, Lund

Introduction

The fluorescence method of FALCK and HILLARP for demonstrating catecholamines and their immediate precursor, dopa, has provided a new tool for histochemical studies of melanocytes. Some of the histological observations with the fluorescence method will be reviewed, and recent chemical studies on identification of catechols in malignant melanoma will be reported.

Fluorescence of Normal and Pathologic Melanocytes after Formaldehyde Treatment

The first evidence of histochemically detectable catechols in melanocytes was presented by FALCK and RORSMAN in 1963. Cells with dendrites in the basal layer of epidermis of human Caucasian skin showed formaldehyde-induced fluorescence when treated according to the method of FALCK and HILLARP. The presence of fluorescence in normal melanocytes has since been confirmed in reports primarily concerned with pathological melanin-forming cells [8, 10]. It was assumed that β-(3,4-dihydroxyphenyl)-alanine, dopa, was the compound responsible for the fluorescence observed in melanocytes. Caucasian epidermal melanocytes produce an increased amount of pigment and exhibit an increased fluorescence intensity when stimulated to increased functional activity [1, 15].

In contrast, the highly active melanocytes of normal Negro skin show little or no fluorescence [2].

The melanocytes in the hair bulbs of pigmented guinea pigs show a formaldehyde-induced fluorescence, but in albino skin or in the white parts of multicolored animals the hair melanocytes show no fluorescence. Fluorescence has been observed in the melanocytes of black as well as of red hair and is stronger in the latter [14].

Cells of human pigmented nevi and malignant melanomas of the skin have been extensively studied for formaldehyde-induced fluorescence [5, 8, 10, 20]. *Junctional nevi* have an intense green to yellow fluorescence. In *dermal nevi* the fluorescence is less pronounced than in junctional nevi.

Human malignant melanomas of the skin show very strong green-yellow or sometimes yellow fluorescence. The fluorescence of cutaneous metastases of melanoma resembles that of the primary tumors, but in secondary ones in the lymph nodes the fluorescence intensity of the individual malignant cells may differ widely. A single lymph node may contain fluorescent and non-fluorescent melanoma cells.

The fluorescence of melanoma cells is generally localized in the cytoplasm [8, 10]. In cytological preparations of formaldehyde-treated melanoma imprints the granular appearance of the fluorescence is striking [5].

While human skin melanomas show a strong formaldehyde-induced fluorescence, investigation of different *transplanted melanomas* in experimental animals has not revealed any fluorescence [12, 14]. Neither, for example, were three different hamster melanomas, two melanotic and one amelanotic, obtained from Dr. FORTNER at the Sloan-Kettering Institute, New York, fluorescent after formaldehyde-treatment, nor did three types of mouse melanomas, B 16, Harding-Passey and S 91, show fluorescence.

Only four of eleven human *ocular melanomas* studied showed fluorescence, and even then only parts of the tumors fluoresced [6]. The fluorescence of ocular melanomas resembled more that of lymph node metastases of skin melanomas than that of the invariably strongly-fluorescent primary skin tumors.

Identification of Melanoma Catechols which give Fluorescence with Formaldehyde

Of the biogenic substances that condense with formaldehyde to form fluorescent products, dopa seemed to be the most probable, since this amino acid is an intermediary substance in melanin synthesis. Chemical investigations showed that the dopa content of melanomas was such that this substance might be responsible for the fluorescence after formaldehyde-treatment.

The color of the formaldehyde-induced fluorescence of cells synthesizing melanin has been described as green to yellow. Dopa gives a green fluorescence.

Microspectrofluorimetry of melanoma cells has shown the excitation maximum at 425–430 nm and emission maximum at approximately 510 nm [5]. Thus, the emission of melanoma cells did not correspond to the emission of dopa, which has a maximum emission at 480 nm. ROST and POLAK [20] also found that the spectra of dopa were not consistent with those from melanocytes.

Other catechols were assumed, therefore, to be responsible for the formaldehyde-induced yellow fluorescence of human malignant melanomas, and chemical methods were used for the isolation and identification of such catechols.

No adrenaline, dopamine or 5-hydroxytryptamine was found in human malignant melanomas and very small amounts of noradrenaline were detected [9]. Hydrolysis of perchloric acid precipitates of melanomas gave considerable additional amounts of dopa. But also another substance which gave fluorescence with formaldehyde was found in melanoma extracts. This substance had a lower Rf value than dopa when chromatographed in phenol-hydrochloric acid on paper. Oxidized eluates of melanomas showed two distinct emissions; one was that of dopa, the other, at a higher wavelength, of an unidentified substance [11].

It has been shown that B 16, Harding-Passey and S 91 mouse melanomas and different hamster melanomas contain dopa [11, 16, 21], but they do not fluoresce on treatment with formaldehyde [12, 14].

The poor correlation between the dopa content and the fluorescence characteristics of different melanomas further motivated attention to the second compound in human melanomas, found by us in 1966. This compound seemed to be a catechol and we assumed that it could be a peptide with dopa in N-terminal position, since dopa must have a free amino group to give fluorescence with formaldehyde. In order to test this hypothesis we synthetized enzymatically a dopa peptide by incubating L-tyrosyl-glycyl-glycine with polyphenoloxidase for 2 h at room temperature. Adsorption to Al_2O_3 and elution with 0.1 N HCl gave a catechol which could be separated from dopa by chromatography in butanol/acetic acid/water. The dopa peptide formed moved more slowly than dopa, gave fluorescence with formaldehyde and became red in visible light when sprayed with potassium ferricyanide. Spectrophotofluorimetry after oxidation according to Anton and Sayre [3] showed that the oxidized dopa peptide had the same fluorescence as dopa. Thus, the compound observed in human melanomas with emission at a longer wavelength than dopa after oxidation was probably not a dopa peptide. Since dopa in N-terminal position in peptides gives the same fluorescence as dopa after oxidation, determination of dopa using the method of Anton and Sayre does not give information on the amount of free dopa. The occurrence of free dopa in human melanomas, however, has been proved by the finding of dopamine formation after incubation of tissue extracts with dopa decarboxylase [9, 11].

The studies on the dopa peptide had shown that dopa in N-terminal position could give formaldehyde-induced fluorescence, but the human melanoma compound with formaldehyde-induced fluorescence, which fluoresced at a longer wavelength than dopa after oxidation, seemed to be of another chemical nature.

We made the assumption that the melanoma catechol could be a thioether of dopa or of a dopa peptide [18, 19]. Incubation of L-tyrosine or L-tyrosyl-glycyl-glycine with cysteine or glutathione in the presence of polyphenoloxidase gave rise to the formation of compounds which, after adsorption to Al_2O_3 and subsequent elution, were separable from dopa and the dopa peptide by chromatography (table I). The compounds formed fluoresced yellow after formaldehyde treatment and the fluorescence after oxidation according to Anton and Sayre differed from the fluorescence of dopa and the dopa peptide (fig. 1).

Table I. Rf values of catechols[1]

Substances incubated	Rf value
Dopa (standard)	0.30
Tyrosyl-glycyl-glycine	0.18
Tyrosine + cysteine	0.10
Tyrosine + glutathione	0.09
Tyrosyl-glycyl-glycine + cysteine	0.08
Tyrosyl-glycyl-glycine + glutathione	0.09

1 Rf values of catechols formed by incubation of tyrosine or tyrosyl-glycyl-glycine with polyphenoloxidase and cysteine or glutathione (n-butanol/acetic acid/water, 60:15:25, ascending chromatography). Dopa and the dopa peptide fluoresced green, and the other compounds yellow, after formaldehyde treatment.

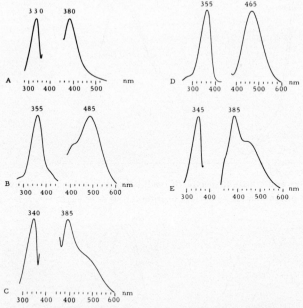

Fig. 1. Excitation and emission spectra for fluorophores formed at oxidation of dopa (A), and of catechols formed at incubation with polyphenoloxydase of tyrosine + cysteine (B), tyrosine + glutathione (C), tyrosyl-glycyl-glycine + cysteine (D), or tyrosyl-glycyl-glycine + glutathione (E). (Courtesy of RORSMAN *et al.* [19].)

The fluorescence characteristics of dopa conjugated with cysteine were very similar to those of the catechol previously described in human melanoma [11].

Last year we had the opportunity to study more extensively a melanoma with large amounts of a catechol similar to that previously described in melanomas [4]. The melanoma tissue examined consisted of skin metastases in a red-haired male with a primary skin melanoma. Chromatography and electrophoresis of the catechol adsorbed to and eluted from Al_2O_3 showed that it moved like cysteinyl dopa. Fluorometry of the Al_2O_3 eluate after oxidation showed a compound with fluorescence emission maximum at 485 nm and excitation maximum at 355 nm corresponding to the maxima of cysteinyl dopa (fig. 2) [19].

Fluorimetry after formaldehyde treatment of the chromatographed catechol from melanoma showed excitation maximum at 420 nm after alkaline treatment and at 375 nm after acid treatment. The corresponding emission maxima were at 530 nm and 510 nm. All maxima of the melanoma catechol corresponded to those of cysteinyl dopa (fig. 3).

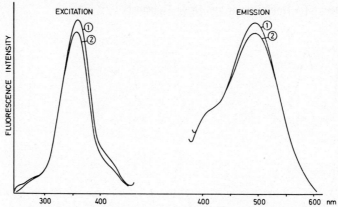

Fig. 2. Excitation and emission spectra of oxidized cysteinyl dopa (1) and of oxidized materials from melanoma (2). (Courtesy of BJÖRKLUND *et al.* [4].)

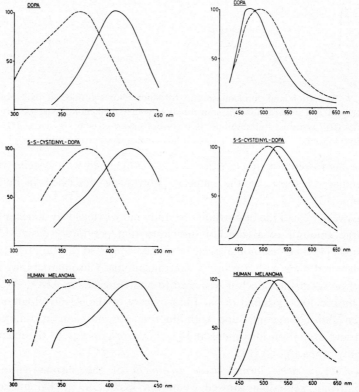

Fig. 3. Exci ation and emission spectra of dopa, cysteinyl dopa and of materials from melanoma after formaldehyde-treatment. Solid lines represent fluorescence in alkaline medium; broken lines represent fluorescence in acid medium. (Courtesy of BJÖRKLUND *et al.* [4].)

UV *spectrophotometry* of the Al_2O_3 eluates of the melanoma extracts showed absorption maxima at 255 nm and 292 nm and a minimum at 275 nm (fig. 4). The maxima and the minimum corresponded to those of cysteinyl dopa [17] or of a cysteinyl dopa peptide [18]. The amount of cysteinyl dopa was calculated to be about 50 µg per g wet weight melanoma tissue, using the molar extinction coefficient of cysteinyl dopa at 292 nm as given by PROTA et al. [17].

Heating of Al_2O_3 eluate in 2 N HCl gave a violet color. Photometry showed absorption maxima at 587 and 330 nm and a slope at 550 nm. After alkalinization, the solution turned yellow-brown and the absorption maximum was at 482 nm. These colors are characteristic for the reaction product of cysteinyl dopa [NICOLAUS and PROTA, personal communication].

All characteristics of the catechol extracted from the melanoma metastases were identical to those of cysteinyl dopa. Cysteinyl dopa is considered to be the precursor of the pigment in red hair [17]. The fact that the melanoma studied was obtained from a patient with red hair may suggest that the presence of cysteinyl dopa in a melanoma reflects the original pigment type of the individual. However, it is also possible that the presence of cysteinyl dopa is due to an aberation of melanin synthesis in some melanomas, independent of the original pigmentation of the patient.

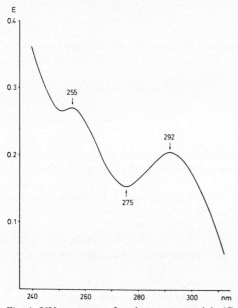

Fig. 4. UV-spectrum of melanoma material. (Courtesy of BJÖRKLUND et al. [4].)

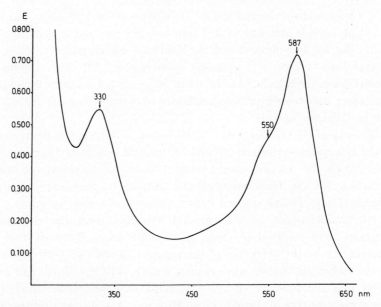

Fig. 5. Absorption spectrum of melanoma material heated in 2 N HCl. (Courtesy of BJÖRKLUND *et al.* [4].)

Acknowledgements

This investigation has been supported by grants from The Swedish Medical Research Council (B71-14X-712-06A and B71-14X-56-07A) and from The Swedish Cancer Society (67-111).

References

1 AGRUP, G.; FALCK, B. and RORSMAN, H.: Formaldehyde-induced fluorescence of epidermal melanocytes after a single dose of ultraviolet irradiation. Acta derm.-vener. *51:* 353 (1971).
2 AGRUP, G.; FALCK, B.; OLIVECRONA, H. and RORSMAN, H.: Formaldehyde-induced fluorescence of epidermal melanocytes of Caucasian and Negro skin. Acta derm.-vener. *51:* 350 (1971).
3 ANTON, Aa. H. and SAYRE, D. F.: The distribution of dopamine and dopa in various animals and a method for their determination in diverse biological material. J. Pharmacol. exp. Ther. *145:* 326 (1964).
4 BJÖRKLUND, A.; FALCK, B.; JACOBSSON, S.; RORSMAN, H.; ROSENGREN, A.-M. and ROSENGREN, E.: Cysteinyl dopa in a human malignant melanoma. Acta derm.-vener. (In press).

5 EHINGER, B.; FALCK, B.; JACOBSSON, S. and RORSMAN, H.: Formaldehyde-induced fluorescence of intranuclear bodies in melanoma cells. Brit. J. Derm. *81:* 115 (1969).
6 EHINGER, B.; OLIVECRONA, H. and RORSMAN, H.: Malignant melanomas of the eye as studied with a specific fluorescence method. Acta path. microbiol. scand. *69:* 179 (1967).
7 FALCK, B. and RORSMAN, H.: Observation on the adrenergic innervation of the skin. Experientia *19:* 205 (1963).
8 FALCK, B.; JACOBSSON, S.; OLIVECRONA, H. and RORSMAN, H.: Pigmented nevi and malignant melanomas as studied with a specific fluorescence method. Science *149:* 439 (1965).
9 FALCK, B.; JACOBSSON, S.; OLIVECRONA, H.; OLSEN, G.; RORSMAN, H. and ROSENGREN, E.: Determination of catecholamines, 5-hydroxytryptamine and 3,4-dihydroxyphenylalanine (dopa) in human malignant melanomas. Acta derm.-vener. *46:* 65 (1966).
10 FALCK, B.; JACOBSSON, S.; OLIVECRONA, H. and RORSMAN, H.: Fluorescent dopa reaction of nevi and melanomas. Arch. Derm. *94:* 363 (1966).
11 FALCK, B.; JACOBSSON, S.; OLIVECRONA, H.; RORSMAN, H.; ROSENGREN, A. M. and ROSENGREN, E.: On the occurrence of catechol derivatives in malignant melanomas. Comm. Dept. of Anatomy, University of Lund, Sweden, No. 5 (1966).
12 FALCK, B.; JACOBSSON, S.; OLIVECRONA, H.; RORSMAN, H.; ROSENGREN, A. M. and ROSENGREN, E.: Dopa in melanin-producing cells. G. ital. Derm. *44–110:* 493 (1969).
13 OLIVECRONA, H. and RORSMAN, H.: Specific fluorescence in guinea-pig melanocytes. Acta derm.-vener. *46:* 497 (1966).
14 OLIVECRONA, H. and RORSMAN, H.: Fluorescence microscopy of malignant melanomas in the Syrian golden hamster. Acta derm.-vener. *46:* 401 (1966).
15 OLIVECRONA, H. and RORSMAN, H.: The effect of roentgen irradiation on the specific fluorescence of epidermal melanocytes. Acta derm.-vener. *46:* 403 (1966).
16 POMERANTZ, H. S. and WARNER, C. M.: Identification of 3,4-dihydroxyphenylalanine as tyrosinase cofactor in melanoma. Biochem. biophys. Res. Comm. *24:* 25 (1966).
17 PROTA, G.; SCHERILLO, G. e NICOLAUS, R. A.: Struttura e biogenesi delle feomelanine. IV. Sintesi e proprietà della 5-S-cisteinildopa. Gazz. chim. ital. *98:* 495 (1968).
18 RORSMAN, H.; ROSENGREN, A. M. and ROSENGREN, E.: Fluorometry of a dopa peptide and its thioether. Acta derm.-vener. *51:* 179 (1971).
19 RORSMAN, H.; ROSENGREN, A. M. and ROSENGREN, E.: Fluorometry of catecholthioethers. Acta derm.-vener. (in press).
20 ROST, F. W. D. and POLAK, J. M.: Fluorescence microscopy and microspectrofluorimetry of malignant melanomas, naevi and normal melanocytes. Virchows Arch. Path. Anat., Abt. A. *347:* 321 (1969).
21 TAKAHASHI, H. and FITZPATRICK, T. B.: Large amounts of deoxyphenylalanine in the hydrolysate of melanosomes from Harding-Passey mouse melanoma. Nature, Lond. *209:* 888 (1966).

Author's address: Dr. H. RORSMAN, Department of Dermatology, University of Lund, *Lund* (Sweden)

Control of Pigmentation

A Quantal Bioassay for Melatonin

B. C. FINNIN and B. L. REED

Victorian College of Pharmacy, Parkville, Vic.

Introduction

Melatonin was first isolated from mammalian pineal glands by LERNER et al. [6] but its physiological role is not yet completely understood. One of the main difficulties is that only extremely small quantities of melatonin are present in the pineal gland, and much more might be learned about its storage and disposition by the use of a sensitive and specific assay method.

Both instrumental and biological assay methods are available for estimating melatonin concentrations, but all of them have some disadvantages. Gas-liquid chromatography (GLC) [2, 3, 5] and spectrofluorimetry [1, 7, 8, 11] are the two main instrumental methods. Some GLC methods have the advantage of being specific, but their sensitivity is low. The sensitivity of GLC methods can be improved dramatically by converting melatonin to its heptafluorobutyryl derivative [3]. Spectrofluorimetric methods can be made highly sensitive [7, 8], but indole alkylamines closely related chemically to melatonin, and usually found in association with it, have similar fluorescence spectra, so that these methods lack specificity.

Melatonin produces melanosome aggregation within the melanophores of amphibians and this reaction forms the basis of two bioassay methods. The method of RALPH and LYNCH [12] involves assessment of the pigment aggregation produced in tadpoles by immersing them in melatonin-containing solutions. The relatively high specificity and the lack of interference by biological materials are important advantages of this method. However, since a tadpole culturing program is essential and each test animal must be viewed under the microscope, this method is costly in terms of time and facilities.

The bioassay method of MORI and LERNER [10] involves photometric measurement of light reflected from, or transmitted through, isolated pieces of frog skin immersed in test solution. Since noradrenaline, acetylcholine and 5-hydroxytryptamine also cause paling of frog skin [9, 10], interference is likely in biological samples unless a separation scheme is used first. This method overcomes the subjectiveness of methods necessitating the visual assessment of melanophore index, but it involves the use of considerably more equipment. The sensitivity of some of the assay methods that have been used is shown in table I.

The bioassay method that we have developed is based upon the action of melatonin on the freshwater pencil fish, *Nannostomus beckfordi anomalus*. The pencil fish undergoes a circadian colour change. During the day this fish displays a horizontal black band on each side along the entire length of the body and extending into the caudal fin (the day band, fig. 1). At night this band disappears from the body and is replaced by two black spots (the night spots, fig. 2).

Melatonin administered to the fish induced the change from day to night colouration [13]. To produce this change, melatonin must cause a fading of the day band and a darkening of the night spots. We have investigated over 100 indoles for their ability to cause this reaction [4, 14]. Melatonin was 1000 times as active as any other of these indoles. Many other agents have an effect on pencil fish melanophores, however, none of these can induce the change from day to night colouration [15]. The specificity and sensitivity of the change from day to night colouration suggest the use of this animal for the bioassay of melatonin.

Table I. Sensitivity of methods for melatonin assay

Method	Sensitivity (ng)
Spectrofluorimetry	15
Spectrofluorimetry (0-phthalaldehyde-)	5
Gas chromatography	10,000
Gas chromatography (heptafluorobutyryl-)	0.02
Tadpole bioassay	0.10
Frog skin bioassay	0.01

Fig. 1. The pencil fish showing the day band.
Fig. 2. The pencil fish showing night spots.

Method and Results

Young adult specimens of *Nannostomus beckfordi anomalus* of either sex and of average weight 155 mg (range 130 to 200 mg) were used. The fish were kept at 24° to 25 °C on a white background for at least two weeks before experiments. The experiments were always conducted during the afternoon.

Intraperitoneal injections were made with a 50 µl syringe fitted with a 30 gauge needle. The volume injected was 10 µl. Melatonin (Sigma Chemical Co.) was dissolved in 90% alcohol and diluted with Young's Teleost Ringer to give a final concentration of alcohol of 0.02% or less. After injection, fish were kept in 1 l beakers, 5 fish to a beaker and on a dark background, for observation. When kept on a light background the pencil fish colouration may fade and this makes the melatonin-induced change less distinct. However, all fish clearly display the full day colouration shown in figure 1 after less than 5 min on a black background, and this makes the response to melatonin clearer and easier to see. The maximum response to

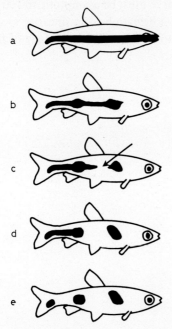

Fig. 3. A diagram of an untreated pencil fish (a), and fish showing increasing degrees of response to melatonin (b, c, d and e). The endpoint colour is shown by fish (c) the break in the band (indicated by the arrow) is an important feature.

any dose of melatonin occurred within 10 min after injection, and its duration ranged from 5 min to several hours depending on the dose.

An untreated fish (a), together with 4 fish showing increasing degrees of response to melatonin (b, c, d and e), is shown in figure 3. In some fish, either at night or after injection of high doses of melatonin, the band in the region of the caudal peduncle does not fade, the maximum response to melatonin in these fish is that shown by fish (d) in figure 3. The most suitable endpoint was found to be the occurrence of a break in the day band in the area indicated by the arrow in fish (c) which shows the endpoint colouration. Any fish in which the day band showed a break in this position, accompanied by development of the anterior night spot, was recorded as having a positive response to melatonin.

A plot of the percentage of fish showing a positive response against the logarithm of the dose of melatonin (fig. 4) appears linear over the range from 0.05 to 0.15 ng. At least 100 fish were used to determine each point. The percentage of fish showing a positive response at a dose level of 0.2 ng was 92%. This fell outside the 95% confidence limits drawn about the regression line. Since the dose-response curve must be asymptotic to 100%, it is likely that the result above indicates a deviation from linearity at the 0.2 ng dose level. No attempt was made to estimate the percentage of fish

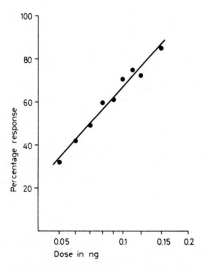

Fig. 4. A plot of the percentage of fish showing a positive response to melatonin against the log of the dose. At least 100 fish were used to determine each point. The regression line is shown.

Table II. Estimation of melatonin concentrations using a (2+2) method

Test No.	Dilutions selected by screening		Assay percentage response[1]		Standard[2] percentage response[1]		Calc. Potency ng/ml	True Potency ng/ml	Fiducial Limits 0.01<p<0.05
	T_1	T_2	T_1	T_2	S_1	S_2			
1	1:2	Undil.	44	72	28	78	10.8	11	8.4–14.2
2	1:200	1:100	40	68	28	78	1,010	1,000	777–1,300
3	1:2	Undil.	52	88	28	78	13.4	15	10.5–16.8

1 Per cent response of 50 fish.
2 The same results from 50 fish at 5 ng/ml (S_1) and 50 fish at 10 ng/ml (S_2) were used as standards for all three determinations.

responding at doses lower than 0.05 ng, since the assessment of the endpoint at low doses becomes more difficult. It is probable that the useful range of the assay will be from 0.05 to 0.15 ng.

The sensitivity of different batches of pencil fish to melatonin has proved uniform over the 9 months of this study. Bartlett's test for comparing the variance of the batches shows that they could be considered to come from the same population.

The assay of unknowns can be handled using a (2+2) method (two test dilutions plus two standard dilutions). Three test solutions of melatonin, the concentrations of which were unknown to the experimenter, were screened by injection each into 5 fish. If all fish showed a positive response, dilutions were made in Teleost Ringer until a dilution was found which, when injected into 5 fish, produced the endpoint in some but not all fish. Fifty fish were then injected at this dilution and on the basis of the result obtained, another dilution chosen and a further 50 fish injected. Two standard solutions were chosen, each for injection into 50 fish, to give suitable reference responses.

The potency of each of the test solutions was calculated and confidence limits obtained using standard statistical methods. The results obtained from these estimations are shown in table II.

Discussion

We have used the pencil fish method to identify melatonin in single pineal glands of the rat, rabbit and guinea-pig and in pooled glands of

5 mice. Melatonin has also been detected in extracts from 20 pooled fish retinas. These preliminary experiments involving only crude extraction procedures indicate that the assay is not affected by the presence of protein or other biological materials, since melatonin concentrations found agree well with published data. Since it is highly specific, this reaction of pencil fish can be used simultaneously to identify and assay melatonin with little separation or purification of extracts.

This test is less subjective than the other methods requiring visual assessment of response, since there is no need to grade the response. Since microscopic examination of each test animal is not necessary, an assay can be completed in 3 hours. Because pencil fish can be readily obtained from tropical fish distributors and can be re-used after resting for one week, there is no need for a breeding program. Thus only a small amount of equipment is needed.

Intraperitoneal injection, instead of administration *via* the swimming water, was used to avoid interference with absorption by traces of surfactants or organic solvents which can modify transfer across the gills. A black background eliminates the need for darkening agents which add a further variable to be controlled in some methods [10].

It is hoped that application of this bioassay may lead to a better understanding of the role of melatonin in physiological control of pigment cells in fishes and amphibia. The sensitivity, specificity and simplicity of this method commend it for use until a more sensitive and specific instrumental method is developed.

Acknowledgements

Gratitude is expressed to Dr. ANNE STAFFORD and members of the Pharmaceutical Society of Victoria for the support that made this investigation possible.

This work was supported by a grant from the Australian Research Grants Committee.

References

1 BALEMANS, M. G. M.; EBELS, I. and VONK-VISSER, D. M. A.: Separation of pineal extracts on Sephadex G-10 I. A spectrofluorimetric study of indoles in a cockerel pineal extract. J. neuro-visc. Relat. *32:* 65–73 (1970).
2 BROOKS, C. J. W. and HORNING, E. C.: Gas chromatographic studies of catecholamines, tryptamines, and other biological amines. I. Catecholamines and related compounds. Analyt. Chem. *36:* 1540–1545 (1964).

3 DEGEN, P. H. and BARCHAS, J. D.: Gas chromatographic assay for melatonin. Proc. Western Pharmacol. Soc. *13:* 34 (1970).
4 FINNIN, B. C.; LANGDON, P. W. and REED, B. L.: Unpubl. observations (1971).
5 GREER, M. and WILLIAMS, C. M.: Gas-chromatographic determination of melatonin and 6-hydroxymelatonin. Clin. chim. Acta *15:* 165–168 (1967).
6 LERNER, A. B.; CASE, J. D.; TAKAHASHI, Y.; LEE, T. H. and MORI, W.: Isolation of melatonin, the pineal gland factor that lightens melanocytes. J. Amer. chem. Soc. *80:* 2587 (1958).
7 MAICKEL, R. P. and MILLER, F. P.: Fluorescent products formed by reaction of indole derivatives and o-phthalaldehyde. Analyt. Chem. *38:* 1937–1938 (1966).
8 MILLER, F. P. and MAICKEL, R. P.: Fluorometric determination of indole derivatives. Life Sci. *9 (1):* 747–752 (1970).
9 MOLLER, H. and LERNER, A. B.: Melanocyte stimulating hormone inhibition by acetylcholine and noradrenaline in the frog skin bioassay. Acta endocrin., Kbh. *51:* 149–160 (1966).
10 MORI, W. and LERNER, A. B.: A microscopic bioassay for melatonin. Endocrinology *67:* 443–450 (1960).
11 QUAY, W. B.: Differential extractions for the spectrophotofluorometric measurement of diverse 5-hydroxy- and 5-methoxyindoles. Analyt. Biochem. *5:* 51–59 (1963).
12 RALPH, C. L. and LYNCH, H. J.: A quantitative melatonin bioassay. Gener. comp. Endocr. *15:* 334–338 (1970).
13 REED, B. L.: The control of circadian pigment changes in the pencil fish: a proposed role for melatonin. Life Sci. *7:* 961–973 (1968).
14 RUFFIN, N. E.; REED, B. L. and FINNIN, B. C.: The specificity of melatonin as a melanophore controlling factor in the pencil fish. Life. Sci. *8 (2):* 1167–1174 (1969).
15 RUFFIN, N. E.; REED, B. L. and FINNIN, B. C.: Pharmacological studies on teleost melanophores. Europ. J. Pharmacol. *8:* 114–118 (1969).

Author's address: BARRIE C. FINNIN, Victorian College of Pharmacy, *Parkville, Victoria 3052* (Australia)

The Effect of Various Drugs on the Response of Isolated Frog Skin Melanophores to Melanocyte-Stimulating Hormone (MSH) and Adenosine 3', 5' – Monophosphate (Cyclic AMP)[1]

R. Novales and Barbara J. Novales

Department of Biological Sciences, Northwestern University, Evanston, Ill.

Introduction

There is now convincing evidence, reviewed by Novales [8], that pituitary MSH brings about melanin dispersion in frog and perhaps other melanophores by increasing the cyclic AMP content of these cells. Although it is known that melanin dispersion in response to MSH or cyclic AMP is inhibited reversibly by hypertonicity [7, 9], the mechanism by which the inhibition is produced is still uncertain. The recent demonstration that cytochalasin B inhibits skin darkening by MSH [6] and causes melanin aggregation in MSH-treated melanophores [5] suggests that some cytochalasin B-sensitive organelles may be important for melanin dispersion. Since McGuire and Moellman [5] found that significantly fewer microfilaments were present in cytochalasin B-treated epidermal melanophores, their results suggest that microfilaments are required for the maintenance of melanin dispersion. In addition, we have found [Novales and Novales, unpublished data] that cytochalasin B inhibits iridophore contraction in response to MSH, suggesting the possible importance of microfilaments in the response of these cells also.

The present study provides results obtained with several drugs well known to inhibit RNA synthesis (actinomycin), protein synthesis (puromycin), or activate cyclic nucleotide phosphodiesterase (imidazole), in an attempt to further elucidate the possible role of these factors in producing melanin dispersion. In addition, it has been found that local anesthetics have a significant action on frog skin chromatophores that may provide new information bearing on their mechanism of action.

[1] This investigation was supported by N.S.F. Grants Nos. GB-4956X and GB-23065.

Materials and Methods

The thigh and leg skins of adult *Rana pipiens* frogs were employed, using the technique developed by SHIZUME *et al.* [14]. Skins were removed from double-pithed animals and paled by 4 rinses in Ringer's solution. During this time, the melanin granules aggregate in the melanophores and the iridophores expand. The skins were then mounted on aluminum frames and placed, epidermal side down, in 30 ml beakers, containing 20 ml of Ringer's solution. Darkening agents (MSH or cyclic AMP) were present in the presence or absence of the drugs which were dissolved in advance in the Ringer. Darkening was measured as a decrease in reflectance with a Photovolt Photoelectric Reflection Meter, Model 610. After darkening, skins were routinely examined with a binocular dissecting microscope to assess the chromatophore changes which underlay the darkening. Sometimes, they were fixed and whole mounts prepared for compound microscopy. In these cases, brightfield microscopy was used for the melanophores and darkfield for the iridophores. The MSH used was a synthetic α-MSH (Hofmann) and the cyclic AMP was obtained from either Mann Research Laboratories, Inc. or from Sigma Chemical Co. Caffeine (Nutritional Biochemicals Corp.), imidazole (Sigma), procaine (Sigma), puromycin dihydrochloride (Nutritional), and tetracaine (Winthrop Laboratories) were all commercially obtained. However, actinomycin D was a gift of Dr. H. B. WOODRUFF of Merck, Sharpe and Dohme. It was first dissolved in ethanol and then properly diluted, with an appropriate ethanol control.

Results

Table I shows that actinomycin D significantly inhibited the response to MSH up to 33% over the effective concentration range that was studied

Table I. The effect of actinomycin D on the response of the melanophores of isolated frog skin to MSH[1]

Actinomycin D (μg/ml)	Decrease in reflectance (1 h G.U.)	% Inhibition[2]	P[2]
0	42 ± 2 (10)[3]	–	–
12.5	33 ± 3 (3)	21.4	<.05<.01
25	35 ± 1 (3)	16.7	<.01>.005
50	28 ± 2 (3)	33.3	<.005
100	28 ± 3 (10)	33.3	<.005
100[4]	30 ± 5 (4)	28.6	<.05>.005
100[5]	30 ± 3 (4)	28.6	<.05>.01

1 MSH concentration: 10 U/ml.
2 Compared to MSH in the absence of actinomycin D.
3 Figures are means ± standard errors; numbers of skins in parentheses.
4 15 min pretreatment.
5 30 min pretreatment.

(12.5–50 µg/ml). Attempts to increase the inhibition by increasing the actinomycin D concentration to 100 µg/ml were ineffective, even when 15 or 30 min preincubations were employed (table I). Such skins darkened normally with MSH when placed in Ringer in the absence of the actinomycin. Actinomycin D also inhibited the response to cyclic AMP to the same degree (fig. 1). However, since actinomycin D in the ethanol had a significant darkening in the absence of either MSH or cyclic AMP (fig. 1), the degree of inhibition may have been even greater than the measured one-third. On the other hand, puromycin at a concentration of 100 µg/ml had no inhibitory action on the darkening of frog skin by either MSH or cyclic AMP.

Imidazole, an activator of phosphodiesterase, had a small darkening action (table II). This darkening was additive with the darkening produced by other darkening agents, e.g., MSH, caffeine or isoproterenol (table II).

Local anesthetics also had a significant darkening action, procaine being less effective than tetracaine in this regard (table III). Furthermore, the results obtained with the local anesthetics show that melanin dispersion is

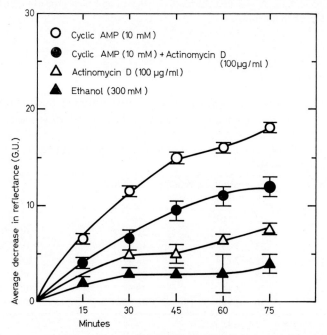

Fig. 1. The effect of actinomycin D on the response of isolated frog skin to cyclic AMP. The effects of actinomycin in the alcohol diluent, as well as the alcohol diluent alone are also shown. Skins had been previously paled in Ringer's solution. All agents in Ringer's solution.

Table II. The effect of imidazole on the responses of the chromatophores of isolated, paled frog skin[1]

Imidazole (mM)	MSH (U/ml)	Caffeine (mM)	Isoproterenol (μg/ml)	Δ60 min (G.U.)
10	–	–	–	9 ± 1 (2)[2]
40	–	–	–	9 ± 1 (4)
–	0.1	–	–	7 ± 2 (4)
–	0.5	–	–	3 ± 3 (5)
10	0.5	–	–	375 ± 5 (4)
40	0.1	–	–	12 ± 1 (4)
–	–	2.6	–	18 ± 2 (12)
40	–	2.6	–	25 ± 3 (9)
–	–	–	50	9 ± 3 (6)
40	–	–	50	15 ± 2 (9)

1 All skins paled by rinses in Ringer's solution.
2 Figures are means ± standard errors; numbers of skins in parentheses.

Table III. The effect of local anesthetics on the responses of the chromatophores of isolated, paled frog skin[1]

Drug	Concentration (mM)	Δ60 min[2] (G.U.)	Chromatophores[3,4,5]	
			Melanophores	Iridophores
Procaine	3.7	2 ± 1 (4)[6]	o	x
	7.3	4 ± 0.5 (9)	o	o
	11.0	8 ± 0.4 (7)	x	o
Tetracaine	0.1	2 ± 0.5 (4)	o	x
	0.5	11 ± 2 (4)	x	x
	1.0	13 ± 1 (4)	x	o
	3.7	22 ± 2 (5)	*	o
	7.3	24 ± 3 (3)	*	o
None	–	1 ± 1 (21)	o	*

1 All agents in Ringer's solution for a 60-min period.
2 Decrease in reflectance over the 60-min period in galvanometer units.
3 o aggregated state of melanin or iridophore platelets.
4 x = partially dispersed state of melanin or iridophore platelets.
5 * = dispersed state of melanin or iridophore platelets.
6 Figures are means ± standard errors; numbers of skins in parentheses.

much more important than iridophore contraction in causing reflectance decreases in isolated frog skin. Thus, 7.3 mM procaine had no effect on the melanophores, but fully contracted the iridophores, causing an average reflectance decrease of only 4 galvanometer units (G.U.). However, 7.3 mM tetracaine produced *both* melanin dispersion and iridophore contraction, causing an average reflectance decrease of 24 G.U. Melanin dispersion thus accounted for a decrease of 20 G.U., much greater than that associated with iridophore contraction.

Discussion

Cyclic AMP and MSH were each inhibited to approximately the same degree by actinomycin D, further supporting the concept that MSH action is mediated by cyclic AMP. However, in the absence of more information, it would be unwise to draw any further conclusions, e.g., that DNA-dependent RNA synthesis is required for the response. PRASAD [11] found that actinomycin D injections paled intact salamanders *(Amphiuma tridactylum)*. He concluded that the effect probably resulted from an inhibition of MSH secretion by the pars intermedia, because the effect could be counteracted by exogenous administration of MSH. Furthermore, the rate of paling by actinomycin was similar to the rate at which paling occurs after hypophysectomy in this species, further supporting the concept that it was due to a decrease in MSH levels, rather than an inhibition of MSH action on the melanophores. The present study has shown that although MSH action is slightly inhibited, significant melanin dispersion can occur in the presence of actinomycin D. The data obtained with puromycin support the concept that protein synthesis may not be a necessary concomitant for melanin dispersion in response to MSH or cyclic AMP. An attempt was made to enhance the actinomycin D inhibition with acridine orange, in view of the recent report [13] that acridine orange enhances the uptake of actinomycin D into lymphocytes. However, no increase in the present actinomycin D inhibition was obtained with acridine orange. The relative lack of effect of actinomycin D or puromycin on the action of MSH recalls their lack of effect on vasopressin stimulation of sodium transport by toad bladder [2], another cyclic AMP-mediated peptide hormone response [10].

The effect of imidazole was opposite to the expected effect, in view of the established ability of imidazole to stimulate cyclic nucleotide phosphodiesterase [1] and thus, presumably, to reduce cellular cyclic AMP. A number of imidazole antagonisms of cyclic AMP-mediated hormonal responses have been tabulated by ROBISON *et al.* [12]. Among these are the actions of ACTH on adrenal steroidogenesis and fat cell lipolysis. However, no such antagonism was found to the response of frog skin to MSH, caffeine or isoproterenol. It is unlikely that the MSH concentration was too high, because imidazole (40 mM) had no effect on the response to a low concentration of MSH (0.5 U/ml). The darkening action of caffeine [7] and of isoproterenol [4] have been previously reported. Neither of these agents was inhibited by imidazole. The mode of action of imidazole in darkening frog skin is unknown. It may be acting to increase cyclic AMP levels. However, propranolol (0.1 mM) had no effect on the imidazole response, indicating that β-adrenergic stimulation is probably not involved. Propranolol can block the response of *Xenopus laevis* melanophores to epinephrine [9], and there is abundant evidence that β-adrenergic stimulation increases cellular cyclic AMP content [16]. If imidazole darkening were mediated by cyclic AMP, some potentiation of the imidazole response by caffeine might have been expected, and none was found (table II). Finally, the results of WIGGINS and HARCLERODE [17] are similar to the present results. They found that imidazole mobilized ^{45}Ca from sparrow bone, *in vitro*, the opposite of the expected effect, in view of its hypocalcemic action *in vivo* in the rat.

The darkening action of the local anesthetics was of the greatest interest, because it suggests that cyclic AMP may be involved in some of their other actions. TERCAFS [15] earlier reported that cocaine (0.5 mM) had no effect on the melanophores of *R. temporaria* skin. Tetracaine was more potent than procaine in darkening frog skin and this is also true of its local anesthetic potency [18]. The data obtained with these agents support the conclusion that melanin dispersion has a greater effect on skin reflectance than does iridophore contraction. It has been known since the discovery of the pars intermedia hormone that iridophore contraction occurs in response to this hormone. Most recently, HADLEY and BAGNARA [3] have called renewed attention to the importance of iridophore contraction in the reflectance change produced by MSH. The present study shows that iridophore contraction does have some effect on reflectance, but it is far less than the effect produced by melanin dispersion. Thus, the implicit assumption of most investigators that they have mainly been measuring melanin dispersion by the reflectance measurements has been warranted, to a large degree. Furthermore, most investigators have been accompanying reflectance measurements with light microscopy of responding skin.

Acknowledgements

The authors wish to thank Prof. KLAUS HOFMANN of the University of Pittsburgh for the kind gift of synthetic α-MSH and Dr. H. B. WOODRUFF of Merck, Sharpe and Dohme Research Laboratories for the gift of the actinomycin D.

References

1 BUTCHER, R. W. and SUTHERLAND, E. W.: Adenosine 3', 5'-phosphate in biological materials. I. Purification and properties of cyclic 3', 5'-nucleotide phosphodiesterase and use of this enzyme to characterize adenosine 3', 5'-phosphate in human urine. J. biol. Chem. *237:* 1244–1250 (1962).
2 EDELMAN, I. S.; BOGOROCH, R. and PORTER, G. A.: On the mechanism of action of aldosterone on sodium transport: the role of protein synthesis. Proc. nat. Acad. Sci., Wash. *50:* 1169–1177 (1963).
3 HADLEY, M. E. and BAGNARA, J. T.: Integrated nature of chromatophore responses in the *in vitro* frog skin bioassay. Endocrinology *84:* 69–82 (1969).
4 MCGUIRE, J.: Adrenergic control of melanocytes. Arch. Derm. *101:* 173–180 (1970).
5 MCGUIRE, J. and MOELLMANN, G.: Cytochalasin B: effects on microfilaments and movement of melanin granules within melanocytes. Science *175:* 642–644 (1972).
6 MALAWISTA, S. E.: Cytochalasin B reversibly inhibits melanin granule movement in melanocytes. Nature, Lond. *234:* 354–355 (1971).
7 NOVALES, R. R.: The effects of osmotic pressure and sodium concentration on the response of melanophores to intermedin. Physiol. Zool. *32:* 15–28 (1959).
8 NOVALES, R. R.: On the role of cyclic AMP in the function of skin melanophores. Ann. N.Y. Acad. Sci. *185:* 494–506 (1971).
9 NOVALES, R. R. and DAVIS, W. J.: Cellular aspects of the control of physiological color changes in amphibians. Amer. Zool. *9:* 479–488 (1969).
10 ORLOFF, J. and HANDLER, J. S.: The cellular mode of action of antidiuretic hormone. Amer. J. Med. *36:* 686–697 (1964).
11 PRASAD, K. N.: The lightening effect of actinomycin D on the skin of *Amphiuma tridactylum*. Proc. Soc. exp. Biol. Med. *124:* 619–623 (1967).
12 ROBISON, G. A.; BUTCHER, R. W. and SUTHERLAND, E. W.: Cyclic AMP (Academic Press, New York 1971).
13 ROTH, E. F., Jr. and KOCHEN, J.: Acridine potentiation of actinomycin D uptake and activity. Science *174:* 696–698 (1971).
14 SHIZUME, K.; LERNER, A. B. and FITZPATRICK, T. B.: *In vitro* bioassay for the melanocyte stimulating hormone. Endocrinology *54:* 553–560 (1954).
15 TERCAFS, R. R.: Chromatophores and permeability characteristics of frog skin. Comp. Biochem. Physiol. *17:* 937–951 (1966).
16 TURTLE, J. R. and KIPNIS, D. M.: An adrenergic receptor mechanism for the control of cyclic 3', 5' adenosine monophosphate synthesis in tissues. Biochem. biophys. Res. Comm. *28:* 797–802 (1967).
17 WIGGINS, J. and HARCLERODE, J.: Effect of imidazole on bone ^{45}Ca release in white-throated sparrow, *Zonotrichia albicollis*. Comp. Biochem. Physiol. *34:* 297–300 (1970).
18 WILSON, A. and SCHILD, H.: Applied pharmacology (CLARK); 9th ed. (Little, Brown, Boston 1959).

Author's address: Dr. RONALD R. NOVALES, Department of Biological Sciences, Northwestern University, *Evanston, IL 60201* (USA)

Effects of Ultraviolet Radiation on Melanophore System of Fish[1]

R. Fujii, T. Nakazawa and Y. Fujii

Department of Biology, Sapporo Medical College, Sapporo, and Division of Biology, National Institute of Radiological Sciences, Chiba

Introduction

Ultraviolet (UV) radiation may be used in studies of the effects of radiant energy on cellular functions and subcellular structures, because UV rays of a certain wavelength are specifically absorbed by certain molecules constituting cell organelles, leading to functional depression. This method has been successfully used with micro-organisms and cells in culture [5, 6].

However, there is as yet very little information about the effects of UV rays on neuro-effector systems. This is especially true of short UV rays. This is due to the fact that most photons are absorbed by the overlying tissue layer so that they cannot reach the cells under study. In an attempt to obtain some information about the mechanism of photoinactivation by biologically active short UV rays in neuro-effector systems, we have adopted an autonomic nerve-pigment cell system of a teleost fish, making use of special *in vitro* preparations with the melanophores and their controlling nerves exposed [1, 3].

Material and Methods

The goby, *Chasmichthys gulosus,* of body length between 55 and 70 mm, was used as the experimental material. Split tail-fin preparations, made according to the method of Fujii [1], were mounted on a cover slip by means of a pair of fine glass needles. They were irradiated in a Petri dish containing physiological solution, which had the following composition: 233 mM NaCl, 8.1 mM KCl, 2.3 mM $CaCl_2$, 3.7 mM $MgCl_2$, 5.6 mM glucose,

[1] This paper is dedicated to Professor H. Kinosita on the occasion of his retirement from the Zoological Institute of the University of Tokyo.

5.0 mM tris-HCl buffer (pH 7.4). A commercial germicidal lamp (Matsushita, GL-15, or Toshiba, GL-15) was used as the source of 254 nm-UV rays. The intensity of each lamp was measured with a germicidal lamp intensity meter (Toshiba, GI-1). In all experiments, the dose rate was kept constant at 36 erg $\cdot$ mm^{-2} $\cdot$ sec^{-1}. During exposure the solution containing the preparations was gently agitated with a small rotator operated by a magnetic stirrer. Unless otherwise stated, the preparations were exposed to UV rays from the dermal side. Usually, the whole preparation was irradiated. In some experiments, however, a narrow UV beam, 0.45–2 mm wide, was applied across the middle part of a preparation. Such beams were obtained by putting a pair of screens just over the proximal and distal parts of the fin portion. For this purpose safety razor blades were employed, their edges placed so as to make the beam parallel to the source lamp.

Norepinephrine and melatonin in physiological saline, were employed as pigment-aggregating substances which act directly on the melanophores. Adrenergic melanin-aggregating nerves were electrically stimulated through a pair of platinum wires, 500 μ in diameter and insulated except at the tips: using an electronic stimulator (Nihon Kohden, MSE-3), biphasic square pulses (10 Hz, 1 msec duration, 30 V) were applied to the proximal part of the preparation. Presynaptic neural elements or nerve endings were selectively stimulated to release adrenergic transmitter by perfusing the preparation with physiological saline in which the total Na ions had been replaced with K ions [1].

Responses of the interradial melanophores about the middle of the preparation were recorded either photomicrographically or photoelectrically [4], using a cadmium sulfide photoconductive cell (Hamamatsu TV, P140) and an electronic paper recorder (Hitachi, QPD-74).

Results and Discussion

The direct responses of pigment cells to photic stimuli have been reported in a variety of animals. Very little, however, is known about the responses to UV rays. As for fish chromatophores, SPAETH [8] noted that melanophores in an isolated *Fundulus* scale aggregated in response to short UV. According to him, the effect was reversible, though the UV intensity was not given. In the present observation a similar response was observed, with melanosomes in cellular processes tending to aggregate near the root of the processes. The magnitude of such response increased until the total dosage reached 1.3×10^5 erg $\cdot$ mm^{-2} and then in higher doses the effect became irreversible. In any case, however, a considerable amount of pigment still remained widely scattered within the cellular projections. In the right half of figure 1, the skin was irradiated with UV rays of 1.3×10^5 ergs, and the characteristic response of the melanophores was seen. A minimal discernible response was observed at the dose as low as 3×10^4 ergs. No response was recorded when the preparations were radiated from the epidermal side.

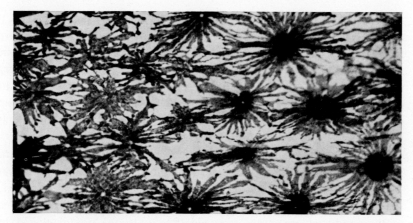

Fig. 1. Photomicrograph of melanophores on the interradial membrane of a split tail-fin preparation of the goby, *Chasmichthys gulosus*. In the right half of the figure the skin was irradiated with 1.3×10^5 erg·mm^{-2} UV rays, 254 nm in wavelength. × 310.

Figure 2 indicates the effects of UV radiation on the responsiveness of melanophores to norepinephrine. The concentration of the catecholamine was 5×10^{-6} M, which is known to be about minimal for arousing the fastest pigment aggregation [2]. A retardation of the response was already observed at 3×10^4 erg·mm^{-2}. Upon increasing the exposure, both the rate and the magnitude of the response decreased until no pigment move-

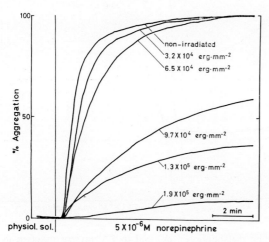

Fig. 2. Responses of goby melanophores to norepinephrine; recorded on split preparations exposed to UV rays of various doses.

ment was detectable in melanophores irradiated with doses above 2×10^5 ergs. The site of the observed photoinactivation may not be adrenergic receptors, but could be linked to the pigment-driving machinery, since melatonin which aggregates melanosomes independently of adrenergic receptors [2] could not forward the response beyond the level attained by norepinephrine. The inactivation might be responsible for the irreversible changes in the cytoplasmic colloidal state, i.e., sol-gel transformation [6], though future studies are naturally needed.

In order to see the effects of UV light on the function of the melanin-aggregating nerve fibers, split-fin preparations irradiated with a UV beam, 2 mm wide, were employed. While the electrical stimulation was applied to the proximal part of the fin portion, the responses of melanophores distal to the irradiated area were recorded (fig. 3). In the non-irradiated preparation, a nervous volley for 1 min elicited a rapid aggregation of melano-

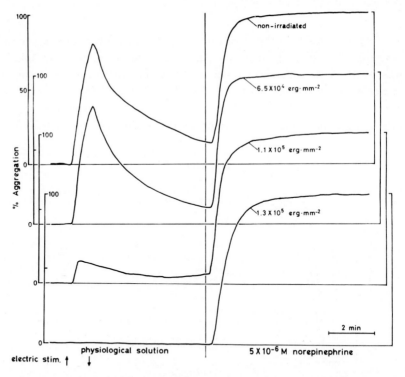

Fig. 3. Responses of goby melanophores, exposed to UV rays of different doses, to electrical stimulation of melanin-aggregating nerves, followed by full aggregation by norepinephrine.

somes. The maximal level of the response was later attained by the addition of norepinephrine to the perfusing saline. As typically shown in figure 3, an exposure less than 7×10^4 erg·mm^{-2} did not affect the axonal conduction. Increasing doses, however, caused a rapid depression of the response. For instance, the deflection seen in the preparation injured by 1.1×10^5 ergs was much smaller than that of the control response. In the case of higher exposures, the conduction of nerves was totally inhibited. So far as we know, the UV sensitivity of autonomic fibers has not been previously reported. It is of interest that the present value for the conduction blockade is comparable with those in motor axons, which GIESE [6] and LIEBERMAN [7] cited in their review articles. Thus, the present results suggest that an analogous mechanism to that currently proposed by LIEBERMAN [7] might be involved in the UV-induced blockade of conduction in the fine autonomic fibers. The structural alterations of UV-absorbing molecular components in the outer and inner leaflets of nerve membranes could be the cause of the photoinactivation. To elucidate this, further work, including spectral sensitivity studies, should be carried out.

It is well known that the increase in the K^+ concentration in perfusing medium induces melanosome aggregation in teleost melanophores. Our earlier studies revealed that K ions act on the presynaptic neural elements or nerve endings to release adrenergic transmitter molecules, which then

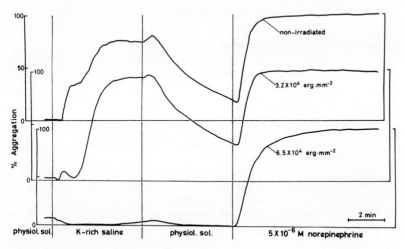

Fig. 4. Responses of goby melanophores, exposed to UV rays of different dosesl to K^+ stimulation of presynaptic neural elements or nerve endings, followed by ful, aggregation by norepinephrine.

bring about the pigment aggregation in the cells [1]. As exemplified by a series of measurements shown in figure 4, we examined the effects of UV radiation on the K-induced responses. Until the total exposure reached 3×10^4 erg $\cdot$ mm^{-2}, no discernible changes took place. Increasing doses over this value, however, brought about a rapid decrease in the height of the response, and practically no deflection was seen in the preparation to which 6.5×10^4 ergs were applied. The easier inactivation might be due to the physico-chemical changes of molecular structures constituting presynaptic membrane through which transmitter molecules pass out. The UV light might otherwise act on the Ca-uptake mechanism, since the transmitter release is coupled with the entry of Ca ions through the presynaptic membrane [4].

The adrenergic nerve-melanophore system existing in split-fin preparations provides a very good model of autonomic effector systems. By making use of these preparations, we have shown in the present study that the UV sensitivity differed among components in this system. The transmitter-releasing function was the most susceptible to UV rays, while conduction of the nerves was rather resistant. The fact that the pigment mobility was more gradually inhibited with increasing exposures, suggests the superadded development of injury in the effector cells. The more accurate determination of sites of UV lesion in neuro-effector elements needs further refined physiological as well as submicroscopic morphological investigation. Studies utilizing UV techniques would help in understanding both the physiological mechanisms as well as the molecular architecture in neuro-effector systems in general.

References

1 FUJII, R.: Mechanism of ionic action in the melanophore system of fish. I. Melanophore-concentrating action of potassium and some other ions. Annot. zool. japon. *32:* 47–58 (1959).
2 FUJII, R.: Demonstration of the adrenergic nature of transmission at the junction between melanophore-concentrating nerve and melanophore in bony fish. J. Fac. Sci. Univ. Tokyo, Sect. IV *9:* 171–196 (1961).
3 FUJII, R.: Chromatophores and pigments; in W. S. HOAR and D. J. RANDALL, Fish physiology, vol. 3, pp. 307–353 (Academic Press, New York 1969).
4 FUJII, R. and NOVALES, R. R.: Cellular aspects of the control of physiological color changes in fishes. Amer. Zool. *9:* 453–463 (1969).
5 GIESE, A. C.: Photophysiology; vol. 2 (Academic Press, New York 1964).
6 GIESE, A. C.: Effects of ultraviolet radiations on some activities of animal cells;

in F. URBACH, The biologic effects of ultraviolet radiation, pp. 61–76 (Pergamon Press, Oxford 1969).
7 LIEBERMAN, E. M.: Ultraviolet radiation effects on isolated nerve fibers. Adv. biol. med. Phys. *13:* 329–350 (1970).
8 SPAETH, R. A.: The physiology of chromatophores of fishes. J. exp. Zool. *15* 527–585 (1913).

Author's address: Dr. RYOZO FUJII, Department of Biology, Sapporo Medical College, *Sapporo 060* (Japan)

Experimental Pharmacology

The Effect of 4-Isopropylcatechol on the Harding-Passey Melanoma

S. S. BLEEHEN

Department of Dermatology, University College Hospital Medical School, London

Introduction

In 1966 CHAVIN and SCHLESINGER found that a number of diverse chemical compounds when injected subcutaneously into black goldfish *(Carassius auratus* L), had a selective destructive effect on melanocytes and melanophores in the skin. The compounds screened included hydroquinone and some of its derivatives, catechol, several derivatives of catechol and a number of mercaptoamines. Hydroquinone was considered to be the most potent 'anti-pigmentary' compound [5]. These workers later observed that hydroquinone caused the destruction of the abnormal melanocytes and melanophores of melanomas occurring on the hybrid swordtail fish *(Xiphophorus helleri)* and produced a regression in the size of the primary tumours and metastases [6]. The discovery that a number of diverse chemical agents had a potent depigmenting effect gave new impetus into research on the mode of action of these compounds and their effect on normal and malignant mammalian melanocytes. Two studies [1, 2] showed that of a total of 54 compounds tested for their depigmenting effect by topical application to the skin of black guinea pigs, 4-isopropylcatechol (4-IPC) was the most potent. Biopsies taken from 4-IPC treated depigmented areas were examined by light and electron microscopy and showed a marked lack of functional melanocytes, indicating that the cutaneous depigmentation resulted from a selective destructive effect of this compound on melanocytes. *In vitro* studies [3, 8] showed that a number of substituted phenols, including 4-hydroxyanisole (monomethyl ether of hydroquinone, MMEHQ), 4-isopropylcatechol and 4-tertiary butylcatechol had a rapid selective toxic effect on guinea pig melanocytes in tissue culture. The effect of the monoethyl ether of hydroquinone (MEEHQ) on a number of experimental transplantable melanomas was studied by FRENK and OTT [7], but this compound did not prolong the survival time of treated animals.

The purpose of this paper is to report the preliminary findings of the effect of 4-IPC on the Harding-Passey melanoma, transplanted in C57 female black mice.

Materials and Methods

Harding-Passey mouse melanoma was obtained from tumour-bearing mice (supplied from the Chester Beatty Research Institute, London). Tumour fragments approximately 2–3 mm in size were inserted by means of a trocar into the subcutaneous tissue in the right flank of C57 black female mice. When the tumours were just palpable, about 2–3 weeks after transplantation, the animals were divided into two groups. One group of animals was treated daily for 5 days a week with intraperitoneal injections of 0.1 ml of 0.1M 4-isopropylcatechol (4-IPC). A similar number of control animals was injected with 0.1 ml of 0.9% saline. Treatment was continued up to 8 weeks or until the animal died. The survival time of each animal was recorded and an attempt was made to assess the size of the tumour at death.

Portions of a 4-IPC-treated Harding-Passey melanoma and similar portions of a saline-treated tumour were processed for light microscopy and sections were cut and stained with haematoxylin and eosin and also with a silver melanin stain. For electron microscopy, portions of 4-IPC and saline-treated tumours were cut into pieces 1 mm^3 and fixed in 3% glutaraldehyde. After embedding in araldite, ultra-thin sections were cut and stained with uranyl acetate and lead hydroxide.

Results

There was an apparent initial retardation of the growth of the Harding-Passey melanoma in mice treated with 4-IPC. However, this was difficult to assess, and measurements of the eventual size of the tumours at death were approximately the same as in saline-treated control animals. Both 4-IPC and saline-treated Harding-Passey melanomas were very melanotic. It was noted that in the 4-IPC-treated animals there was an increased number of white and depigmented hairs, particularly on the flanks.

The survival times of both 4-IPC and saline-treated mice are shown on the graph (fig. 1). The mean survival time of 4-IPC-treated animals was significantly prolonged, being 62.1 ± 4.1 days, compared with 48.5 ± 3.8 days in the saline-treated group. Two animals in the saline-treated group are not included in the data since their deaths were due to accidental causes and occurred soon after the start of treatment.

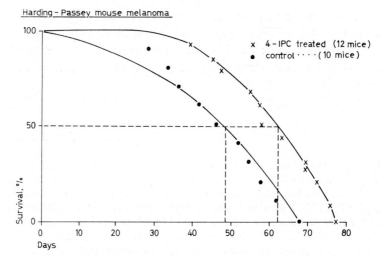

Fig. 1. Graph showing survival times of C57 black mice with Harding-Passey melanoma, treated with 4-IPC or saline.

Table I. Ultrastructural findings 4-IPC/control treated Harding-Passey melanoma

	4-IPC	Control
Melanosomes: early	+	+
late	+ +	+ + +
Vacuoles/lysosomes	+	+
Electron-dense round bodies	+ +	+
Lamellated 'myelin bodies'	+	±

Histological studies did not reveal any significant differences between the 4-IPC and the saline-treated Harding-Passey melanomas. The amount of pigment in melanin containing cells was the same in both. Many mitotic figures were seen in both (fig. 2 and 3). On electron microscopy, minor differences were noted and the ultrastructural cytological findings are summarised in table I. There was a relative increase in the number of electron-dense round bodies and lamellated 'myelin bodies' in the 4-IPC-treated Harding-Passey melanoma cells.

Discussion

This preliminary study of the effect of 4-isopropylcatechol on the Harding-Passey melanoma indicates that treatment with this compound

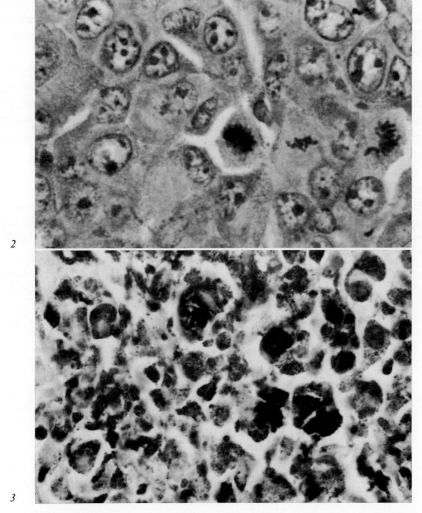

Fig. 2. Harding-Passey melanoma, showing melanoma cells and several mitotic figures. Haematoxylin and eosin. × 1,200.

Fig. 3. Harding-Passey melanoma. Pigment in melanocytes and macrophages. Fontana melanin stain. × 800.

significantly prolongs the mean survival time in C57 black mice bearing this tumour. Although no striking histological changes were noted in the 4-IPC-treated tumours, compared with control saline-treated tumours, the growth appeared to be retarded. Previous studies [1, 3, 8] have shown that

4-IPC has a selective lethal effect on mammalian melanocytes. Heavily pigmented cells with marked tyrosinase activity are more susceptible to damage by a number of substituted phenols, in particular 4-isopropylcatechol, 4-hydroxyanisole (MMEHQ) and 4-tertiary butylcatechol [8].

Although FRENK and OTT [7], were unable to demonstrate any prolongation of survival time in animals bearing a number of experimental transplantable melanomas and treated with MEEHQ, the Harding-Passey melanoma was not studied. Perhaps the greatly increased tyrosinase activity of the subcellular particles of the Harding-Passey melanoma cell, as compared with the B16 and Cloudman S91 melanomas [9], is relevant.

There is now strong evidence to support the view that the melanocytotoxic effect of a number of substituted phenols, such as 4-IPC and MMEHQ, involves the oxidation of these compounds by tyrosinase with the formation of free radical derivatives [8]. These free radicals may initiate a chain reaction of lipid peroxidation that results in the destruction and death of the melanocyte.

It is hoped that further studies on these 'melanocidal' compounds will lead to a new chemotherapeutic approach in the treatment of malignant melanoma in man.

Acknowledgements

This work was supported in part by University College Hospital and by Unilever Research Laboratories, Isleworth who synthesised and supplied 4-isopropylcatechol.

References

1 BLEEHEN, S. S.; PATHAK, M. A.; HORI, Y. and FITZPATRICK, T. B.: Depigmentation of the skin with 4-isopropylcatechol, mercaptoamines and other compounds. J. Invest. Derm. 50: 103–117 (1968).
2 BLEEHEN, S. S.: Cutaneous depigmentation by chemical agents. Abstr. VIIth Int. Pigment Cell Conf., Seattle, Wash., 1969.
3 BLEEHEN, S. S. and RILEY, P. A.: The in vitro effect of substituted phenols on melanocytes. Abstr. Brit. Ass. Derm. Winter Meet., London 1970.
4 CHAVIN, W. and SCHLESINGER, W.: A new series of depigmentational agents in black goldfish. Naturwissenschaften 53: 163 (1966).
5 CHAVIN, W. and SCHLESINGER, W.: Some potent depigmenting agents in the black goldfish. Naturwissenschaften 53: 413–414 (1966).
6 CHAVIN, W. and SCHLESINGER, W.: Effects of melanin depigmentational agents upon normal pigment cells, melanoma and tyrosinase activity; in W. MONTAGNA

and F. Hu Advances in biology of the skin, vol. 8, pp. 421–445 (Pergamon Press, Oxford 1967).
7 FRENK, E. and OTT, F.: Evaluation of the toxicity of the monoethyl ether of hydroquinone for mammalian melanocytes and melanoma cells. J. Invest. Derm. *56:* 287–293 (1971).
8 RILEY, P. A.: Mechanism of pigment-cell toxicity produced by hydroxyanisole. J. Path. Bact. *101:* 163–169 (1970).
9 SEIJI, M.: Subcellular particles and melanin formation in melanocytes; in W. MONTAGNA and F. Hu Advances in biology of the skin, vol. 8, pp. 189–222 (Pergamon Press, Oxford 1967).

Author's address: Dr. STANLEY S. BLEEHEN, Rupert Hallam Department of Dermatology, Sheffield University, Hallamshire Hospital, *Sheffield S10 2JF* (England)

Studies on the Mechanism of the Anti-Melanoma Effect of Polyinosinic-Polycytidylic Acid (PIC)

Does Exogenous Interferon Mimic the Anti-Melanoma Effect of PIC? Does Thymectomy Plus Irradiation Abrogate the Effect of PIC?

R. S. BART and A. W. KOPF

Oncology Section, Skin and Cancer Unit, New York University Medical Center, and Department of Dermatology, New York University School of Medicine, New York, N.Y.

Introduction

We have shown that the double-stranded synthetic RNA, polyinosinic-polycytidylic acid (PIC), inhibits the growth of B-16 malignant melanomas, and prolongs the survival of C-57 black mice bearing them [2, 3, 15]. One possible mechanism of this inhibition of growth of malignant melanoma is through induction of interferon, since PIC is a known interferon inducer [1, 8, 11] and interferon itself has been reported to inhibit some murine tumors [5, 9, 10]. Another possible mechanism of melanoma inhibition by PIC may be by heightening anti-tumor immunologic activity, since immunologic reactivity to B-16 melanoma has been demonstrated [4, 14] and since PIC is known to boost immunologic responses [19].

The following two experiments were performed with B16 melanoma in C57 mice to find out a) if the administration of exogenous interferon would mimic PIC in its anti-melanoma effect, and b) if PIC would lose its anti-tumor effect in thymectomized-irradiated (i.e., presumably immunodeficient) mice.

Materials and Methods

In both experiments, B-16 malignant melanoma was transplanted as follows. Excised tumors were pressed through a metal mesh to remove stromal elements. The melanomas were then mixed with a small amount of minimal essential medium (about 15% of final volume). A 13-gauge trochar was used to implant 0.05 ml of the above mixture subcutaneously into the axilla of each C-57/6J black mouse. Female mice were used for both experiments.

Interferon Experiment

Serum interferon was purchased from Abbott Laboratories (Chicago, Illinois). It was induced in 18 to 20 g mice by intravenous inoculation of egg-embryo-grown Newcastle disease virus (NDV) followed by exsanguination in 5 to 6 hours. The serum was inactivated by lowering the pH to 2.0 with HCl. Control serum was processed in the identical manner except that the mice had not received NDV. Interferon potency was assayed by the 50%-plaque-inhibition technique by Dr. JAN VILCEK. The concentration was 19,000 U/ml in the 'interferon serum' and 40 U/ml in the control serum. Twenty-nine mice, 5-weeks old, were transplanted on day 0 with malignant melanoma (fig. 1). Each of 14 mice was then injected intraperitoneally with a total of 76,000 U of interferon in 8 equal doses between days 2 and 11 following tumor implantation. Each of 15 control mice was treated with an equal volume of control serum on the same days. All mice were killed on day 14 and the volumes of their tumors were measured by liquid displacement.

Thymectomy-Irradiation Experiment

Eighteen mice were thymectomized at 4 weeks of age. Nine days later they were irradiated with 220 r to their dorsal surfaces and 220 r to their ventral surfaces (100 kV peak; 1.9 ml aluminum half-value-layer). This schedule was selected to deliver an average of 350 r to the tissues. The mice were transplanted with B16 melanoma 10 days after irradiation. The date of transplant was considered day 0 (fig. 2). Each of eight mice received eight injections of 150 μg polyinosinic-polycytidylic acid (PIC) in phosphate buffered saline (PBS) intraperitoneally between days 2 and 11. The 10 control animals received PBS alone on the same days. All mice were sacrificed on day 14 and the volumes of their tumors measured. Twenty-three additional mice were sham-operated, divided into two groups, and treated with PIC or PBS.

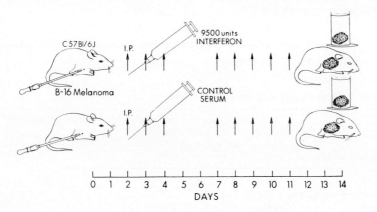

Fig. 1. Protocol for interferon experiment. Note similar tumor volumes at end of experiment.

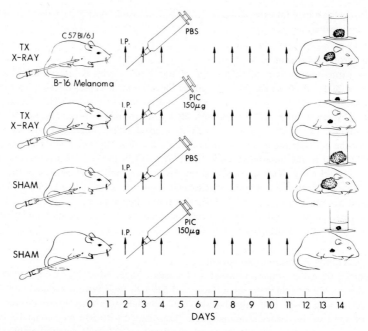

Fig. 2. Protocol for thymectomy plus x-ray experiment. Note smaller tumor volume in PIC-treated group despite thymectomy and irradiation.

Results

Interferon Experiment

It can be seen from table I that exogenous murine serum interferon had no effect on tumor volumes.

Table I. Effect of interferon on tumor volume

	Number of mice	Average tumor volume (ml)	Probability[1]
Control serum	15	0.99	
Interferon-containing serum	14	0.96	>0.8

[1] Probability, using t-test, that difference in average volumes was due to chance.

Table II. Effect of PIC on tumor volume in thymectomized-irradiated mice

	Treatment	Number of mice	Average tumor volume (ml)	Probability[1]
Sham-operated non-irradiated controls	PBS	10	1.60	
				<0.001
Sham-operated non-irradiated controls	PIC	13	0.28	
Thymectomized-irradiated	PBS	10	1.08	<0.001
Thymectomized-irradiated	PIC	8	0.27	

1 Probability, using t-test, that difference in average volumes was due to chance.

Thymectomy-Irradiation Experiment

The data in table II indicates that PIC inhibits the growth of B-16 malignant melanoma despite previous thymectomy and irradiation of the mice.

Discussion

The mechanism of inhibition of the growth of B-16 malignant melanoma by polyinosinic-polycytidylic acid (PIC) is unknown. It has been suggested that PIC may exert this inhibition by either a direct effect on the tumor cells or via an indirect effect. The above experiments were designed to study two possible indirect effects: a) whether interferon, which is induced by PIC, inhibits the growth of the melanoma, and b) whether PIC augments the host's anti-tumor immunologic response.

The possibility that PIC has its anti-melanoma effect via interferon is suggested by the fact that interferon itself has been shown to inhibit the induction or growth of various murine cancers [5, 9, 10, 17] and also by the fact that PIC is known to be a potent inducer of interferon [1,8]. In our experiment we were unable to demonstrate an anti-melanoma effect by the administration of interferon. However, the failure to demonstrate an anti-melanoma effect by *exogenous* interferon does not entirely rule out the possibility that PIC works via interferon induction. One could theorize that interferon has anti-melanoma properties but that it must be present in sufficient amounts *within* the tumor cells to have its anti-tumor effect. It could be that PIC induces such intracellular levels of interferon whereas exogenous interferon, in the amounts we used, does not. WEINSTEIN et al. [20] and LEVY et al. [17] found that exogenous interferon, unlike PIC, did not decrease the rate of induction of Maloney sarcoma virus (MSV) tumors in mice; nor did it cause regression of the tumors. However, interferon did have an inhibitory effect on MSV tumors in tissue culture [17].

In seeking an explanation for why we failed to demonstrate inhibition of melanoma growth with interferon whereas GRESSER et al. [10] succeeded in prolonging the survival of mice bearing other tumors, we considered, among other factors, possible differences in the doses of interferon used. GRESSER and colleagues used 20,000 U daily, beginning 24 h after tumor implant; we used 9500 U daily beginning 48 h after transplant. However, because biologic assays are used to titer interferon, equating potencies unit-for-unit may be misleading. Another factor we considered is possible differences in the cell surfaces of B-16 melanoma as compared to those of tumors which respond to exogenous interferon. It is possible that interferon is less able to penetrate into melanoma cells. Of interest is the report that human ocular melanoma cells have a thick acid mucopolysaccharide coat [16].

Polyinosinic-polycytidylic acid is known to augment both cell-mediated and humoral immunologic responses [6, 12, 13, 19]. Since it has also been shown that immunologic reactivity may play a role in the control of the growth of B-16 melanoma in mice [4, 21], a possible mechanism of anti-melanoma action by PIC could be by augmenting this reactivity. Theoretically then, PIC should have a reduced effect in immuno-deficient mice if it works by this mechanism.

The method of thymectomy-irradiation which we used has been shown to delay homograft rejection in mice, indicating immunosuppression [18]. However, despite thymectomy and irradiation of our mice, PIC retained its anti-melanoma effect. Several explanations were considered. First, the animals may not have been sufficiently immunosuppressed by the method used. In contrast to the findings that immunosuppression by anti-lymphocyte serum results in increased tumor growth [21], animals immuno-suppressed by the method of thymectomy-irradiation we used did not show such increased tumor growth (table II). We are currently planning experiments using neonatal thymectomy plus x-irradiation of C-57 mice to see if more profound immunologic deficiency can be achieved. Secondly, PIC may restore immune competence in thymectomized mice [7]. Thirdly, PIC may not have its anti-melanoma effect via an immune response.

In conclusion, our findings do not support the hypotheses that the mechanism of the anti-melanoma effect of PIC is via interferon induction or via augmented immunologic reactivity.

Acknowledgment

We thank Messrs. PAUL GLUCK, JAY COOPER and STEVEN LAM for their technical assistance as well as Dr. JAN VILCEK for his valuable suggestions and for assaying the interferon potency.

This work was supported by a grant from the National Cancer Institute (No. 1 R01-CA 1205-01).

References

1 BARON, S.; DUBUY, H.; BUCKLER, C. E.; JOHNSON, M. D. and WORTHINGTON, M.: Factors affecting the interferon response of mice to polynucleotides; in R. F. BEERS and W. BRAUN Biological effects of polynucleotides, pp. 45–54 (Springer, New York 1971).
2 BART, R. S. and KOPF, A. W.: Inhibition of growth of murine malignant melanoma with synthetic double-stranded ribonucleic acid. Nature, Lond. *224:* 372–373 (1969).
3 BART, R. S.; KOPF, A. W. and SILAGI, S.: Inhibition of the growth of murine malignant melanoma by polyinosinic-polycytidylic acid. J. Invest. Derm. *56:* 33–38 (1971).
4 BARTLETT, P. C.: Changes in the immunogenetic relationship of the B-16 melanoma to the C57B1/6 mouse. Bull. Tulane med. Fac. *26:* 199–207 (1967).
5 CAME, P. E. and MOORE, D. H.: Inhibition of the spontaneous mammary carcinoma of mice by treatment with interferon and poly I:C. Proc. exp. Biol. Med. *137:* 304–305 (1971).
6 CAMPBELL, P. A. and KIND, P.: Bone marrow-derived cells as target cells for polynucleotide adjuvants. J. Immunol. *107:* 1419–1423 (1971).
7 CONE, R. E. and JOHNSON, A. G.: Regulation of the immune system by synthetic polynucleotides. J. exp. Med. *133:* 665–676 (1971).
8 FIELD, A. K.; TYTELL, A. A.; LAMPSON, G. P. and HILLEMAN, M. R.: Inducers of interferon and host resistance. Proc. nat. Acad. Sci., Wash. *58:* 1004–1010 (1967).
9 GRESSER, I.; COPPY, J.; FONTAIN-BROUTY-BOYE, D.; FALCOFF, R.; FALCOFF, E. and ZAJDELA, A.: Interferon and murine leukaemia. III. Efficacy of interferon pre-

parations administered after inoculation of Friend virus. Nature, Lond. *215:* 174 (1967).

10 GRESSER, I.; BOURALI, J. P.; LEVY, D.; FONTAIN-BROUTY-BOYE, D. et THOMAS, M.T.: Prolongation de la survie de souris inoculées avec des cellules tumorales et traitées avec des préparations d'interférons. C.R. Acad. Sci., Paris (D) *269:* 994–997 (1969).

11 HILLEMAN, M. R.; LAMPSON, G. P.; TYTELL, A. A.; FIELD, A. K.; NEMES, M. M.; KRAKOFF, I. H. and YOUNG, C. W.: Double-stranded RNA's in relation to interferon induction and adjuvant activity; in R. F. BEERS and W. BRAUN Biological effects of polynucleotides, pp. 27–44 (Springer, New York 1971).

12 ISHIZUKA, M.; BRAUN, W. and MATSUMOTO, T.: Cyclic AMP and immune responses. I. Influence of poly A:U and cAMP on antibody formation in vitro. J.Immunol. *107:* 1027–1035 (1971).

13 JOHNSON, H. G. and JOHNSON, A. G.: Regulation of the immune system by synthetic polynucleotides. J. exp. Med. *133:* 649–664 (1971).

14 KATO, K. H. and MARCUS, S.: Mouse malignant melanoma antibody characterized by immunofluorescence. Fed. Proc. *30:* 245 (1971).

15 KOPF, A. W. and BART, R. S.: Prolongation of survival of mice bearing malignant melanoma and inhibition of tumor growth with polyinosinic-polycytidylic acid; in Pigmentation: its genesis and control, pp. 497–501 (Appleton-Century-Crofts, New York 1972).

16 LAPIS, K. and RADNOT, M.: Surface properties of human melanoma cells. Amer. J. Ophthal. *71:* 740–750 (1971).

17 LEVY, H. B.; ADAMSON, R.; CARBONE, P.; DE VITA, V.; GAZDAR, A.; RHIM, J.; WEINSTEIN, A. and RILEY, F.: Studies on the anti-tumor action of poly I: poly C; in R. F. BEERS and W. BRAUN Biological effects of polynucleotides, pp. 55–65 (Springer, New York 1971).

18 MILLER, J. F. A. P.: Effect of thymic ablation and replacement; in R. A. GOOD and A. E. GABRIELSEN The thymus in immunobiology, p.p. 436–464 (Harper and Row, New York 1964).

19 TURNER, W.; CHAN, S. P. and CHIRIGOS, M. A.: Stimulation of humoral and cellular antibody formation in mice stimulated by poly I:C. Proc.exp.Biol. Med. *133:* 334–338 (1970).

20 WEINSTEIN, A. J.; GAZDAR, A. F.; SIMS, H. L. and LEVY, H. B.: Lack of correlation between interferon induction and antitumor effect of poly I: Poly C. Nature, New Biol., Lond. *231:* 53–54 (1971).

21 WOLF-JÜRGENSEN, P.; KOPF, A. W.; LIPKIN, G. and BART, R. S.: Influence of anti-lymphocyte serum on malignant melanoma. J. Invest. Derm. *51:* 441–444 (1968).

Author's address: Prof. ROBERT S. BART, Oncology Section, Skin and Cancer Unit, New York University Medical Center, 562 First Avenue, *New York, NY 10016* (USA)

Neutron Capture Treatment of Malignant Melanoma Using ^{10}B-Chlorpromazine Compound

Y. Mishima

Department of Dermatology, Wakayama Medical University, Wakayama-City.

Malignant melanoma is the cancerous growth of the pigment cells which have the specific function to synthesize melanin. As malignant transformation occurs in pigment cells, they usually acquire a greater melanin-synthesizing ability. Therefore, biological and non-surgical treatment or control of malignant melanoma could be achieved by utilizing this specific function. I should like to present here our accumulated findings [6, 7], together with new, unpublished data on neutron capture treatment of malignant melanoma using ^{10}B-chlorpromazine compounds.

^{35}S-labeled chlorpromazine has been shown by Blois [1] to bind selectively with melanin, presumably by forming a charge transfer complex as a result of the π-electron interaction between an indole nucleus on the surface of melanin particles and a molecule of chlorpromazine. The distribution of ^{35}S-chlorpromazine 12 h after administration is present not only in malignant melanoma tissue but also in various internal organs, such as the adrenal glands, spleen, liver, and kidneys. However, the tissue distribution of chlorpromazine after 3 days was found to be highly localized in melanin-containing tissues, melanoma and the eyes. The concentration is 10 to 20 times that of other tissues. This same phenomenon has been observed in the tissue 7 days after systemic administration.

It has been shown [3, 4] that thermal neutrons are well absorbed by the non radioactive boron-10 isotope and this absorption is known to result in the emission of alpha particles and lithium atoms which share between them an energy of 2.79 MeV, transferred totally to tissue (fig. 1). Since these particles are relatively large, the primary radiation injury is confined within a distance of 10 to 14 μ from the point of the activated boron atom, a phenomenon which has been used in neurosurgery. This distance of 10 to 14 μ ap-

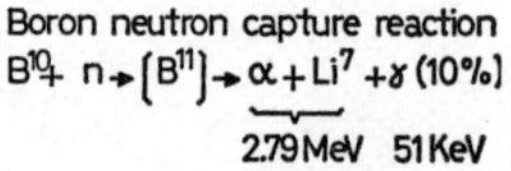

Boron neutron capture reaction
$B^{10} + n \rightarrow [B^{11}] \rightarrow \alpha + Li^7 + \gamma$ (10%)
         $\underbrace{\qquad\qquad}$
         2.79 MeV   51 KeV

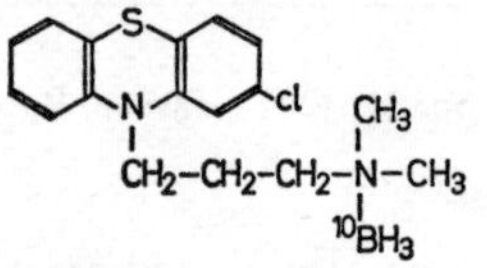

Fig. 1. Absorption of thermal neutrons by ^{10}B resulting in α-particle reaction and release of energy.

Fig. 2. Chemical structure of our synthesized ^{10}B-chlorpromazine borane.

proximates the diameter of individual melanoma cells. At this point, COOPER and I [2] considered synthesizing a molecular hybrid compound for the selective eradication of malignant melanoma cells, using their specific melanin-synthesizing functions. In cooperation with a boron chemist, TOSHIO NAKAGAWA, I have been able to synthesize ^{10}B-chlorpromazine borane with 96% ^{10}B as transparent needle crystals having a melting point of 96.5°C (fig. 2).

Prior to the neutron-irradiation experiment on malignant melanoma, the distribution of irradiation energy of our Thermal Column of Kyoto University Nuclear Reactor Heavy Water Facility had been investigated by SATO [8] of our study group, using a human body phantom. The distribution of irradiated energy within a human body phantom containing Rossi's tissue equivalent solution, which is composed of carbon, hydrogen, oxygen and other elements, similar to the soft tissue of the human body, has been measured after irradiation with KUR Thermal Column of Nuclear Reactor operated at 1000 kW for 30 min, resulting in a dosage of 5×10^8 nvt/cm²/min on the surface of the phantom using a 10×10 cm size collimator. In this human phantom, an acrylic, cube-shaped model of melanoma ($5 \times 5 \times 5$ cm) was placed which contained, in addition to Rossi's solution, ^{10}B-compound of 70 µg/g concentration, to measure the energy distribution inside and outside the melanoma model. The energy distribution is shown by rem/min and clearly reveals that the irradiated energy is highly concentrated within this melanoma model as compared with surrounding normal tissue equivalent solution.

Based on the above findings, we have carried out eleven experiments, irradiating Fortner's and Green's melanotic malignant melanoma in a total of about 500 Syrian golden hamsters, using KUR Thermal Column of Nuclear Reactor with and without local administration of ^{10}B-compound. In cooperation with KANDA, ONO and SHIBATA [5], we found that a dosage of 2×10^{12} nvt/cm² to 1.2×10^{13} nvt/cm² by 3 MW for 17 min to 5 MW for 40 min respectively, is the range of optimal dosage in the present experimen-

Table I. Neutron dosage measured on the surfaces of experimental cage and hamster (3MW, 17 min)

Location	Neutron dosage	Gamma ray dosage
Acrylic cage surface	$\sim 2.5 \times 10^{12}$ nvt	
Hamster's front surface	$\sim 2 \times 10^{12}$ nvt	~ 480r
Hamster's back surface	$\sim 0.5 \times 10^{12}$ nvt	~ 400r

tal system. Melanoma-bearing hamsters are held without anesthesia within a specially designed acrylic cage. These are placed directly in front of a bismuth plate in order to avoid an increase in gamma rays and to maintain constant neutron dosage. Table I shows a neutron flux on the surface of an acrylic cage, and the hamster skin surface of both sides.

On the basis of these phantom experiments, we have carried out first *in vivo* thermal neutron irradiation experiments, using local administration of simple ^{10}B-compound, $^{10}BF_3CaF_2$ having a concentration of 8.9 mg/0.1 ml physiological saline. Growth curves of control and irradiated Fortner's melanotic malignant melanoma expressed by individual tumor volume (calculated by VAN WOERT and PALMER's equation [9] and measured every 12 h after transplantation) show that in Fortner's melanotic melanoma there is a steep linear increase in tumor volume between 6 and 7 days following implantation. Tumor growth curves of neutron-irradiated Fortner's melanomas without B-compound administration show an apparent suppression of melanoma growth after neutron irradiation, but this suppression effect tends to be lost 15 to 16 days after transplantation. Melanoma growth curves of neutron-irradiated Fortner's melanomas, combined with local administration of $^{10}BF_3CaF_2$ similar to the melanoma-phantom model experiment, show that tumor growth suppression remains longer and, in most cases, is still seen 17 to 20 days after transplantation and 13 to 16 days after irradiation. Melanoma growth curves of $^{10}BF_3CaF_2$ administration without neutron irradiation reveal no suppression of melanoma growth. A summary of these inhibition rates of melanoma growth observed 8 days after irradiation in the four groups (fig. 3) shows that thermal neutron irradiation combined with $^{10}BF_3CaF_2$ produced maximum suppression of melanoma growth.

Not only *in vivo* experiments as just described, but also *in vitro* neutron-irradiation experiments have revealed a distinct difference in suppression effect. *In vitro* thermal neutron irradiation at a dosage of 1.2×10^{13} nvt/cm^2

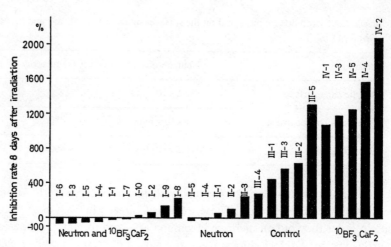

Fig. 3. Suppression effect of thermal neutron irradiation observed in individual cases of the four experimental groups.

has been aplied to melanoma cut into small strips of approximately 1 × 1 × 2 mm, which were suspended in saline containing antibiotics. Only Group I received the local administration of $^{10}BF_3CaF_2$ at 8.9 mg/0.1 ml physiological saline 1 h prior to the excision. Figure 4 shows the melanoma growth

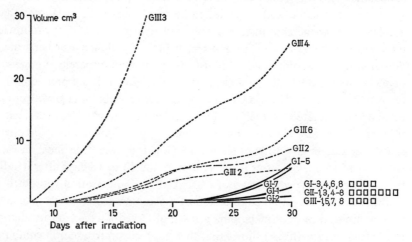

Fig 4. Growth of Fortner's malignant melanomas observed in groups of *in vitro* neutron irradiation experiments: a) treated with both $^{10}BF_3CaF_2$ and thermal neutron irradiation (GI); b) with thermal neutron irradiation only (GII), and c) with no treatment (GIII). The squares on the base line represent individual cases which did not grow melanomas.

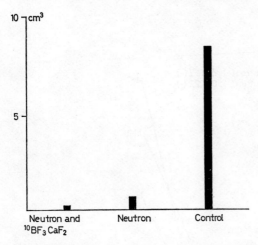

Fig. 5. The average of melanoma volume based on data of eight cases per group, 18 days after the *in vitro* irradiation.

curves observed for 30 days after *in vitro* thermal neutron irradiation. Melanoma receiving neutron irradiation with a ^{10}B-compound are most strongly suppressed. The average results of each group (fig. 5) show the difference in suppression more clearly.

As a next step, we carried out the thermal neutron irradiation experiment, using ^{10}B-chlorpromazine borane. The following is one of the representative results of this experiment. Melanoma growth curves of the control group again show a steep linear increase in tumor volume between 6 and 7 days following implantation, while growth curves of melanoma receiving only thermal neutron irradiation show a distinct suppression of tumor growth which, however, tends to be lost 36 h after irradiation (fig. 6). In contrast to the preceding groups receiving only thermal neutron irradiation, suppression of tumor growth curves of neutron irradiated Fortner's melanomas combine with the local adminstration of ^{10}B-chlorpromazine borane is retained longer and is still present 48 h after irradiation. As shown in figure 7, the inhibition rate between the first and second day following thermal neutron irradiation, obtained for each case of malignant melanoma of these three groups, shows maximum suppression effect in melanoma groups treated with the combination of thermal neutron irradiation and ^{10}B-chlorpromazine borane.

It may be concluded that the combination: thermal neutron irradiation – boron compound most strongly depressed the growth rate of malignant

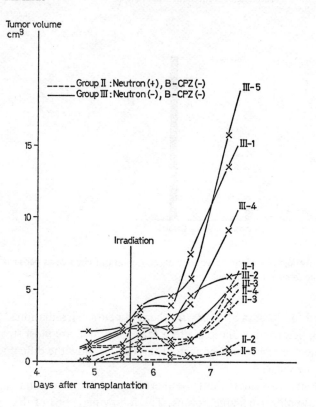

Fig. 6. Tumor growth curves of control group and neutron-irradiated group in the *in vivo* ^{10}B-chlorpromazine borane-neutron irradiation experiment.

melanoma, as compared with the control group and the group with neutron irradiation only, and that the eradication of melanoma may be achieved by finding further optimum methods of boron administration and successive neutron irradiation.

Acknowledgements

Dr. K. KANDA, Professor S. SHIBATA and Mr. K. ONO of the Nuclear Reactor Institute, Kyoto University, generously cooperated in the nuclear physics part of this study. Mr. AKIRA MATUMOTO and Mr. MASANORI YOSHIDA of our laboratory gave technical assistance.

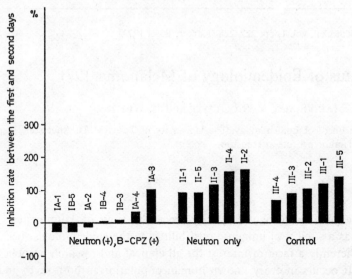

Fig. 7. Inhibition rate of melanoma growth between the first and second day after thermal neutron irradiation on Fortner's melanotic malignant melanoma, with and without ^{10}B-chlorpromazine borane, revealed that the combined therapy was most effective.

References

1 BLOIS, M. S.: On chlorpromazine binding *in vivo*. J. Invest. Derm. *45:* 475–481 (1965).
2 COOPER, M. and MISHIMA, Y.: Unpubl. data
3 ENTZIAN, W.; SOLOWAY, A. H.; RAJU, R.; SWEET, W. H. and BROWNELL, G. L.: Effect of neutron capture irradiation upon malignant brain tumors in mice. Acta radiol., Stockh. *5:* 95–100 (1966).
4 HATANAKA, H.; SOLOWAY, A. H. and SWEET, W. H.: Incorporation of pyrimidines and boron analogues into brain tumor and brain tissue of mice. Neurochirurgia *10:* 87–95 (1967).
5 KANDA, K.; ONO, K. and SHIBATA, S.: Measurement of neutron flux and spectrum in the neutron capture treatment of malignant melanoma. (In preparation).
6 MISHIMA, Y.: The effect of neutron capture treatment using ^{10}B-chlorpromazine compound on malignant melanoma; in M. AKABORI Proc. IInd Meet. Biomed. Use of Nuclear Reactors, pp. 52–60 (University of Kyoto Press, Kumatori 1971).
7 MISHIMA, Y.: Precancerosis of pigment cells. 18th Gener. Assembly Japan Medical Congress, Tokyo, April 5–7, 1971.
8 SATO, T.: Dose distribution curve after irradiation with thermal column of nuclear reactor. 10th Meet. Atomic Energy Soc. Saporo, Japan 1971.
9 VAN WOERT, M. H. and PALMER, S. H.: Inhibition of the growth of mouse melanoma by chlorpromazine. Cancer Res. *29:* 1952–1955 (1969).

Author's address: Professor YUTAKA MISHIMA, Department of Dermatology, Wakayama Medical University, *Wakayama-City 640* (Japan)

Epidemiology of Melanoma

Status of Epidemiology of Melanoma 1971

E. J. MACDONALD, V. McGUFFEE and E. WHITE

Department of Epidemiology, The University of Texas M. D. Anderson Hospital and Tumor Institute at Houston, Tex.

Melanoma, a recognized entity since the time of Hippocrates [9], presents as a subject of immense possibilities to the epidemiologist because it has inherently a facet of interest for all clinical and research branches of science. It occurs in every known human population in both sexes and in all age groups, but in varying incidences. It accounts for less than 2% of total cancer. About 90% originate in the skin, most of the other 10% originate in the eye. Rare cases of melanoma have been reported in the meninges, esophagus, nasopharynx, gallbladder, vagina or anus. Two out of three arise in a preexisting nevus. One-sixth of the total arise in a recognizable congenital nevus. Sometimes there are several melanomas within a single nevus and sometimes they arise in nevi in different parts of the body of the same individual. Of the total, 25% arise in skin where no nevi are present, and are called *de novo* melanomas.

Skin nevi are found most frequently on the trunk, followed by the upper extremities. The distribution of melanoma is different. It occurs most frequently on the head, neck and lower extremities, followed by the trunk. There is, therefore, a greater possibility of nevi of the head, neck and lower extremities transitioning to malignant melanoma, than for those on the trunk. Trauma is considered an exciting factor in those melanomas which arise in an irritated mole.

Two-thirds of the melanomas of the skin arise in some form of nevus, which lends support to the mutant cell theory of origin. Other supporting evidence is the demonstration by electron microscopy of cholinesterase activity in nevi, and also positive dopa and tyrosinase reactions and tyrosinase-active melanosome systems. Similar cholinesterase and tyrosinase activity can be demonstrated in melanotic and amelanotic melanomas, lending credence to the theory of a common origin [10].

Studies of Dubreuilh's melanosis, a precancerous skin pigmentation, support the neural crest theory of melanocarcinogenesis. Dubreuilh's melanosis has many characteristics at odds with the usual behavior of premalignant nevi, such as radiosensitivity, frequent regression and the absence of the amelanotic form. MISHIMA [7] suggests that two developmental pathways may occur. The first he called melanocytic, in which the cells from the neural crest evolve into melanocytes, which may then assume malignant changes. These are the cells which form lentigo senilis or Dubreuilh's precancerous melanosis which may then progress to melanoma. The second pathway he called nevocytic, which postulated the formation of either junctional nevus cells or true melanoma cells originating from bipotential neural crest nevoblasts.

A certain proportion of melanomas have a genetic component. The occurrence of melanoma follows no classic hereditary pattern, however, ANDERSON [2] of our Institution, working with the pedigrees of 106 patients and their relatives, and with the remaining 2,128 patients with melanoma, who were called the non-familial cases, postulated that the underlying genetic mechanism was the involvement of several autosomal gene loci, in addition to the cytoplasmic component transmitted by an affected or carrier female. When melanoma does occur in kindreds, it shares some of the characteristics of other hereditable tumors, such as retinoblastoma. ANDERSON found a distinctly younger age distribution, thus an earlier onset and a high frequency of multiple primary melanomas. This was not due to an unduly long period of risk, for the age of second primary averaged 39 years. Familial melanomas have a high survival rate. In ANDERSON's data, incidence was higher among individuals with affected mothers (17%), than with affected fathers (9%). This finding, he felt, suggests that the inheritance of melanoma involves not only a nuclear component, but a maternal component as well. Further analysis indicated that the latter appeared to be associated more with a cytoplasmic effect than with pre- or post-natal maternal effects. Awareness of this genetic variety and its clinical characteristics gives the alert clinician and the epidemiologist an opportunity to find at an earlier stage of their disease this high incidence group with a better-than-average prognosis.

Reference to theories of origin of melanoma cells is to call attention to the variation possible in melanogenesis, with the consequent problems of specific and reported diagnosis. Pigmented nevi and melanoma in children are cases in point. According to J. LESLIE SMITH [8], the potential for malignant change in congenital giant pigmented nevi is sufficiently high to receive serious consideration. These lesions have usually been considered more a cosmetic problem, but malignant melanoma can develop in them in infancy or years later.

SMITH stated that of reported cases of malignant melanoma in children, 23 to 40% had arisen in giant pigmented nevi. Since these are obvious lesions, early detection should be routine.

The rarity of *de novo,* amelanoma, or melanoma arising from junctional nevus cells being serious in children before puberty has caused such lesions to be considered benign or 'juvenile melanoma'. In the Connecticut registry in 1946, the first one reported in a prepubertal child was a confirmed melanoma with inguinal metastases in a four-year-old boy, but these, though increasing in reported number, are rare.

Reported deaths show a fairly uniform mortality rate in different geographic situations. The literature on melanoma presupposes relatively equal ratio of deaths to cases. Studies on deaths from melanoma in Texas by 254 counties showed very little variation. When the opportunity came to collect every case for a population of four million in 72 counties from every source, including dermatologists' offices, pathology laboratories and tuberculosis hospitals as well as all other institutions, the true occurrence of melanoma was found to bear a differing relationship to reported deaths from melanoma in different regions. It was found that melanoma cases exceeded deaths in the following ratios, generally, as the equator was approached. In El Paso the incidence rate was three times the death rate and this is the usual figure quoted, except in New Zealand and Australia. In San Antonio it was four times, in Corpus Christi five times, in Laredo six times, and in Harlingen seven times the mortality rate. These great differences and improving survival rates with increased incidence suggest that the lesions represented are being detected earlier, or that they are less aggressive than those reported from cancer institutions.

The theory first suggested by MCGOVERN [6] in 1952, that melanoma of exposed areas of the skin was caused by prolonged exposure to the sun, has received much circumstantial evidence to validate it in the work of other Australians, LANCASTER, NELSON, DAVIS, HERRON and McLEOD, and also in my studies in Texas.

The location of most of the lesions in Texas (the lower legs, arms, and face for females and face, arms, shoulders and back for males) suggests that sun exposure plays some part in melanocarcinogenesis.

LEE [5], using the classification suggested in 1969 by CLARK *et al.* [3], divided malignant melanoma in order of decreasingly good prognosis into lentigo-maligna melanoma, superficial spreading melanoma, and nodular melanoma. It is fully appreciated that this is a working hypothesis, subject to change in the future, based on clinical and histological appearances and on variations in the subcellular features.

LEE takes as a starting point the relationship between malignant melanoma and exposure to sunlight, and attempts to bring together into a simple framework the different observations on anatomical distribution and histologic type.

Lentigo maligna begins in the middle aged, but usually is diagnosed first when the patient is elderly. The excess incidence and mortality of all malignant melanoma in low latitudes is found in young adults as well as in the elderly. So the superficial spreading melanoma, most common in lower extremities of females may be the entity involved in the increase, such as that noted in Texas, where the prognosis seems better though the incidence rate is high. The third group, nodular melanomas, are uniformly invasive with a poor prognosis and not site specific.

LEE suggests the possibility of a 'solar circulating factor' and to test his hypothesis expressed the need for publication of a larger series of cases by site, sex and age from different latitudes. As a first step, we have assembled from 69 population segments from around the world, the sex and ethnic data for total melanoma of the skin.

LEE suggested testing the reality of his solar circulating theory by studying melanomas of the eye, and of the covered portions of the body.

ACKERMAN and DEL REGATO [1] describe melanoma of the conjunctiva as relatively rare and arising from three different sources: preexisting nevi, areas of acquired melanosis or *de novo*. Familial occurrence of intraocular melanoma is rare, but follows an autosomal dominant mode of inheritance when it does occur. They describe melanoma of the choroid as the most common site of ocular melanoma. These melanomas arise from the pigmented cells of the uveal tract, and virtually all melanomas of the uveal tract appear to develop from preexisting nevi. Eyes with excess in conjunctival pigmentation are particularly prone to develop melanoma.

We have 69 well-documented cases of melanoma of the eye in the ANDERSON series, and 159 additional in the six survey regions. Sixteen of the Anderson cases, or 16.5%, arose in the conjunctiva, the others in the choroid. LEE theorizes that the conjunctival ones are exposed to scattered ultraviolet light and trauma, while the internal ones are protected from these factors and would be subject to the solar circulating factor.

He feels that tumors of the conjunctiva should resemble lentigo maligna and superficial spreading melanoma, while those of the internal ocular tissues should resemble the nodular types with some of the superficial spreading types. The survival in years in the lesions originating in the choroid or the conjunctiva suggests a difference in etiology. Of the 52 ocular melanomas originating in the choroid, over five years ago, 10 or 19.2% are still alive and few have long survival times. Of the 10 conjunctival primary melanomas, 6 are alive from 6 to 20 years. Of the conjunctival melanomas who died, the durations of survival except for one, of 7 years, were short. Of those who survived over 5 years with ocular melanoma of the conjunctiva, the median age at onset was 47 years; of those who died in less than 5 years, it was 67 years. The median age at onset of individuals with melanoma of the choroid was older in both living and dead groups. An ongoing pathological study of these lesions will give us definitive answers to these descriptive observations.

Melanomas of the eye and of unexposed skin were studied for the last 5 years for 6 regions in Texas. There was a total of 159 eye cases and 691 of unexposed skin. The average five-year crude rate of eye melanoma for this population upward of four million was 0.2 per 100,000 and for unexposed skin 0.9 per 100,000. There was a range in incidence from region to region with high rates for one matching high rates for the other, and *vice versa*. Houston, the most urban region, had the highest rates, 1.4 per 100,000 for melanoma of unexposed skin and 0.3 per 100,000 for eye.

From the volume *Cancer Incidence in Five Continents* [4], the total cancer malignant melanoma cases and age-adjusted rates have been tabulated by sex, together with the four separated, populous regions in Texas, not included in this volume, but adjusted to the same world standard population and covering the same five-year average time interval.

Rates for these 69 population segments have been placed in descending order by latitude degrees north and south by Caucasian except Latin, Latin, Oriental and Negro. Proportion of melanoma of total cancer is shown for each population group for males and females.

For Caucasian males except Latins the variation in age-adjusted incidence rates for total cancer is very great. Seven population groups have total cancer incidence rates in excess of 300 per 100,000; 9 have rates between 250 and 300, 14 between 200 and 250, 9 between 150 and 200 and 6 under 150 per 100,000. This tremendous variation probably reflects, to some extent, as well as true differences in incidence, different methods of data collection. The Texas registries are the only ones in the United States that include patients from dermatologists' offices as well as from hospitals, clinics and laboratories. For the melanomas in the series, 97% were microscopically proved.

The high and low incidence rates in melanoma are related in many but not all instances to the high and low rates for total cancer. Those regions with long established cancer registries have the highest rates. The rates in four regions in Texas are the highest except for New Zealand or Cape Province in South Africa in this series. Exactly comparable data for Australia were not available. The published reports however, show the rates may be the highest in existence. The higher rates cluster about certain latitudes.

Within countries, as for example, England, Canada or the United States, there is great variation. Norway and Sweden at higher latitudes than England or Canada, have much higher rates than most regions of either country. Saskatchewan has as high rates as California or Connecticut even though there are ten degrees difference in latitude.

When the various Caucasian population segments were studied by age-specific rates among males, Manitoba and Saskatchewan in Canada; California; Connecticut; Vas, Hungary; Norway; Cracow, Poland; New Zealand; Hawaiian Caucasians; San Antonio and Houston, Texas had high incidence rates reported under 30.

Among White females, Cape Province, South Africa; Saskatchewan, Canada; California; Nevada; Jews born in Israel; Jews born in Europe or America; Norway; Sweden; Oxford, England; Southwest region of England; Yugoslavia; New Zealand; Hawaii; San Antonio, and Houston, Texas have very high rates.

Melanoma constitutes among females, 3.8% of total cancer of Jews born in Israel, and 4.0% of total cancer among Europeans in New Zealand. It constitutes more than 2.0% of total cancer among female Caucasians in Hawaii; San Antonio, Houston, Texas; and Norway. Among Latins, Cali, Columbia reported the highest proportion of total cancer and among males followed closely by Latins in Harlingen and Houston, Texas.

References

1 ACKERMAN, L. V. and DEL REGATO, J. A.: Cancer diagnosis, treatment and prognosis; 4th ed., pp. 956–977 (Mosby, St. Louis, Mo. 1970).
2 ANDERSON, D. E.: Clinical characteristics of the genetic variety of cutaneous melanoma in man. Cancer 28: 721–725 (1971).
3 CLARK, W. H., Jr. and MIHM, M. C., Jr.: Lentigo maligna and lentigo- maligna melanoma. Amer. J. Path. 55: 39–67 (1969).
4 DOLL, R.; MUIR, C. and WATERHOUSE, J. (eds.): Cancer incidence in five continents Vol. II (UICC, Geneva, 1970).

5 LEE, J.A.H. and MERRILL, J.M.: Sunlight and the aetiology of malignant melanoma: a synthesis. Med. J. Austr. *2:* 846–851 (1970).
6 McGOVERN, V.J.: Melanoblastoma. Med. J. Austr. *1:* 139–142 (1952).
7 MISHIMA, Y.: Melanocytic and nevocytic malignant melanomas. Cancer *20:* 632–649 (1967).
8 SMITH, J. L., Jr.: Problems related to pigmented nevi and melanoma in children. Cancer Bull. *24:* 22–26 (1972).
9 URTEAGA, O. and PACK, G. T.: On the antiquity of melanoma. Cancer *19:* 607–610 (1966).
10 WILKINSON. T. S. and PALETTA, F. X.: Malignant melanoma: current concepts. Amer. Surg. *35:* 301–309 (1969).

Author's address: Prof. ELEANOR J. MACDONALD, Department of Epidemiology, The University of Texas M. D. Anderson Hospital and Tumor Institute, *Houston, TX* (USA)

'Celticity' and Cutaneous Malignant Melanoma in Massachusetts

M. M. LANE-BROWN and D. F. MELIA

Department of Dermatology, Harvard Medical School and the Massachusetts General Hospital, Boston, Mass., and Department of Celtic Languages and Literature, Harvard University, Cambridge, Mass.

Introduction

The Celtic component of the Australian population is disproportionately susceptible to the development of skin cancers, including malignant melanoma [14]. Similarly, the incidence of skin cancers, basal cell carcinoma and squamous cell carcinoma, exclusive of malignant melanoma, is abnormally high among those individuals in Boston, Massachusetts with Celtic names in their lineage [15].

During the latter part of the nineteenth century, Irish immigration to the United States was marked [28]. This period of settlement favorably allows a determination of European origin of some Americans along the lines of ASHLEY and DAVIES [1] who used surnames as a marker for the Welshness in the heterogeneous population of Wales. The Australian report [14] that 'celticity' (coined to imply a person's degree of Celt) contributed to a genetic diathesis for malignant melanoma prompted this study in Massachusetts.

Since primary cutaneous malignant melanoma can be classified into three differing histologic types which relate to prognosis [5, 6, 22–24] a comparison between 'celticity' and histopathologic type was made. This short epidemiological survey can make no claim to explain the processes responsible but the material involved can feasibly illustrate a trend.

Methods

221 melanoma patients were seen at the Pigmented Lesion Clinic of the Massachusetts General Hospital over a two-year period. Of these, 109 had primary tumors which were

histologically classified by Dr. WALLACE H. CLARK, Jr., and/or Dr. MARTIN C. MIHM, Jr. Because of the variety of terminologies used for malignant melanoma, the following terms have been used: type I: melanoma with an adjacent intra-epidermal component of Hutchinson's melanotic freckle type; type II: melanoma with an adjacent intra-epidermal component of superficial spreading type; type III: melanoma without an adjacent intra-epidermal component. The patients were age/sex matched with an initial control group of in-patients at the same hospital. Seventeen of these patients had histologic evidence of skin cancer other than melanoma and a further eleven exhibited actinic keratoses clinically. A second age/sex matched control group with no histologic or clinical evidence of skin cancer was then chosen from more than 500 white patients seen at the Dermatology Clinic of the Massachusetts General Hospital. No Negroid or Mongoloid patients with melanoma were studied and no Negroid or Mongoloid patients were included in the control series.

The degree of 'celticity' of the subjects was determined in the following way:
Each subject (or relative) was asked questions related to ancestry: a) surname (maiden name if female); b) mother's maiden name; c) paternal grandmother's maiden name; d) maternal grandmother's maiden name; e) origin in Europe; f) religion.

A name was judged Celtic if it appeared in the acknowledged lists of Celtic names [3, 7, 8, 11, 13, 16, 17, 18, 19, 20, 21, 26, 31]. Collateral information was often obtainable from the religion, most of those with Irish names being Roman Catholics. Consequently, a Catholic from Ireland named Cox was classified Celtic while a Protestant of the same name was regarded as English, i.e., not Celtic. Yet some apparent English names are Anglicized Irish names, e.g., Collins from O'Cullane; or translations, e.g., Smith for Gowan. Non-Celtic names such as Wilde and Smith were judged Celtic only if the subjects knew that their ancestors were Catholics who came from a Celtic area. When there was doubt they were regarded as non-Celtic. 4+ 'celticity' was used to represent three successive, generations of Celtic surnames (fig. 1); 3+ represents three Celtic names out of four; 2+, two out of four (fig. 2); and 1+, one out of four (fig. 3). Where names were not known they were not counted.

Results

The questionnaire confirmed the presence of similar family names in the Boston area and in a geographically restricted population in the United Kingdom and Ireland – the accepted Celtic areas.

Of 109 primary cutaneous malignant melanomas, 11 were of type I, 65 of type II and 33 of type III [24]. It was found that 63% of the melanoma patients had 1+ 'celticity' or more compared with 36% of the control group in which no skin cancer patients were included ($p < 0.01$). The first control group, which included some patients with skin cancer, had 45% with 1+ 'celticity' or more. 'Celticity' was highest in those patients with type III melanoma (66% were 1+ or more) compared with type I (64% were 1+ or

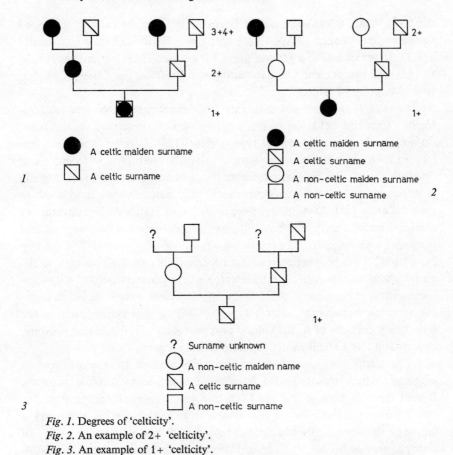

Fig. 1. Degrees of 'celticity'.
Fig. 2. An example of 2+ 'celticity'.
Fig. 3. An example of 1+ 'celticity'.

more) and type II (58 % were 1+ or more). There was no significant difference between the type of melanoma and the degree of 'celticity'.

As collateral information on the 'celticity' of the population in general, a random sample of about 0.5 % (1186) of the names listed in the Boston City Directory [4] was made. The first personal name in columns one and three on each page was used disregarding businesses, proprietary names, vacants, and names preceded by Mrs. 31 % of the 1186 names were Celtic.

Discussion

The Oxford English Dictionary states that the word 'celt' (pronounced selt and also kelt) is derived in English from the French 'celte' (1607), thence the Latin 'celta, celtae', thence the Greek 'keltoi'. Historically, it is applied to the ancient peoples of Western Europe, the Gauls and their continental kin. However, it is also a general name for peoples speaking languages akin to those of the ancient Galli including the Bretons, the Cornish, Welsh, Irish, Manx and Gaelic.

Language and race are disparate since language can be learned arbitrarily. That the Celts, or even a single group of them such as the Irish, do not represent a single racial type is clear [12]. However, under circumstances such as isolation or pressure to adapt, a given group of people can develop a particular genetic characteristic. For instance, sickle-cell anemia in Africans is a relative genetic advantage for heterozygous individuals in malarial areas [25]. Also, the proportion of 'rapid' and 'slow' inactivators of isoniazid varies widely in different human populations. Among Negroes and Europeans about half the population are 'slow' inactivators [9]. Among Japanese [27] the proportion is about 10%. This is a striking contrast to the more typical situation of an inborn error where genes associated with analogous enzyme deficiencies are infrequent. Sufficient genetic isolation alone can preserve apparently neutral types of genetic distinction such as the relative proportions of A, B, O blood groups. ASHLEY [2] found that prostatic hyperplasia has a significantly higher incidence among Celtic people.

The Celtic languages form a distinct branch of the Indo-European languages which are descended from a common reconstructable ancestor. Recent theories suggest that the Indo-Europeans entered Europe from the east sometime around the middle of the third millenium BC [10], with a language that eventually supplanted most of the pre-existing languages of Europe, and which began to break into separate linguistic groups (proto-Celtic, proto Germanic, etc.) in the middle of the second millenium B.C. [10].

By the time we have any linguistic remains the population of Ireland must have spoken a Celtic language almost exclusively. Since Irish remained the primary language of Ireland from at least the fifth century to the nineteenth, virtually 100% of Irish family names surviving to this day must have developed during that period – the majority between the 11th and 14th centuries as in England and France. They can thus serve as a useful tag in studying the movements of the Irish as individuals in the present. The same is true of Scotland and of Wales which had a parallel development of surnames [3, 8].

Although Ireland has been invaded several times since the coming of the Celts there has never been a really large scale population movement into the country in historical times. Thus we have a continuing pool of genetic material in Ireland from at least 400 A.D. which, while absorbing some outside genes, has never been overwhelmed by them. Because this is a very restricted situation in terms of language as well as population, it seems in the broad spectrum quite legitimate to equate possession of a Celtic family name with membership in this genetic subgroup. There is strong evidence that even within Ireland there is little movement of population as evidenced by family name [21].

This situation also obtains in other Celtic-speaking areas which, because of relative geographical and economic isolation, remained quite stable in terms of population movement from the fourteenth century onward. This generalization is confirmed by the distribution of the ABO blood types in Europe; type O is of abnormally high frequency in Celtic speaking areas [25].

Celtic surnames are easily recognizable. Welsh family names are very few in number and usually formed from Christian names (e.g., Williams, Davies) or the word for 'son' plus the Christian name (e.g., Pugh from ap Hugh). Irish and Scottish surnames are generally either patronymics or forms indicating descent from a more distant ancestor or hero. There are also several names originating with 12th century Norman and English invaders which can now be considered distinctively Irish, such as Burke and Barry and a few descriptive names like Walsh (Welshman). There are more Irish surnames in use today than Welsh, but the numbers are still smallcompared to English or French surnames [16, 26].

URBACH [30] used the 1960 decennial census for greater Philadelphia to show prevalence of skin cancer by ethnic origin. The 'foreign born' category is for immigrants and children of immigrants. 'Native' to the U.S. implies 2 or more generations in the U.S. These criteria could introduce a skew in the figures; since a genetic predisposition is postulated, whether a patient is foreign born or American born is not relevant. 'Celticity' allows a more rigorous investigation of the skin cancer and melanoma hypothesis.

The use of Celtic-type names as a tag for a particular genetic mix is not a prejudgement of the case. The gene-pool reflected here may have its base in the Neolithic peoples who preceded the Celts to Europe or in some peculiar combination of other factors. The usefulness of the linguistic data is due to the fact that we still adhere to a system of personal identification developed in Europe during the Middle Ages.

In this study, then, a Celtic name indicates residence in a Celtic country of at least one recent ancestor of the subject. This residence is indicative of membership in an historically stable gene-pool which is for historical reasons associated with Celtic language.

We conclude that the incidence of malignant melanoma is abnormally high among those individuals in Massachusetts with Celtic names in their lineage. 'Celticity' does not significantly relate to the histologic type of the primary cutaneous malignant melanoma.

References

1 ASHLEY D. J. B. and DAVIES, H. D.: The use of the surname as a genetic marker in Wales. J. med. Genet. *3:* 203 (1966).
2 ASHLEY, D. J. B.: Observations on the epidemiology of prostatic hyperplasia in Wales. Brit. J. Urol. *38:* 567 (1966).
3 BLACK, G. F.: The surnames of Scotland (Public Library, New York, 1946).
4 Boston City Directory, Vol. 167 (R.L. Polk & Co., Boston 1970).
5 CLARK, W. H., Jr.: A classification of malignant melanoma in man correlated with histogenesis and biological behavior; in W. MONTAGNA and F. HU. Advances in biology of skin, vol. 8: The pigmentary system, pp. 621 (Pergamon Press, New York 1967).
6 CLARK, W. H., Jr.; FROM, L.; BERNARDINO, E. A. and MIHM, M.C., Jr.: The histogenesis and biologic behavior of primary human malignant melanomas of the skin. Cancer Res. *29:* 705 (1969).
7 DAUZAT, A.: Dictionnaire Etymologique des Noms de Famille et Prénoms de France. (Larousse, Paris 1951)
8 DAVIES, T.: A book of Welsh names, pp. 10–11 (Sheppard Press, London 1952).
9 EVANS, D. A. P.; MANLEY, K. E. and MCKUSICK, V. A.: Genetic control of isoniazid metabolism in man. Brit. med. J. *ii:* 485 (1960).
10 GIMBUTAS, M.: Bronze age cultures in central and eastern Europe (Mouton, The Hague 1965).
11 GOURVIL, F.: Noms de famille de Basse-Bretagne (Editions D'artrey, Paris 1966).
12 HOOTON, E. A. and DUPERTUIS, C. W.: The physical anthropology of Ireland. Papers of the Peabody Museum of Archaeology and Ethnology, Harvard University, vol. 30, pp. 244 (1955).
13 KNEEN, J.J.: The personal names of the Isle of Man (Oxford University Press, London 1937).
14 LANE-BROWN, M. M.; SHARPE, C. A. B.; MACMILLAN, D. S. and MCGOVERN, V. J.: Genetic predisposition to melanoma and other skin cancers in Australians. Med. J. Austr. *1:* 852 (1971).
15 LANE-BROWN, M. M. and MELIA, D. F.: A genetic diathesis to skin cancer. (Abstract) Clin. Res. *19:* 358 (1971)
16 MACLYSAGHT, E.: Irish families, their names, arms and origins, pp. 28 (Hodges, Figgis, Dublin 1957).
17 MACLYSAGHT, E.: More Irish families (O'Gorman, Galway 1960).
18 MACLYSAGHT, E.: Supplement to Irish families (Helicon, Dublin 1964).
19 MACLYSAGHT, E.: Guide to Irish surnames (Hodges, Figgis, Dublin 1964).
20 MACGIOLLA-DOMHNAIGH: Some anglicized surnames in Ireland (Privately printed, New York 1923).

21 MATHESON, R.: Special report on the surnames of Ireland. His Majesty's Stationary Office, pp. 15–24 (Genealogical Publishing Co. Baltimore 1968).
22 MCGOVERN, V.J. and LANE-BROWN, M. M.: The nature of melanoma (Thomas, Springfield, Ill. 1969).
23 MIHM, M. C., Jr.; CLARK, W. H., Jr. and FROM, L.: The clinical diagnosis, classification and histogenetic concepts of the early stages of cutaneous malignant melanomas. New Engl. J. Med. *284:* 1078 (1971).
24 MIHM, M. C., Jr.; FITZPATRICK, T. B.; LANE-BROWN, M. M. and RAKER, J.W.: A colour atlas of primary cutaneous malignant melanoma as an aid to early clinical diagnosis. Proc. (Karger, Basel/New York 1972).
25 MOURANT, A.E.: The distribution of human blood groups; pp. 44–53 (Thomas, Springfield 1954).
26 REANEY, P.H.: A dictionary of British surnames, p. 8 (Routledge and Kegan Paul, London 1958).
27 SUNAHARA, S.; URANO, M. and OGAWA, M.: Genetical and geographical studies on isoniazid inactivation. Science *134:* 1530 (1961).
28 TAEUBER, C. and TAEUBER, A.: The changing population of the United States, p. 56 (Wiley, New York 1958).
29 The Oxford English Dictionary (C.T. Onions, Oxford University Press).
30 URBACH, F.: in F. URBACH. Geographic pathology of skin cancer, the biologic effects of ultraviolet radiation, pp. 635 (Pergamon Press, London 1969).
31 WOULFE, P.: Sloinnte Gaedhael is Gall (Gill, Dublin 1923).

Author's address: Dr. MALCOLM M. LANE-BROWN, Department of Dermatology, Massachusetts General Hospital, *Boston, MA 02114* (USA)

Pigment Variation in Relation to Protection and Susceptibility to Cancer

E. F. Rose

Bantu Cancer Registry, East London, South Africa

Pigmentation in man and animals serves as a light screen and for radiation protection. It is, therefore, only in unusual circumstances that albino animals survive in the wild. The susceptibility of light-coloured animals to skin cancer was well demonstrated by Professor J. Bonsma of Pretoria University, who found a high incidence of ophthalmia and cancer round the eyes of white-faced Hereford cattle. This was related to the amount of pigment. By selective breeding he increased the pigment round the eyes and reduced his disease rate (table I).

Depth of pigmentation is inherited to a different degree in different human races, the Negroes of Africa where the light intensity is highest having the deepest pigmentation. South Africa, by virtue of its geographical situation at the base of the continent, has become an ethnological cul-de-sac for all the invading tribes from the north who had either to wipe out their predecessors or intermingle with them forming hybrid races and cultures [6]. The Transkeian people at the base of the continent show great variation in skin colour through their intermingling, from the pale yellow skins with hamitic features inherited from the Bushmen and Hottentot to the dark skin of the Negro.

Dr. Burrell [3] checked on the skin colour of some 13,000 Transkeian Bantu and found 3 distinctly recognizable types (table II).

The brown colour of the majority of these people distinguishes the Bantu from the Negro of west and central Africa, who has a black skin. The features of the 'black' people of the Transkei are typically negroid, and those of the 'yellowbrown' people resemble the hamite. The 'yellowbrown' people can again be differentiated into those with smooth skins and those with 'dusty' lustreless skins (1 in 18) similar to the freckles found on the albinoid skin.

Table I. Variation in Pigmentation about the eyes of 560 Hereford cattle[1]

Herd[2]	Classification according to amount of pigmentation			Incidence of ophthalmia, other affliction and cancer			
	++	+	−	++	+	−	
A	19	252	59	0	16	25	1 cancer
B	9	63	98	0	0	26	2 cancer
C	3	48	27	0	5	21	
D	16	137	73	0	17	28	
Total	47	500	257	0	38	100	(3 cancers)
% of total (804)	5.8	62.2	32.0	0	4.7	12.4	
Mara	55	159	102	0	17 (1 cancer 5.4)	22	(4 cancers)
% of total (316)	17.4	50.3	32.3	0	5.4	7.0	
Grand total (1120 eyes)	102	659	359	0	55 (1 cancer)	122	(7 cancers)
grand total	9.1	58.8	32.1	0	4.9	10.9	
Proportion and % affected in each class	$0/102$	$55/659$	$122/359$	0	8.3 .02 cancer	34.0 1.9	cancer

[1] In a survey of the amount of pigmentation on the eyelids and on the surroundins hair of Hereford cattle, three groupings were made, namely, those where pigment wag almost or entirely absent (−), those with pigment or broken pigment on the eyelids accompanied by a small, often interrupted ring of hair less than ½ in. around the eyes (+), and those with a large ring at least ½ in. wide around the eyes (++).
[2] Herds A, B, C, D, are maintained in the temperate zone while the Mara herd is in a sub-tropical environment.

Table II. Types of skin in the Bantu

Skin colour	Males/% (N = 9038)		Females/% (N = 4527)		Persons/% (N = 13,565)	
Black	481	5.3%	163	3.6%	644	4.75%
Yellowbrown	419	4.6%	308	6.8%	727	5.36%
'Dark' brown	8138	90.0%	4056	89.6%	12,194	89.89%
Total persons	9038		4527		13,565	

Cancer Rates in Various Races of South Africa

Apart from the colour variation of the people of the Transkei the various races of South Africa present interesting differences of colour as they do in their cancer incidence. Mortality figures for South African whites, Asiatics and coloureds (table III) are interesting in that deaths from skin cancers are greatest in the lightest-skinned people, namely, the whites, and lowest in the darkest-skinned people, namely, the Asiatics, who are composed mainly of dark-skinned Indians. (The coloured are a mixed race whose skin complexion can vary from a very pale yellow to a light brown.)

Because of the treatability of most skin cancers the mortality figures do not give a true indication of incidence of cancer. Table IV is an analysis of biopsy reports of the Frere Hospital Pathology Laboratory, East London. It demonstrates the site of the different cancers reported for the 3 racial groups treated in the Frere Hospital between 1966 and 1967, namely, white, Bantu and coloured. It is obvious from these figures that skin cancer is exceedingly high in the white-skinned people, twice as high as in the other 2 races.

Table V shows a breakdown of histological types of 235 skin cancers seen in East London in an 18-month period between 1967 and 1968, from a total of 798 cancers. Although squamous carcinoma is the most common type of skin cancer in all groups (50%), basal cell carcinoma, with one exception, is confined exclusively to the white-skinned race.

Melanomas

Melanomas comprise 8.1% of the total skin cancers. Seventy-five per cent were in the dark-skinned races. The site of the melanomas in Bantu was on the sole of the foot, except for one on the gum. The only melanoma in a coloured man was on his chin, and the 5 melanomas in the white race were on the leg, neck and 3 unspecified sites. It is usual for melanomas to be found on the sole of the feet in Bantu, an unusual site in white-skinned persons. Lewis, reporting from Uganda in 1967 [5], stated that the site of melanoma on the feet of Africans is coincidental with discrete areas of pigmentation on the soles of the feet. The most significant type of pigment was a small, black, well circumscribed 'spot' which histologically resembled a junctional naevus or lentigo, so that no fundamental difference exists between Negro and Caucasian in the pathogenesis of malignant melanoma which invariably results from a junctional naevus. It is possible that these

Table III. South African mortality figures (1966–1967) from Bureau of Census and Statistics

Number of cases per million	White	Asiatics	Coloured
Primary Lung and bronchus	443	31	244
Primary liver	120	50	157
Oesophagus	63	51	126
Liver (secondary)	0.9	0	18
Lung (secondary)	–	–	–
Cervix	85	55	145
Breast	14	–	19
Skin (melanoma)	45	1	4
Skin (other neoplasm of)	19	7	11

Table IV. Frere Hospital pathology records (1966–1967), showing commonest cancers in prodeo cases in 3 racial groups

	Caucasian (white) (Total cases 621)		Bantu (black) (Total cases 780)		Coloured (brown) (Total cases 50)	
	Cases	% of total	Cases	% of total	Cases	% of total
Mouth	10	1.6	30	3.8	1	2
Tongue	13	2.0	35	4.4	3	6
Oesophagus	6	0.97	126	16.1	8	16
Stomach	7	1.1	18	2.3	–	–
Large bowel	28	4.5	6	0.8	–	–
Bronchus	18	2.8	9	1.1	4	8
Lung	9	1.4	5	0.6	2	4
Liver	10	1.6	41	5.2	2	4
Skin	224	36.0	57	7.3	2	4
Wart (malignant)	5	0.8	–	–	–	–
Mole (malignant)	23	3.7	16	2.0	1	2
Ulcer (malignant)	6	0.97	7	0.9	–	–
Lip	25	4.0	1	0.1	–	–
Breast	76	12.2	53	6.79	5	10
Uterus	17	2.7	8	1.03	2	4
Cervix	24	2.8	157	20.0	12	24
Penis	–	–	5	0.6	1	2
Prostate	20	3.2	17	2.2	2	4
Bladder	17	2.7	4	0.5	–	–

Table V. Histological breakdown of skin malignancies in 3 racial groups

Total cancers (prodeo): 567			Total cancers (private): 231		
Total skin cancer: 124			Total skin cancer 111 (103 persons)		

	Frere Hospital records			Private laboratory		
	Jan. 1967	Jan. 1967–Dec. 1968		Jan.–Dec. 1967	Total	Percentage
	Bantu	Coloured	White	White		
	44% Males	75% Males	51% Males	56% Males		
	55% Females	25% Females	49% Females	33% Females		
Squamous cell cancer	44	3	42	28	117	50%
Basal cell	1	0	15	72	88	37%
Basic squamous	–	–	–	6	6	2.5%
Melanoma (malignant)	13	1	0	5	19	8.1%
Epithelioma	1				1	
Kaposi	2				2	
Skin ulcer differentiated cancer	1	1			1	
Total	62	5	57	111	235	

pigmented naevi on the junction between pigmented and unpigmented areas of the feet are subject to trauma in these usually barefooted persons.

Albinism

The high incidence of skin cancer in the white races of South Africa tends to substantiate the theory that pigmentation plays a protective role against the carcinogenic effects of solar radiation. It is generally accepted in South Africa that albinos are prone to skin cancer. There is a high incidence of albinism in the Transkei (1 in 3,500).

These are the factors that prompted me to investigate albinism in the Transkei with a view to determining their skin cancer rate.

Findings

Dr. BURRELL registered 458 albinos in 1960. In 1971 I investigated 41 families in which there were 71 albinos, 11 with vitiligo, 2 with piebaldingi 18 had a 'red' colour (partial albinism), and 3 with Waardenburg's syndrome.

On the basis of his crude observations regarding skin colour (table II) Dr. BURRELL examined the skins of albino relatives at 42 homes, and for comparison he examined the skins of members of unrelated neighbouring families who denied ever producing albinos (sampling 1 in 10 by statistical round numbers) (table VI).

From these observations there appeared to be considerably less black-skinned relatives in the families of albinos, than in the controls which approximated that of the general population. There also was a marked reversal of dusty to smooth ratios (see table II) in respect of albino relatives and neighbours.

The skin of the albino is white, soft in unexposed parts, but thick and dry on the exposed parts. It is sensitive to light, becoming erythematous from exposure to sun, and crusty sores are evident on face, lips and other exposed parts. A number of these people, as they grow older, develop splotchy dark brown freckles. The hair is yellow and the eyes vary from a light brown (khaki colour) to a very pale blue. Dr. BURRELL claims to have seen the classical signs of photophobia and nystagmus in less than half of the 309 albinos he examined. He describes eye colour as follows:

Pink (red)	39
Limpid green	91
Bluish green	28
Nordic blue	52
Sepia brown	85
Unclassified	14
	309

Sensitivity to light and nystagmus seems to be related to eye colour.

In 27 albino families (71 albinos) investigated I found no pink eyes but I did get a red reflex in a number of persons with light-coloured eyes. Most of those with blue and green eyes had yellow in varying amounts extending from the pupillary borders of the iris, giving the green eyes a khaki ap-

Table VI. Skins of 276 relatives of 46 albinos and skins of 632 unrelated neighbours

Skins called		Relatives of albinos No.	%	Unrelated neighbours No.	%
Black		1	0.3	29	4.6
	Dusty	13	4.7	2	0.3
Yellowbrown.					
	Smooth	3	1.1	34	5.4
Dark brown		259	93.8	567	89.7
Total persons		276		632	

pearance. Nystagmus and photophobia were evident to a varying degree, being absent in brown eyes.

During this investigation, some interesting skin variations were seen. Amongst some of the albino families a type of skin colour occurred which can only be described as 'red'. These 'red' people have the blond yellow hair of the albino, but a red skin which is sensitive to light, and especially in children evinced the crusty sores so evident in the albino skin. The eye colour is brown and I found none with nystagmus (fig. 1 and 2).

Two cases of piebalding were encountered in 2 different families with no history of albinism or of ancestors with piebalding. These children were born with a white forelock and patches of depigmentation on the body. One boy had a 'dalmation type' appearance of the body except for the hands and face. The hair was black apart from the white forelock. This white forelock on its own is not uncommon and is recognized as a familial trait. It is also found in Waardenburg's syndrome [4] which presents with deafness, heterochromia of the eyes with widening of the inner canthi as well as the white forelock. The 3 cases of Waardenburg's syndrome encountered in the Transkei also had areas of lighter pigmentation on the skin.

Apart from these pigment anomalies, depigmentation was found as a result of disease and trauma, e.g., burns, ulcers, lupus erythematosis, and some conditions of tuberculosis of the skin.

Vitiligo is a feature not uncommon to the Transkei and has been found in families with and without a history of albinism.

These last-mentioned conditions of depigmentation are actually aggravated by sunlight as opposed to normal skin which becomes more pigmented under solar radiation.

Pigment Variation in Relation to Protection and Susceptibility to Cancer 243

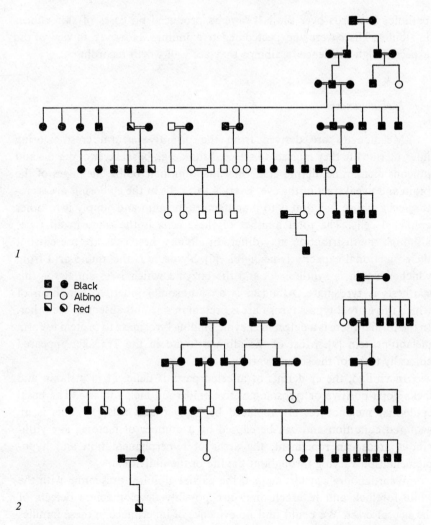

Fig. 1 and 2. The family trees of Bantu albinism and the sporadic occurrence of red-coloured people (partial albinism).

In spite of the fact that most of the albinos manifested with crusty sores on the face, lips, arms and legs, the only skin cancer found in Dr. BURRELL's series was a melanoma at the back of the knee in an albino woman of fifty. Admittedly, this is not a common site for a melanoma in a Bantu. Depth questioning of all my families investigated, as to cause of death in albino relatives produced the same negative results. A check on Frere Hospital pathology records over the last 5 years produced no cases of skin cancer in albinos. These were unexpected negative findings. However, in view of the expected high incidence in albinos they are well worth recording.

Melanogenesis

Melanocytes are derived from the primitive neural crest. During inter-uterine life they migrate to selected sites, e.g., basal layer of dermis and mucous membrane, dermo-epidermal junction, hair bulbs, meninges of the brain and uveal tract of the eye. Every 4 to 5 cells in the epidermis are secretory cells whose function is to manufacture melanin and supply it to other epithelial cells. The total number of these cells is the same in all races although the distribution may differ. In albinos these cells are present but their functional capacity is negligible [6]. Tyrosine is the amino acid from which melanin is synthesized, and the enzyme which is essential for this synthesis is tyrosinase. Albinism is an autosomal inherited condition of which there are 2 types: tyrosinase is present in one and absent in the other. In the former there is a defect in enzyme binding tyrosinase to protein to form melano-protein [9]. Most of the albinos found in the Transkei appeared clinically to be of the second variety.

In vitiligo, the epidermis of affected parts is deficient in melanin and lacks dopa-positive or tyrosine-positive epidermal melanoblasts. The basal epidermal melanocytes are replaced by Langerhans cells [1]. This is an acquired condition and can be caused by a number of factors. The vitiliginous areas can repigment, the areas of hyperpigmentation and hypopigmentation varying throughout the life of the individual.

Waardenburg's syndrome and the partial albinism presenting with the white forelock and heterochromia are possibly developmental defects of the neural crest. We could find no evidence of inheritance in these families with gross defects, although the white forelock alone appeared with regularity in some families.

Conclusion

There is an obvious difference in the types of cancers in different racial groups, and as the function of pigmentation is a protective one against solar radiation, especially that of the U.V. spectrum, a deeper study of the abnormalities of pigment synthesis and the histogenesis of skin cancers in differently pigmented peoples holds promise as a means to improve the understanding and prevention of skin cancer. Our inability to find a high incidence of skin cancer in the albino was unexpected and indicates that other factors besides pigmentation may play a role in the prevention of carcinogenesis of the skin.

Acknowledgements

I am most grateful for the opportunity I have had to use the unpublished material on albinism gathered by the late DR. BURRELL. I would also like to acknowledge the help I have had from my Bantu field worker, MR. BIKITSHA, who personally knew most of the families I visited. I would also like to thank Professor J. BONSMA for allowing me to publish his interesting work on Hereford Breeding.

This research was made possible by a Grant from The National Institute of Health, Bethesda, Md., U.S.A., C.A. 06565, and from the Medical Research Council, South Africa.

References

1 ALLEN, A.C.: The skin. A clinico-pathological treatise; 2nd ed. (Heineman, London 9).
2 BONSMA, J.: Pretoria University Agricultural Department (Personal commun.).
3 BURRELL, R.W.: (Unpublished notes).
4 DI GEORGE, A.M.; OLMSTED, R.W., and HARLEY, R.D.: Waardenburg's syndrome. J. Pediat. 57 (1960).
5 LEWIS, M.G.: Malignant melanoma in Uganda (the role of pigmentation of soles of the feet); in P. CLIFFORD, C.A. LINSELL and G.L. TIMMS. Cancer in Africa; (East African Publishing House, 1968).
6 O'MALLEY, C.K.: The mechanism of skin pigmentation. Sth Afr. med. J. (1960).
7 SCHAPERS, I.: The Khoisan peoples of South Africa (Reprinted) (Routledge and Kegan Paul, London 1963).
8 S.A. Bureau of Census and Statistics: Mortality figures for 1966–1967.
9 WITKOP J. and NANCE, W.: Amer. J. hum. Genet. 22: 55–74 (1970).

Author's address: DR. E.F. ROSE, Bantu Cancer Registry, *East London* (South Africa)

Geographic Pathology of Malignant Melanoma in Japan

W. MORI

Department of Pathology, Tokyo Medical and Dental College, Tokyo

Malignant melanoma has been known as one of the most malignant neoplasms, from the viewpoint of cancer control, on the one hand, and as one of the most interesting tumors, from the viewpoint of geographic pathology, on the other. The latter is to be the main focus of the present discussion.

General Incidence, Sex and Age Distribution

It has long been said that malignant melanoma is a rare disease in Japan, especially when compared with Western countries. Although I agree with this to some extent, the difference seems a little smaller than expected.

As far as the data obtained from autopsy cases in various countries is concerned [5, 10, 13, 14, 16], melanoma incidence in Japan does not seem to be very low, as shown in table I. However, it should be noted that incidence

Table I. Incidence of malignant melanoma among autopsy cases in various areas of the world

	Author	No. of all autopsy cases	No. of malignant melanoma cases	%
California	STEINER	35,293	53	0.15
Canal Zone	ENOS et al.	18,241	7	0.04
England	MORI	5,214	11	0.21
Illinois	STEINER	8,000	21	0.26
Japan	MORI	136,599	212	0.16
Maryland	MACCALLUM	12,000	10	0.08

among autopsy cases does not always reflect general incidence of the disease in a community, since a collection of rather rare diseases may occur in this group.

The material used for this survey was obtained by reviewing the Annual of pathological autopsy cases in Japan, 1958–1969 [7], and all autopsies of malignant melanoma performed in Japan within the 12-year period were included. Thus, a total of 295 cases of malignant melanoma were collected and analyzed. Other procedures concerning materials and methods employed for this study were the same as described in a previous paper [14] in which an analysis of material in a recent 10-year period (212 cases) was reported.

One hundred and seventy-six cases were of males and 119 of females, the former constituting 59.7% of the total. This incidence closely paralleled the 3:2 male/female ratio of the autopsy population during the period under study, i.e., no predominance of this tumor was found in either sex.

Age and sex distributions are shown in figure 1. The highest incidence was seen between 50 and 69 years of age in both sexes.

Primary Sites

Primary sites of the tumors were first divided into 3 major groups: cutaneous, muco-cutaneous-junctional, and non-cutaneous. The last group was further divided into 2 subgroups: mucosal and non-mucosal (table II). Cutaneous neoplasms make up about half, i.e., only 51.5% of all melanomas.

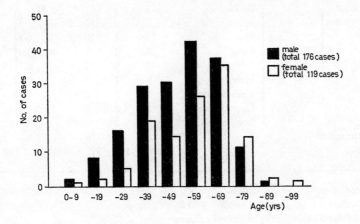

Fig. 1. Total incidence of malignant melanoma among autopsy cases in Japan.

Table II. Primary sites of malignant melanoma in Japan (295 autopsied cases)

	Primary sites	Number of cases (M)	(F)	Total	Total %		
Cutaneous melanoma	Head and face	13	6	19	6.4 } 8.8		51.5
	Neck	5	2	7	2.4		
	Axillary fossa	7	1	8	2.7 } 9.5		
	Arm and hand	3	8	11	3.7		
	Finger[1]	6	3	9	3.1		
	Trunk	10	11	21	7.1		
	Inguinal area[2]	8	1	9	3.1		
	Thigh and leg	13	8	21	7.1		
	Ankle and foot back	11	6	17	5.8 } 25.4		
	Foot sole	12	7	19	6.4		
	Toe	8	1	9	3.1		
	Undetermined	2	0	2	0.7		
Muco-cutaneous junctional melanoma	Mouth lip	2	1	3	1.0		5.8
	Anorectal area	4	4	8	2.7		
	Female external genital area	0	5	5	1.7		
	Glans penis	1	0	1	0.3		
Non-cutaneous melanoma	Mucosal	Nasopharynx[3]	9	5	14	4.7 } 16.9	21.7
		Paranasal cavity	9	5	14	4.7	
		Oral cavity	7	15	22	7.5	
		Esophagus	4	0	4	1.4	
		Rectum	1	5	6	2.0	
		Vagina	0	4	4	1.4	
	Non mucosal	Brain	10	7	17	5.8 } 6.4	15.6
		Spinal cord	2	0	2	0.7	
		Eye	12	6	18	6.1	
		Lung	2	1	3	1.0	
		Adrenal	2	0	2	0.7	
		Liver	0	1	1	0.3	
		Mediastinum	1	0	1	0.3	
		Parotid gland	1	0	1	0.3	
		Achilles tendon	1	0	1	0.3	
Undetermined		10	6	16			5.4
Total		176	119	295			100

1 Seven cases were in the thumb.
2 Including 2 cases from the scrotum and 2 from the penis.
3 Including 3 cases from pharyngeal mucosa.

This is a very low value compared with the results of similar surveys from Western countries [1, 3, 12, 15], and this relative predominance of non-cutaneous over cutaneous melanomas may be one of the causes which gave the mistaken impression that malignant melanoma is very rare in Japan. Mucosal melanoma is said to be rather common among native African negroes [2, 9], and our figure in Japan seems to be similar.

Almost half of cutaneous melanomas, or 24.5% of the total, originated in the lower extremities, the sole of the foot being one of the predisposed locations. As to etiologic factors in the evolution of melanoma, HEWER [6] and others suggested the role of trauma, especially walking barefoot. In modern Japan, walking without shoes or sandals outside the house is extremely rare; yet the sole of the foot is one of the most common sites of the neoplasm. Trauma may be one of the causes, but it seems reasonable to search for other factors, such as an increased susceptibility of the pigment cells in some areas of the skin. Although a considerable number of melanomas occur in the head, face and neck, or extremities, the conclusion that areas exposed to sunlight are sites of predilection does not seem to be supported either in this survey.

There were 17 (5.8% of the total) lesions occurring at the mucocutaneous-junctions (table II). The majority of mucosal melanomas occurred in the oral (22 cases), naso-pharyngeal (14), and paranasal cavities (14), amounting to 7.5%, 4.7% and 4.7%, respectively of all melanomas. The esophagus, rectum, and vagina are the other primary sites listed in this group.

Non-cutaneous, non-mucosal melanomas, originating in the central nervous system and eye, were found 46 times (15.6% of the total). The former seem to have originated mostly from the meninges. The lung, adrenal, and liver, etc., were other (doubtful) primary sites. Melanomas originating in the central nervous system were relatively common in young subjects, while ocular lesions had an age distribution similar to that of cutaneous origin.

In 16 cases (5.4% of the total) the primary site remained unknown, even after autopsy. Some of these tumors may have been cutaneous, the patient not having mentioned a surgical removal years back. But some others may have been non-cutaneous with extensive metastases at the time of autopsy preventing detection of the primary sites.

Figure 2 shows the age and sex distribution of cutaneous, muco-cutaneous-junctional and mucosal melanomas. These 3 diagrams differ somewhat from one another. The high incidence of cutaneous melanoma has a wide range of approximately 30–69 years; muco-cutaneous-junctional

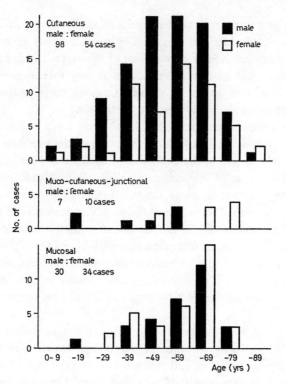

Fig. 2. Age and sex distribution of various malignant melanoma in Japan (comparison of cutaneous, muco-cutaneous-junctional, and mucosal melanomas).

lesions are uncommon in all age groups, while mucosal melanoma seems to be more frequent in females than in males, with a peak incidence in the 7th decade.

Regional Differences in Incidence in Japan

Japan consists of four main islands which form an arc running from northeast to southwest. Genetically, the Japanese are fairly uniform and the country, therefore, is often considered as a region well-suited for the study of problems of geographic pathology. Conditions may be representative not only of Asia in general but also of regional differences, since climate differs markedly from the north end to the south.

Table III. Local difference in incidence of malignant melanoma in Japan

District	Survey with autopsy cases (1958–1967)			Survey with mortality statistics (1960–1967)				
	No. of all autopsy cases (I)	No. of malignant melanoma cases (II)	(II)/(I) ratio %	No. of all death cases (III)	No. of malignant tumor cases (IV)	No. of malignant melanoma cases (V)	(V)/(III) ratio %	(V)/(IV) ratio %
A	7,650	18	0.24	248,464	37,171	27	0.011	0.073
B	19,281	48	0.25	1,304,894	187,470	130	0.010	0.069
C	69,603	96	0.14	1,669,934	259,603	221	0.013	0.085
D	23,693	33	0.14	1,487,130	225,905	136	0.009	0.060
E	16,372	17	0.10	789,977	111,663	67	0.008	0.060
Total	136,599	212	0.16	5,500,399	821,812	581	0.011	0.071

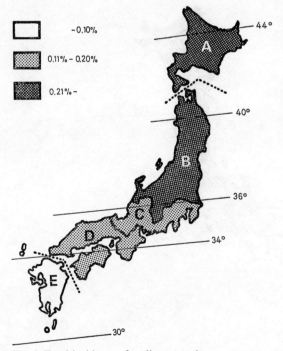

Fig. 3. Total incidence of malignant melanoma among autopsy cases in Japan.

In this study, Japan was divided into 5 districts for comparative work, A B C D and E from the north to south. The incidence of all malignant melanoma, irrespective of their primary site, among autopsy cases was calculated for each district and compared (table III, fig. 3). In the two northern districts, A and B, the incidence among all autopsy cases was 0.24 and 0.25 %, respectively. In the two central districts, C and D, the incidence was 0.14 and 0.14 %, respectively; and in the southernmost district, E, it was 0.10 %. The difference between districts (A, B) and C, D) as well as that between (A, B) and (E) were significant ($P = 0.05$), but that between (C, D) and (E) was not. The dividing line between the districts of high and low incidences of melanoma seemed to lie along the 36th parallel.

The data described above is derived from a previous paper [14]. In order to determine whether the incidence among autopsy cases exactly reflected the true incidence of the disease among the general population, another analysis has been made based on death certificate, with the assistance of Prof. SEGI and his associates at Tohoku University. The result is shown in the right half

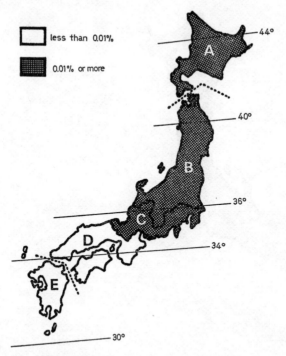

Fig. 4. Total incidence of malignant melanoma in mortality statistics in Japan.

of table III and also in figure 4. Again, significant difference is seen between the north and south but this time the dividing line has moved a little south to the 35th parallel.

Solar carcinogenesis has been thought to have some responsibility for promoting the evolution of melanoma [4, 8, 11]. However, as far as the result of this survey is concerned, melanoma had a higher incidence in the north of Japan where people are exposed to less sunshine than in the south. This tendency was seen with both cutaneous and non-cutaneous melanomas. Moreover, as described above, in Japan non-cutaneous melanomas occurring in sites not exposed to light were as common as cutaneous tumors. These facts seem to suggest that the role of sunlight in the causation of melanoma is still a moot question. A possible hypothesis is that the intrinsic factor of low susceptibility of the skin to melanoma plays a much more important role than the extrinsic factor of sunlight.

Ectopic Pigmentation

The relatively high incidence of malignant melanoma of the nose among Africans might be related to ectopic pigmentation of the nasal cavity, commonly seen with African negroes [9]. We have not yet come to a conclusion whether the relatively high incidence of mucosal melanomas in the Japanese can be attributed to ectopic pigmentation. But, with the Japanese, it is true that we see pigmented spots grossly, or melanocytes microscopically, in the mucosa of oral, nasal, or paranasal cavities with some degree of frequency and, sometimes, of the esophagus, vagina, and even in the wall of a dermoid cyst of the ovary.

TAKAGI and ISHIKAWA [17] reported that ectopic pigmentation in the oral cavity among the Japanese was not rare and sometimes malignant melanoma developed from these sites. This coincides with my experience, and these points need to be further studied.

Acknowledgements

This investigation was supported by a grant from the Japanese Government. The author wishes to thank the Japanese Pathological Society for allowing him to use the data listed in the 'Annual of the pathological autopsy cases in Japan'. Many thanks are due to Prof. M. SEGI, Prof. H. MAEDA, and Miss K. ISHIWATA for their assistance in this work.

References

1 ALLEN, A.C. and SPITZ, S.: Malignant melanoma, a clinicopathological analysis of the criteria for diagnosis and prognosis. Cancer 6: 1–45 (1953).
2 BROOMHALL, C. and LEWIS, M.G.: Malignant melanoma of the oral cavity in Ugandan Africans. Brit. J. Surg. 54: 581–584 (1967).
3 CHAUDHRY, A.P.; HAMPEL, A. and GORLIN, R.J.: Primary malignant melanoma of the oral cavity: a review of 105 cases. Cancer 11: 923–928 (1958).
4 DAVIS, N.C.; HERRON, J.J. and MCLEOD, G.R.: Malignant melanoma in Queensland, analysis of 400 skin lesions. Lancet ii: 407–410 (1966).
5 ENOS, W.F. and HOLMES, R.H.: Malignant melanoma in the tropics. Amer. J. Path. 27: 523–533 (1951).
6 HEWER, T.F.: Malignant melanoma in coloured races: the role of trauma in its causation. J. Path. Bact. 41: 473–477 (1935).
7 Japanese Pathological Society (ed.): Annual of the pathological autopsy cases in Japan (I–XII, 1958–1969), Tokyo, 1959–1970 (in Japanese).
8 LANCASTER, H.O.: Some geographical aspects of the mortality from melanoma in Europeans. Med. J. Austr. i: 1082 (1956). (Abstr., see 8a.)
8a LANCASTER, H.O. and NELSON, J.: Sunlight as a cause of melanoma. Med. J. Austr. i: 452 (1957). (Abstr., quot. MCGOVERN, V.J. and BROWN, M.M.L.: Malignant melanoma; in The nature of melanoma, chapt. 7, pp. 92–116 [Thomas, Springfield 1969]).
9 LEWIS, M.G. and MARTIN, J.A.M.: Malignant melanoma of the nasal cavity in Ugandan Africans. Relationship of ectopic pigmentation. Cancer 20: 1699–1705 (1967).
10 MACCALLUM, W.G.: Tumors (continued); in A textbook of pathology, 7th ed., chapt. 66, pp. 1098–1115 (Saunders, Philadelphia Pa. 1942).
11 MACDONALD, E.J.: The epidemiology of melanoma. Ann. N.Y. Acad. Sci. 100: 4–17 (1963).
12 MILTON, G.W. and BROWN, M.M.L.: Malignant melanoma of the nose and mouth. Brit. J. Surg. 52: 484–493 (1965).
13 MORI, W.: A review of autopsy records at the Department of Pathology, University of Cambridge, 1961–1967 (to be published).
14 MORI, W.: A geo-pathological study of malignant melanoma in Japan. Path. Microbiol. 37: 169–180 (1971).
15 RAVID, J.M. and ESTEVES, J.A.: Malignant melanoma of the nose and paranasal sinuses and juvenile melanoma of the nose. Arch. Otolaryng. 72: 431–444 (1960).
16 STEINER, P.E.: Cancer; race and geography (Williams and Wilkins, Baltimore, Md. 1954)
17 TAKAGI, M. and ISHIKAWA, G.: To be published.

Author's address: DR. WATURU MORI, Department of Pathology, Tokyo Medical and Dental College, *Yushima, Bunkyo-ku, Tokyo* (Japan)

The Incidence of Other Primary Tumors in Patients With Malignant Melanoma[1]

W. S. FLETCHER

Department of Surgery, University of Oregon, Medical School, Portland, Ore.

Introduction

In a recent review of 154 patients with malignant melanoma reported from the University of California at San Francisco [2], a 20% incidence of coexisting malignant neoplasms was found. Considering the fact that 46% of the patients died of their melanoma and were no longer at risk of developing a further cancer, the incidence is striking indeed.

It is surprising that there has been little or no recognition of this high incidence of second primary tumors in patients with malignant melanoma. If it is true, several possible explanations for such a finding exist. First, the observation could be due to chance alone. Second, it might represent a bias in patient-population selection. Third, it might represent some unique environmental difference, and fourth, it could represent the fact that university hospitals tend to have more comprehensive workups and to follow their patients indefinitely. Considering the fact that the American Cancer Society estimates 25% of the population will have cancer, such follow-up might allow identification of a higher proportion of patients with double primaries.

To determine whether or not the observed 20% incidence of coexisting malignant neoplasms was valid and if it could be demonstrated in another series, a review of patients with malignant melanoma was undertaken in a separate geographic location. It encompassed 2 university affiliated hospitals and 5 private hospitals with organized tumor registries.

---

[1] This study was supported in part by the American Cancer Society, Oregon Division, and part in by the Gamma Nu Chapter of Beta Sigma Phi.

Methods and Materials

The seven hospitals surveyed represent all but two of the major medical facilities in the greater Portland, Oregon, area. Collectively, these hospitals serve as the primary care facility for a local population of roughly 1,000,000 and as the referral center for much of the Columbia River basin which adds approximately 2,000,000 patients to that number. Although high rainfall and relatively little sunshine are characteristic of the local climate, four-fifths of the total area is arid with a high number of sunny days. This latter may account for the relatively high occurrence of malignant melanoma here. Some bias may also be introduced by the fact that only the most severely afflicted patients tend to be referred out of their local area to the metropolitan area.

The hospitals surveyed were the University of Oregon Medical School Hospitals and Clinics (UOMS) – 209 patients; the Portland Veterans Administration Hospital – 156 patients; St, Vincent's Hospital – 128 patients; Good Samaritan Hospital – 116 patients; Physicians and Surgeons Hospital – 31 patients, and Emanuel Hospital – 59 patients. Figure 1 depicts graphically the numbers of patients with malignant melanoma and the numbers of patients with malignant melanoma and another primary tumor from each of these hospitals.

Results and Discussion

A predominant number of the secondary primary tumors arose in squamous epithelium but no other consistent pattern of association between types of cancer emerged. For instance, of the 28 patients reported from UOMS there were 13 multiple skin cancers, 1 leukemia, 4 sarcomas of different types, 4 carcinomas of the cervix, 2 carcinomas of the ovary, 1 each of the colon, vagina, and vulva, and 3 carcinomas of the rectum.

Roughly one-half (15) of these 28 patients developed their malignant melanoma prior to the second primary tumor.

Of these 209 patients 45 are alive without disease, 17 are alive with disease, and 15 are lost to follow-up. Five patients are dead without disease, 111 are dead with disease, 10 are dead of unknown causes, and 6 are lost to follow-up, giving a raw survival of only 36.9 %.

MOERTEL et al., in a survey of 37,580 patients in a general tumor clinic population [5] reported an incidence of second primary tumors of 5.1 %. This figure is consistent with other reported series and our own incidence of second primary tumors in a population of over 7,000 patients recorded in the UOMS tumor registry. Analysis of the data with respect to the number of expected second primary tumors using the 5.1 % figure shows that the incidence of second primary tumors in patients with malignant melanoma

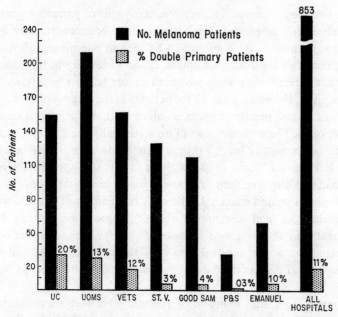

Fig. 1. Numbers of patients with malignant melanoma and those of patients with malignant melanoma + another primary tumor from the 7 hospitals concerned.

is consistently higher than the incidence found in a general tumor registry population. This is significant at the 0.01 level by the X^2 test (table I).

Several facets of this collection of data bear comment: 1. The error in reporting second primary malignant tumors in patients with malignant melanoma can only be on the low side, as the incidence was reported from

Table I

	No. melanoma patients	Expected No.[1] other cancers	Observed No. other cancers	Significance X^2
University of California	154	7.91	31	67.40[2]
University of Oregon	209	10.66	27	25.05[2]
Total of 7 Hospitals	853	43.50	94	58.62[2]

1 5.1 % occurrence of double primary tumors, MOERTEL et al. [5].
2 Highly significant (P<0.01).

documented pathology reports. This is particularly true of patients reported from private hospitals where the population tends to be more transient and there is a strong tendency for a patient to be seen in one hospital for one disease and in another for a different illness. 2. Some bias may be introduced by the fact that patients with more advanced cancer tend to be referred to university hospitals. However, when all the patients in the area are considered, the incidence of second primary tumors is still twice that of a general tumor clinic population. 3. The author knows of no statistically valid technique for projecting the occurrence of second primary tumors in patients who are no longer at risk. However, many of the patients in this study were relatively young and, had they survived their first cancer, would presumably be at great risk of developing a second cancer. 4. The early recognition of patients with malignant melanoma and the application of appropriate therapy has improved greatly in the past two decades. Thus, for stage I melanoma properly treated the 5-year survival rate is of the order of 78%, and for patients with stage II disease the 5-year survival rate is of the order of 38% [3].

Accordingly, increasing numbers of patients who have survived their melanoma are alive and at risk of developing a second tumor. Until such time as further evidence may become available, a first melanoma should be thought of as the red flag of carcinogenesis and these patients should be followed at regular intervals with comprehensive cancer screening examinations.

Figure 2 serves as a useful conceptual model to consider some of the known factors which have to do with whether a given patient with cancer survives his disease or not. As may be seen, the current concept of carcinogenesis implies, first, a genetic predisposition; second, a carcinogenic stimulus, and third, a latent period of undetermined length.

The fact that there are 'melanoma families' has been documented [1]. Whether this is due to truly genetic or environmental influences is not yet ascertained. There is strong circumstantial evidence that malignant melanoma is a tumor of viral etiology [6] as shown by the presence of cross-reacting antibodies in patients with the disease as compared to the presence of these antibodies in normal controls and in the families and close associates of the patients. What the latent period may be for the development of a malignant melanoma is unknown but the fact that the disease is frequently seen in people in their twenties and thirties, suggests that it may not be of long duration.

Factors which influence the course of the disease once established, are slightly better known than factors that influence the etiology. At the present

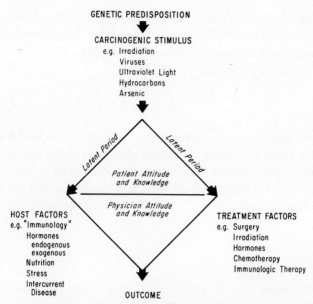

Fig. 2. Conceptual model of factors influencing survival in patients with cancer.

time, surgery offers the best, if not the only, chance of cure. Response rates with chemotherapy or immunotherapy approach 20%, but cures are rarely effected.

Isolated bits of information are beginning to fall into place regarding 'host resistance' factors which affect the outcome of the disease. For instance, it is known that in approximately 8% of the patients with malignant melanoma the primary tumor itself disappears spontaneously [7] and several spontaneous regressions of advanced cancer have been documented [8]. It has further been determined that antitumor antibody titers fall in patients with progressive disease, and that in at least some of these patients 'blocking antibodies' can be demonstrated [4]. The exact role of stress, hormones, pregnancy, irritation, nutrition, and the like is difficult to document scientifically; however, there are few experienced oncologists who do not feel that these factors are of significance in the outcome of this disease in at least some patients.

Although all the pieces of the malignant melanoma puzzle have not yet been identified, enough of them are in place to indicate the general outline of the picture. At this time it appears that whatever genetic, carcinogenic, and host resistance factors are at work to produce the malignant melanoma are apparently also at work to produce other cancers.

Until such time as more specific associations can be identified, all patients with malignant melanoma must be considered to be at high risk of developing other primary cancers and should be followed accordingly.

Conclusion

It is concluded that whatever genetic, environmental, or immunologic factors combine to cause malignant melanoma also cause other malignant neoplasms. Therefore, patients with malignant melanoma should be followed very carefully, not only for recurrence of malignant melanoma, but for the development of other cancers.

References

1 ANDERSON, D. E.; SMITH, J. L., Jr. and MCBRIDE, C. M.: Hereditary aspects oj malignant melanoma. J. Amer. med. Ass. *200:* 741–746 (1967).
2 FRASER, D. G.; BULL, J. G., Jr. and DUNPHY, J. E.: Malignant melanoma and coexisting malignant neoplasms. Amer. J. Surg. *122:* 169–174 (1971).
3 GOLDSMITH, H. S.; SHAH, J. P. and KIM, D.-H.: Prognostic significance of lymph node dissection in the treatment of malignant melanoma. Cancer *26:* 606–609 (1970).
4 HELLSTROM, I.; HELLSTROM, K. E.; EVANS, C. A.; HEPPNER, G. H.; PIERCE, G. E. and YANG, J. P. S.: Serum mediated protection of neoplastic cells from inhibition by lymphocytes immune to their tumor specific antigens. Proc. nat. Acad. Sci. *62:* 362 (1969).
5 MOERTEL, C. G.; DOCKERTY, M. B. and BAGGENSTOSS, A. H.: Multiple primary neoplasms. 1. Introduction and presentation of data. Cancer *14:* 221–230 (1961).
6 MORTON, D. L.; MALMGREN, R. A.; HOLMES, E. C. and KETCHAM, A. S.: Demonstration of antibodies against human malignant melanoma by immunofluorescence. Surgery *64:* 233-240 (1968).
7 SMITH, J. L., Jr. and STEHLIN, J. S., Jr.: Spontaneous regression of primary malignant melanomas with regional metastases. Cancer *18:* 1399–1415 (1965).
8 SUMMER, W. C.: Spontaneous regression of melanoma: Report of a case. Cancer *6:* 1040–1043 (1953).

Author's address: Prof. WILLIAM S. FLETCHER M. D., Department of Surgery, University of Oregon Medical School, *Portland, OR 97201* (USA)

Melanoma Biology

The Research and Clinical Approach to Melanoma at The University of Texas M. D. Anderson Hospital and Tumor Institute

R. Lee Clark

The University of Texas M. D. Anderson Hospital and Tumor Institute, Houston, Tex.

From 1944 through August 1971, 2991 patients with malignant melanoma have been seen at The University of Texas M.D. Anderson Hospital and Tumor Institute. Presently six basic science and three clinical departments are conducting cooperative and independent studies on malignant melanoma.

Medical Genetics
Dr. David E. Anderson

Familial melanoma has been diagnosed in over 3% of melanoma cases at Anderson Hospital since 1968. Thirty-seven kindreds are being investigated, one with ocular and 36 with cutaneous melanoma. Significant differences have been demonstrated between familial and non-familial melanoma, familial showing a distinctly younger age distribution, significantly earlier average age at diagnosis (fig. 1a), a high frequency of multiple primary melanomas, and a significantly higher survival rate (fig. 1b), probably because it is a more superficial type of disease, and therefore, more amenable to treatment.

Genetic mechanisms appear to be complex and to involve several autosomal gene loci, some degree of dominance although not a simple mendelian trait (fig. 1c), and an extranuclear component. In these kindreds, melanoma frequently develops in pre-existing nevi of light complexions and more frequently in offspring of carrier mothers and grandmothers (14%), than from carrier fathers (7%), or from transmitting mothers whose fathers were transmitters (7%). We interpret this as transmission of a nuclear component by a male parent and possibly also a cytoplasmic factor when the transmitting parent is female, provided her mother was also a transmitter. The cytoplasmic component, although believed to enhance tumor formation, has not been identified, partially due to the dearth of melanoma specimens from patients with affected mothers as opposed to those with affected fathers.

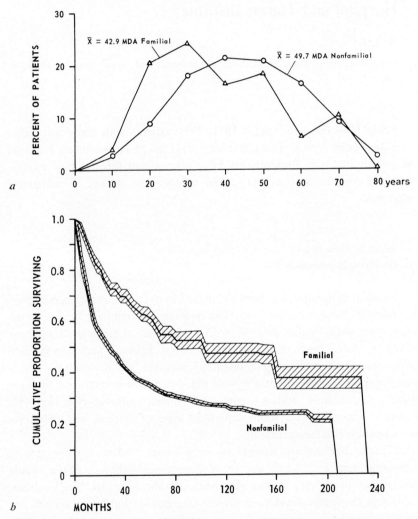

Fig. 1. a Age at first diagnosis; familial and nonfamilial melanoma; *b* Survival after diagnosis in months; familial and nonfamilial; *c* Pedigree of family with hereditary malignant melanoma.

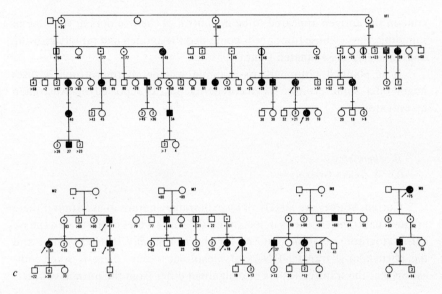

Virology

Dr. JAMES M. BOWEN and Dr. CHARLES M. MCBRIDE
(Collaborators: Dr. LEON DMOCHOWSKI and Dr. EVAN HERSH)

Preliminary studies with electron microscopy have thus far failed to reveal evidence of virus particles in malignant melanoma. The indirect immunofluorescence test on sera of melanoma patients, however, reveals that the sera of some patients contain antibodies reacting to an antigen associated with autologous or homologous tumor cell nucleoli.

A frozen section of melanoma tissue treated with normal control serum before staining with conjugate demonstrates a lack of dye in nucleoli. A frozen section from the same melanoma stained with the patient's own serum and conjugate demonstrates intense staining of abnormally large nucleoli in virtually all tumor cells, indicating a positive reaction.

After 10 subcultures, cells from a different melanoma demonstrate that the nucleolar antigen persists. Approximately 40 to 50% of those patients tested have the antigen but do not produce serum antibodies against it; 15 to 20% of patients have antibodies against their own tumor cell nucleoli. The nucleolar antigen appears to be a protein complexed in some manner with nucleolar RNA.

There is a striking correlation between the presence of the nucleolar antigen and the clinical status of the patient. Patients whose cells did not contain the antigen appeared to be in control of their tumors at the time of sampling; those whose tumor cells contained the antigen had rapidly growing and/or widely disseminated tumor.

It is planned ultimately to examine whether dermatotropic viruses might offer some promise as oncolytic agents, as reported by Australian investigators.

Biochemistry
(Dr. A. CLARK GRIFFIN)

Current studies are largely of an exploratory nature to ascertain if there are unusual transfer-RNAs in melanomas. From studies of the reverse-phase chromatographic elution patterns for the isotopically labeled tyrosyl and the phenylalanyl transfer-RNAs of melanoma tissue, there is some indication that the transfer-RNAs of melanoma differ from the normal counter-

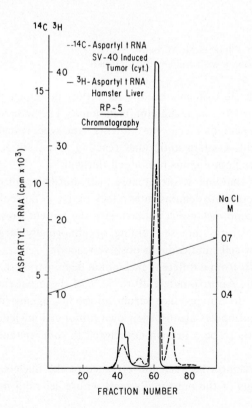

a

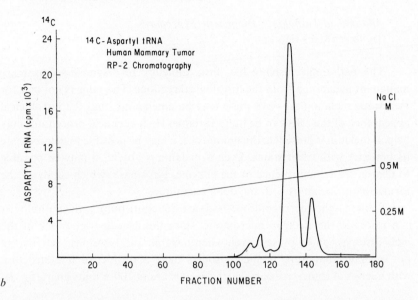

Fig. 2. a Aspartyl t-RNA chromatography curves in normal hamster liver tissue and in hamster SV-40 induced tumor; *b* Aspartyl t-RNA chromatography curves in human mammary tumor.

parts. These 2 amino acids were selected because of their role in the formation of melanin. It is difficult, however to obtain sufficient quantities of homogenous melanoma tissue to produce the quantity of transfer-RNA required for the 7 to 8-foot chromatography columns. Consequently, these studies are considered to be preliminary.

During the past few months, however, a new chromatography column, only 18 in. in length, has been developed, requiring only 10 to 20% of the previously required quantity of transfer-RNA, and is a more sensitive testing procedure. Studies may now progress at a more rapid pace.

In many cell lines, altered transfer-RNAs may result from viral infection. For normal hamster liver tissue and experimentally induced tumors of chemical origin, our reverse-phase chromatography studies reveal only 2 peaks of radioactive aspartyl transfer-RNA. There appears to be a third peak for aspartyl transfer-RNA from virally induced hamster tumors (SV-40) (fig. 2a). Tentative conclusions are that this third peak may have some direct relation to viral oncogenesis.

The two human tumors tested thus far – a mammary carcinoma (fig. 2b) and a malignant melanoma – have each shown this third peak.

Anatomical Pathology: Diagnostic Procedures
(Dr. Bruce Mackay)

The pathologist usually has little difficulty in diagnosing a primary malignant melanoma from the histological sections. The same is not true for metastatic melanomas, since these may be amelanotic, and the histological appearance of the cells can be quite variable. Histochemical procedures may help. If melanin is present in the tumor cells, it may be possible to demonstrate its presence with silver stains. Even if melanin is absent, it may be possible to demonstrate the presence of the enzyme, tyrosinase (which catalyzes the formation of melanin) with the dopa technique.

If these light microscopic methods are not contributory, the pathologist can resort to the electron microscope, since the protein framework of the melanosomes can be present even when melanin and tyrosinase activity are not evident. Immature premelanosomes have a distinctive fine structure with a series of transverse bars which are only about 100 Å units apart (fig. 3).

Anatomical Pathology
(Dr. J. Leslie Smith, Jr.)

Anatomical pathologists and clinicians have designed a study to investigate the possible prognostic value of the correlation of cytological and histopathological changes of primary malignant melanoma with biological behavior. One result of the study has been documentation of a series of histological changes which occur during spontaneous regression of primary lesions. Evidence indicates that patients with metastatic melanoma, in the absence of a primary lesion, have usually had spontaneous regression of the primary tumor. To date, no consistent relationship between biological behavior and cell type, degree of pleomorphism, nuclear or nucleolar characteristics, degree of mitotic activity, and degree of pigmentation has been demonstrated with light microscopy.

The degree of chronic inflammatory response invoked by the primary lesion is believed to be a reflection of host-tumor immune relationship and is probably responsible for differences in biological behavior of lesions. The inflammatory response is the histological feature of importance in tumor regression and is related to tumor cell degeneration.

Further studies leading from these observations are: 1. whether the degree of inflammation can be used to predict future behavior of the disease

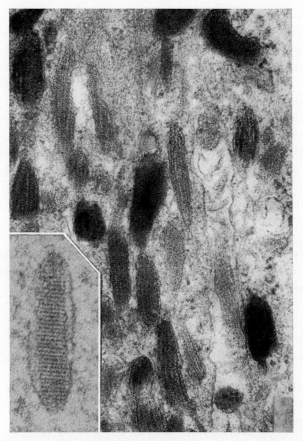

Fig. 3. Electron microscopic photo of premelanosomes in metastatic melanoma tissue. x 16,000 (detail x 115,000).

in individual patients; 2. establishment of a workable and uniformly applicable staging classification to correlate macroscopic morphology with behavior.

Surgery (Experimental)
(Dr. MARVIN M. ROMSDAHL)

Basic biological and physiological studies of several strains from one human malignant melanoma cell line involve: a) the inhibition of tyrosinase activity of highly pigmented strains with low pigment-producing cell lines, appearing to indicate that pigmentation may be more closely correlated with the presence or absence of a specific soluble inhibitor protein than with tyrosinase. Preliminary studies indicate this is a low molecular weight protein with relatively rapid electrophoretic mobility; b) cell kinetic studies that reveal greatly reduced pigment production when melanoma cells are in exponential growth. This and other evidence suggests 1. cells restrict lesser priority functions in favor of propagation requirements, 2. pigmentation probably occurs primarily in the stationary phase of cell division (G_1 and G_0); c) tyrosinase activity can be demonstrated in human serum by a highly sensitive technique using acrylamide gel disc electrophoresis. Serum values are significantly elevated in a majority of patients with advanced and/or disseminated malignant melanoma, although serum from patients with other types of cancer have values in the same range as that of normal individuals (fig. 4). Studies are being done to confirm the belief that the failure of the serum of some patients with advanced disease to show elevated values is probably due to differences in the soluble and insoluble fractions of tyrosinase *in vivo*.

Surgery (Clinical)
(Dr. CHARLES M. MCBRIDE and DR. A. J. BALLANTYNE)

Treatment of head and neck melanoma is still largely surgical because of the difficulties of perfusion and infusion. Cooperative studies with the

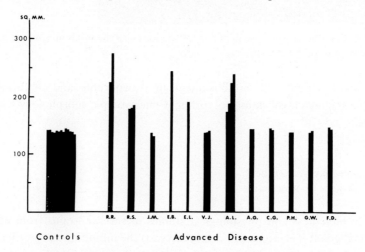

Fig. 4. Serum tyrosinase in patients with malignant melanoma.

Department of Developmental Therapeutics in the effectiveness of antitumor agents for recurrence or distant metastases and in immunological and immunotherapeutic techniques are being done.

Isolation perfusion has been the primary treatment for melanoma of the extremities since 1958. Until 1962, phenylalanine mustards, prepared in prophylene glycol, were used. In 1962, water-soluble 1-phenylalanine mustard dihydrochloride (PAM), or Melphalan, became available and has been used on 240 patients for the treatment of 301 malignant melanomas of the extremities. Comparative figures for isolation perfusion and traditional surgery reveal an increase of 5-year survival rates of approximately 11 % in the perfusion group (table I). In addition, the number of skin grafts and amputations has been reduced, and the development of intransit metastases was markedly reduced. Operative mortality was less than 1 % and the average hospital stay was 8 days.

Developmental Therapeutics
(Cell Kinetics, Chemotherapy and Pharmacology)
(Dr. JEFFERY A. GOTTLIEB)

Intensive chemotherapy of melanoma is currently underway. Pilot studies with imidazole carboxamide, alone and in combination with BCNU (bis-chloroethyl-nitrosourea) and vincristine, were further pursued in cooperation with the Southwest Cancer Chemotherapy Study Group, where a 16 to 25 % complete and partial remission rate was seen. Pilot studies with intra-arterial imidazole carboxamide have been encouraging, particularly in the control of local disease. Camptothecin, imidazole carboxamide mustard, and procarbazine have also come under recent investigation. Hydroxyurea, bleomycin, and adriamycin have initially proven ineffectual

Table I. Raw 5-year survival rates for patients having conventional surgical therapy as compared with patients having perfusion therapy

Stage	Conventional %	Perfused %
I	71	86
II	38	57
III	18	24
IV	10	8
Over-all	51 %	62 %

Table II. Chemotherapeutic agents used in treatment of metastatic malignant melanoma

Agents	Patients studied	% response
Imidazole carboxamide (DIC)	165	19
BCNU, vincristine and DIC	133	23
Procarbazine and DIC	64	33
DIC, intra-arterial (local only)	20	55
Hydroxyurea	21	14
Camptothecin	15	13
Cytosine arabinoside	15	13
Imidazole mustard	20	10
Adriamycin	14	7
Bleomycin	11	0

and no current studies are planned. Pharmacological and chemotherapeutic studies in conjunction with studies of the biological behavior of melanoma are currently underway. Numerous new antitumor agents are being explored and treatment for melanoma metastatic to the central nervous system is receiving intensive investigation. Cell-cycle times in disseminated melanoma have been determined and alterations of these by chemotherapeutic agents are being studied (table II).

Developmental Therapeutics (Immunology)
(Dr. EVAN HERSH)

In close cooperation with the Department of Surgery, a multifaceted approach to immunology and immunotherapy is being conducted. The overall objectives are: 1. to define the relationship between the immunocompetence of the patient and his prognosis; 2. to determine the number of patients who have specific tumor immunity and the relationship of this immunity to prognosis; and 3. to develop methods for the immunotherapy of patients with malignant melanoma.

Immunocompetence can be determined in several ways:

1. The *in vitro* blastogenic response of the patient's lymphocytes to the plant mitogen, phytohemagglutinin, can be measured by H^3-thymidine incorporation after 3 to 5 days of culture and correlates with cell-mediated immunocompetence. The blastogenic response is suppressed by chemotherapy. Response was measured before, 1 day, and 9 days after a 5-day course of chemotherapy. Those patients who had a recovery, rebound and overshoot of this lymphocyte reactivity had a relatively good prognosis.

Those patients whose lymphocyte reactivity remained suppressed after chemotherapy had a poor prognosis (fig. 5a).

2. Delayed-hypersensitivity skin testing can be used to evaluate cell-mediated general immunocompetence, as effective tumor immunity is also

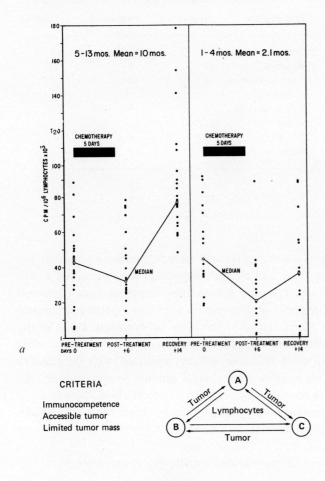

Fig. 5. a In vitro blastic responses to phytohemagglutinin and survival after chemotherapy; b Design and results of adoptive immunotherapy.

Table III. Relationship between chemotherapy, delayed hypersensitivity and response to therapy in cancer patients

Skin test results		Response to therapy		
Pre-therapy	Post-therapy	Progression	Regression	No change
Negative	Positive	0	11	1
Positive	Negative	19	0	1
Positive	>100% increase in diameter	0	10	17
Positive	>50% decrease in diameter	2	0	3
Positive	No change	1	4	6
Negative	Negative	15	1	0
All improved		0	21	18
All not improved		37	5	10

cell mediated. Antigens which can be used include candida, dermatophytin, dermatophytin-O, varidase, and mumps. Responses were measured before and after chemotherapy. Pre- and post-therapy responses were compared (table III). Delayed-hypersensitivity responses of some patients improved after therapy, while others remained the same or worsened. Post-therapy positive responses and tumor regression paralleled in some patients, while non-response or worsened response usually coincided with progression of tumor growth. This indicates that general immunocompetence for cell-mediated immunity is necessary for response to cancer chemotherapy.

So far, two approaches to immunotherapy have been used. Each shows promising preliminary results.

Cross-Grafting and Cross-Transfusion (Adoptive Immunotherapy) (fig. 5b)

Criteria for selection of patients were a) immunocompetence; b) accessibility of tumor for excision; c) limited tumor mass. Patients were treated in sets of three. Tumor grafts were implanted from A to C; C to B; and B to A. Three weeks later, immune lymphocytes were transfused from graft recipient to donor. The procedure was well tolerated. There was 1 objective and 4 subjective responses in 12 patients (one was not evaluable).

Table IV. Local immunotherapy with activated lymphocytes in metastatic melanoma tumor nodules

Data category	Tumors injected with			
	Activated Lymphocytes	Non-activated Lymphocytes	Saline	None
Number tumors injected	29	18	13	15
% tumors regressing	93	39	7	7
% tumors progressing	3.5	45	72	86
% stable	3.5	16	21	7
Mean % volume change	−64	+95	+166	+135

Local Immunotherapy (table IV)

Patients' lymphocytes were activated *in vitro* with phytohemagglutinin, then inoculated directly into the metastatic tumor nodules. In almost every instance, striking tumor regression was observed in the injected nodules. Tumor progression was observed after injection of non-activated lymphocytes, saline, or in those nodules to which nothing was done. This demonstrates that lymphocytes of patients with advanced disease are still capable of exerting antitumor effect.

Plans for the future in immunotherapy include a) active specific immunization with tumor antigens and tumor cells; b) non-specific immunotherapy with BCG; c) combination surgery-immunotherapy and chemotherapy-immunotherapy.

Epidemiology
(Miss ELEANOR J. MACDONALD)

Cooperative studies with Drs. DOUGLAS GORDON and NEVILLE DAVIS of Queensland, Australia, are in progress to determine the influence of humidity and climate by latitude and sun exposure on skin melanoma incidence.

The data tabulated from 6 large areas of the State of Texas, covering 7 degrees of latitude and 4 million people, indicate an increase in skin melanoma incidence in Caucasians as one approaches the equator. This increased incidence appears to be more related to latitude and high humidity than to the number of sunshine hours. Figure 6 demonstrates the incidence to be 3.9/100,000 population in El Paso, where there are the highest possible

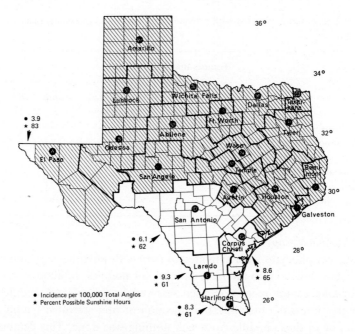

Fig. 6. Melanoma in Texas by latitude and possible sunshine hours.

sunshine hours and zero humidity, and 9.3/100,000 population in Laredo, where there are fewer possible sunshine hours but higher humidity. The latter incidence figures are similar to the intracoastal incidence rates of Queensland, Australia.

Pediatrics
(Dr. JORDAN R. WILBUR)

The current treatment of primary retinoblastoma is basically unchanged. In most cancer treatment centers, enucleation of the first affected eye is done; if tumor appears in the other eye, radiotherapy is given in an attempt to eradicate the tumor and preserve useful vision. However, in some centers, radiotherapy is being used on the initially involved eye. Newer techniques of local control, used when advisable, include photo-coagulation therapy and cryotherapy.

Metastatic retinoblastoma, particularly to the central nervous system, has been uniformly fatal. We are currently attempting to treat a patient with a complex combination chemotherapy and radiotherapy program. Prior to this treatment, there were a large number of retinoblastoma cells in the spinal fluid. Presently, the patient is without evidence of disease four months after treatment.

Author's address: Dr. R. LEE CLARK, The University of Texas M. D. Anderson Hospital and Tumor Institute, *Houston, TX 77025* (USA)

Naevi of Childhood

A. J. Cochran and K. M. Cochran

University Department of Pathology, Western Infirmary and Division of Medicine, Victoria Infirmary, Glasgow

Introduction

The greatest part of the extensive literature on pigmented tumours is understandably devoted to malignant melanoma. However, the infrequency of this tumour prior to puberty [1, 5, 6] and the relative frequency of the enigmatic juvenile melanoma [1, 5, 7] in this age group suggest that further studies of the pigmented tumours of childhood may be rewarding. Two main components might be involved in the low aggressiveness of these tumours in children; factors inherent in the tumour and factors concerned with the individual's reactivity to the tumour. We have examined the histology of naevi from children in an attempt to identify promising areas for further study of the biology of naevi.

Materials

The histology of all pigmented skin tumours received at the Department of Pathology, The Royal Hospital for Sick Children, Glasgow between 1963 and 1968 was reviewed. A *proforma* listing features of potential interest was prepared and all slides examined for these points. Clinical and follow-up information was obtained and related to the pathological features. Counts of basal layer clear cells in the epidermis over and adjacent to tumours were performed as described previously [3].

Results

A. Clinical Features

There were 60 patients in the group from whom 62 tumours had been removed. Detailed clinical information was obtained for 50 patients, 24

Table I. The age of children with different types of pigmented tumour

Diagnosis	Total	Age			
		0–3	3–6	6–9	9–12
Compound naevus	45[1]	6	9	14	14
Juvenile melanoma	11[1]	3	3	1	2
Junctional naevus	3	2	0	0	1
Intradermal naevus	3	1	1	1	0
All tumours	62	12	13	16	17

1 Exact age of 2 CN and 2 JM patients not known.

Table II. The site distribution of the pigmented tumours

Lesion		Site of tumours			
	No.	Head and neck	Trunk	Upper limb	Lower limb
Compound naevus	42	10%	23%	3%	6%
Juvenile melanoma	11	5%	3%	1%	2%
Junctional naevus	3	0%	2%	0%	1%
Intradermal naevus	3	1%	2%	0%	0%

males and 26 females. Table I shows the age at presentation of the group as a whole and subdivided by histological diagnosis. The frequency of compound naevi increased with age. By contrast the majority of juvenile melanomas occurred before 6 years of age. Compound naevi were most common on the trunk and the area of the head and neck, but for juvenile melanoma the order was reversed (table II). The most frequent reasons for presentation were concern about the lesion itself (22 out of 50) and real or imagined growth of the tumour (22 out of 50). By contrast with the situation in malignant melanoma, bleeding was an infrequent presenting feature (2 out of 50).

B. Histological Features

Sixty two tumours were available. There were 45 compound naevi, 11 juvenile melanomas, 3 junctional naevi and 3 simple intradermal naevi. The juvenile melanomas could be subdivided into 6 with spindle cell morphology

Table III. A comparison of histological features of the cells of compound naevi and juvenile melanomas

Frequency of features / total number of tumours	Compound naevi	Juvenile melanoma
Nuclear and cytoplasmic pleomorphism	2/45	7/11
Delayed maturation	10/45	11/11
Giant cells	9/45	5/11
Mitotic figures	3/45	3/11

and 5 with epithelioid cell morphology. In view of the small number of junctional naevi and intradermal naevi available these will not be considered further. We have subdivided the histological features into those inherent to the tumour cells and those relating to the interaction of tumour cells and surrounding tissues.

a) Factors Inherent to the Tumour Cells (Table III)
1. Pleomorphism

The cells of compound naevi were with only two exceptions very uniform. By contrast, 7 out of 11 juvenile melanomas had cells which showed marked nuclear and cytoplasmic pleomorphism.

2. Delayed Maturation

In the majority of compound naevi, once naevus cells had lost contact with the nests of junctional activity they rapidly altered in morphology and 'matured'. They altered from moderate sized round to oval cells with a clear 'cytoplasm' on conventional staining to a rather unvarying population of small uniform cells with a limited amount of pale eosinophilic or ambophilic cytoplasm and a small dense basophilic nucleus (fig. 1).

The 'normal' maturation pattern was seen in 35 out of 45 compound naevi. In the remaining 10 maturation did occur but there was a band of immature cells of intermediate morphology between the epidermis and the deeper mature layer. These naevi we have called delayed maturation naevi (fig. 2). Maturation was not seen in any of the 11 juvenile melanomas.

3. Giant Cells

Multinucleate giant cells were seen in 5 out of 11 juvenile melanomas and 9 out of 45 compound naevi.

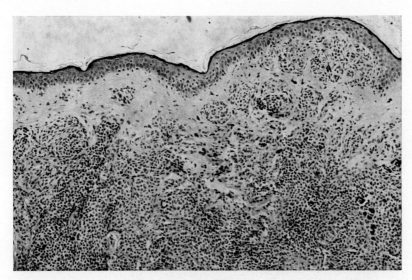

Fig. 1. 'Normal' maturation of compound naevus. Cells convert to naevus cell morphology immediately below nests of junctional activity. Haematoxylin and eosin. x 125.

4. Mitoses

Mitoses were distinctly uncommon, occurring in 3 out of 45 compound naevi, 3 out of 11 juvenile melanomas and in 1 out of 3 junctional naevi. All were normal bipolar mitoses.

b) Interaction of Tumour Cells and Adjacent Tissues (Table IV)

1. Epidermal Ulceration and Invasion

Ulceration was present in 3 out of 11 juvenile melanomas but in none of the 45 compound naevi. A small increase in suprabasal clear cells was seen in 8 out of 45 compound naevi and a considerable increase in these cells was present in the 3 ulcerated juvenile melanomas.

2. Abnormalities of Vascular Pattern

Telangiectatic blood vessels were noted in 5 out of 11 juvenile melanomas, usually in the zone between the intradermal part of the tumour and the epidermis. Similar vessels were present in only 2 out of 45 compound naevi.

3. Lymphoid Cell Infiltration

Moderate to considerable numbers of small lymphocytes were seen in association with 10 out of 11 juvenile melanomas but around only 8 out of 45

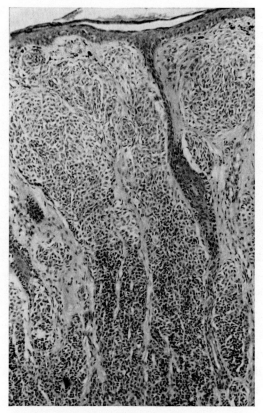

Fig. 2. Delayed maturation naevus. There is a zone of cells of intermediate size and morphology between the level of the nests of junctional activity and the fully developed naevus cells. Haematoxylin and eosin. x 125.

compound naevi. These cells were confined to the deep periphery of the compound naevi, but had infiltrated between tumour cells in the juvenile melanoma although still maximally concentrated in the tumour periphery. Plasma cells were less frequent but occurred in 6 out of 11 juvenile melanomas and 1 out of 45 compound naevi. The plasma cells were always associated with lymphocytes. While lymphoid cells were present in the 3 ulcerated juvenile melanomas, they were also prominent in 7 tumours which had intact covering epidermis.

4. Tumour Edge

The tumour edge was sharply defined in 15 out of 45 compound naevi and 5 out of 11 juvenile melanomas but 'infiltrating' tumour cells were

Table IV. A comparison of some histological features of compound naevi and juvenile melanomas

Frequency of feature / total number of tumours	Compound naevi	Juvenile melanomas
Epidermal ulceration	0/45	3/11
Suprabasal clear cells	8/45	3/11
Abnormal vascular patterns	4/45	5/11
Lymphocyte infiltration	8/45	10/11
Plasma cell infiltration	1/45	6/11
Tumour edge:		
a) sharply defined	15/45	5/11
b) infiltrative	30/45	6/11
Lateral junctional activity	8/45	1/11
Melanocyte pattern A	2/45	0/11
B	25/45	11/11
C	8/45	0/11

present separate from the main tumour mass in 30 out of 45 compound naevi and 6 out of 11 juvenile melanomas.

5. Lateral Junctional Activity and Melanocyte Patterns

Nests of junctional activity were seen peripheral to the epidermis over the main tumour mass in 8 out of 45 compound naevi but in only 1 out of 11 juvenile melanomas.

We have previously described three patterns of basal layer clear cell distributions in the epidermis around melanocytic tumours [4]: pattern A, symmetrical increase of melanocytes; pattern B, no increase, and pattern C, asymmetrical increase. In this study 26% of compound naevi were pattern A, 57% pattern B and 17% pattern C. The juvenile melanomas all had a normal frequency [3] of melanocytes in the surrounding epidermis.

C. Follow-up Studies

Thirty patients were followed for periods which varied from several months to nine years. None of these patients developed local recurrences or metastatic tumours.

Discussion

The distribution of the different histological categories of tumour is similar to that described previously in children [5]. The high frequency of lesions showing junctional activity (95%) is identical to the figure cited by McWhorter and Woolner and reflects the growth of these tumours during childhood.

The characteristic appearance of juvenile melanomas is mainly due to nuclear and cytoplasmic pleomorphism, the formation of giant cells and a total absence of maturation to the normal naevus cell morphology. We consider the concept of maturation failure worthy of further study. In most compound naevi the change from melanocyte morphology and dimensions to those of the naevus cell occurs rapidly when contact with the zone of junctional activity is lost. In a minority of compound naevi there is a delay in this process which leaves a zone of cells intermediate in morphology and position between melanocytes and naevus cells, the delayed-maturation naevus. McWhorter and Woolner indicated that naevi of this type in their series had been previously misinterpreted as malignant. It is possible that the spectrum of appearances from compound naevus through delayed maturation naevus to juvenile melanoma is due to the presence or absence of agents influencing maturation of the melanocytes to naevus cells. A consideration of the nature of factors active in such a system is beyond the scope of this paper but agents akin to the melanocyte chalone [2] might be relevant. It is difficult to know whether malignant melanoma is explicable on this basis, but in view of the widely different clinical behaviour of juvenile melanoma and malignant melanoma, factors other than lack of maturation must be important in determining whether or not malignancy supervenes.

In addition to the obvious variations in tumour cell morphology between compound naevi and juvenile melanoma, there are a number of other histological differences between the tumours. Ulceration was uniformly absent in compound naevi but occurred in association with a considerable increase in suprabasal clear cells in 3 out of 11 juvenile melanomas. An increase in suprabasal clear cells was noted in association with 8 out of 45 compound naevi but the increase was never more than moderate. Abnormal dilated vascular channels which have been described previously as typical of juvenile melanoma [7] were noted in 5 of 11 such tumours in this series but were also seen in a few compound naevi.

Histological evidence of a host attack on tumour cells might include lymphocyte and plasma cell infiltration, peritumoral fibrosis, evidence of cell necrosis and free pigment around the tumour. We sought, but found no evidence of the last three features. Lymphocyte infiltration was seen in almost all juvenile melanomas and also in a few compound naevi. Plasma cells were less frequently seen but occurred mainly in juvenile melanomas and were always accompanied by lymphocytes. The antigens which may have evoked this accumulation of lymphoid cells are not known but in the absence of infection it is arguable that they are reacting to tumour-associated antigens. If this is the case, then it appears that the juvenile melanoma which is morphologically so much more like malignant melanoma may be more 'antigenic' than the compound naevis.

While dense peritumoral fibrosis was not seen, a majority of both compound naevi and juvenile melanomas showed tumour cells separated from the main tumour by collagenous strands, producing an appearance of 'infiltration'. It would be interesting to know whether this is a result of tumour cell motility or perhaps represents a slow attrition of the periphery of the tumour by fibrosis. It is of interest that in end-stage intradermal naevi individual cells and groups of cells are often sequestrated by dense bands of collagen.

The occurrence of lateral junctional activity in some compound naevi and the rarity of abnormalities of melanocyte frequency, morphology and distribution in the epidermis peripheral to juvenile melanomas confirms previous observations [4]. The frequency of the different patterns of melanocyte distribution lateral to compound naevi is very similar to that encountered in a previous study of compound naevi of adults [4]. Compound naevi of childhood are apparently associated with abnormalities of melanocyte distribution in the surrounding epidermis which are virtually identical to those associated with compound naevi in adults.

Acknowledgements

We are obliged to Drs. A. M. McDonald and A. A. M. Gibson, Pathology Department, Royal Hospital for Sick Children, Glasgow for encouragement and practical assistance. Alistair J. Cochran was Peel Medical Research Trust Travelling Fellow during part of the study.

References

1 Allen, A.C. and Spitz, S.: Juvenile melanomas of children and adults and melanocarcinomas of children. Arch. Derm. 82: 325–335 (1960).

2 BULLOUGH, W. S. and LAURENCE, E. B.: Melanocyte chalone and mitotic control in melanomata. Nature, Lond. *220:* 137–138 (1968).
3 COCHRAN, A. J.: The incidence of melanocytes in normal human skin. J. Invest. Derm. *55:* 65–70 (1970).
4 COCHRAN, A. J.: Studies of the melanocytes of the epidermis adjacent to tumours. J. Invest. Derm. *57:* 38–43 (1971).
5 MCWHORTER, H. E. and WOOLNER, L. B.: Pigmented naevi, juvenile melanomas and malignant melanomas in children. Cancer *7:* 564–585 (1954).
6 SKOV-JENSEN, T.; HASTROP, J. and LAMBRETHSEN, E.: Malignant melanoma in children. Cancer *19:* 620–625 (1966).
7 SPITZ, SOPHIE: Melanomas of childhood. Amer. J. Path. *24:* 591–609 (1948).

Author's address: Dr. ALISTAIR J. COCHRAN, University Department of Pathology, Western Infirmary, *Glasgow* (Scotland)

Relationship of Lymphocytic Infiltration to Prognosis in Primary Malignant Melanoma of Skin

P. G. THOMPSON

Department of Plastic and Jaw Surgery, Frenchay Hospital, Bristol.

Introduction

It is well recognised that in virtually every primary malignant melanoma of skin the malignant cells are accompanied by a round-cell infiltration consisting predominantly of lymphocytes [6]. This infiltrate occurs beneath even the earliest lesions, usually in the basal region of the tumour, and in advanced lesions is generally more marked at the margins. It is variable in intensity; some melanomas show an intense lymphocyte infiltration extending right through the lesion, others a moderate infiltrate of the basal layer alone, whilst in some the zone of lymphocyte reaction is confined to the margins of the main tumour mass while the invasive base of the lesion has no obvious reaction. In 1907 SAMPSON HANDLEY [10] commented on this lymphocyte infiltration and noted that it seemed to be related to areas of regression in a malignant melanoma.

In the past decade there has been a rapid increase in the understanding of the mechanisms of host defence to malignant disease, and of the mechanisms. The cellular immune response, mediated by thymus dependent lymphocytes, appears to play a key role.

Regression of malignant melanoma is well known [3]. It may be seen as regression of part or whole of a primary lesion, or as regression of some of the skin nodules in slowly progressive cases with multiple skin metastases. Regression also occurs in benign melanocytic lesions, e.g. Sutton's halo naevus and benign juvenile melanoma of Spitz. In each of these instances regression is accompanied and appears to be mediated by round-cell infiltration of the tumour [12].

Tissue culture experiments clearly demonstrate that lymphocytes from the peripheral blood of patients with malignant diseases are cytotoxic to cells from their own tumour [7]. If a very large ratio of lymphocytes to target cells is cultured together a cross reaction is demonstrated against cells from tumours of the same histological type arising in different patients [11].

Lymphoid infiltration of the primary lesion has already been correlated with improved prognosis in several types of malignant tumour, e.g. carcinoma of the breast (2), carcinoma of the stomach [1], seminoma [8], neuroblastoma [13], and choriocarcinoma [9]. It would therefore seem reasonable to see whether the same relationship exists in malignant melanoma.

Method

In order to determine the influence of the degree of lymphocyte infiltration on the prognosis in melanoma a review was made of 219 consecutive cases of primary malignant melanoma of the skin entered on the Malignant Melanoma Registry of the Department of Plastic Surgery at Frenchay Hospital, Bristol during the years 1960 to 1964.

Of these cases 131 were analysed. They were all lesions which had advanced beyond early invasion of the dermis, but which had not spread beyond the regional lymph nodes on presentation. All the cases included had received adequate surgical treatment, the histology was available for review, and follow-up was complete for at least 5 years after primary treatment.

Eighty eight of the original 219 cases were excluded from the analysis for the reasons shown in table I.

The histological preparations of the primary lesions were examined and classified according to degree of lymphocyte infiltration before the outcome of each case was determined from the records.

The cases were divided into three groups according to lymphocytic response:

1. *Marked:* those cases in which there was an intense lymphocyte infiltration of the whole lesion or of the whole of its base forming a complete basal zone of infiltration.

2. *Intermediate:* those cases with a moderate and readily recognisable lymphocyte infiltration, but in which a group of cells in the tumour or an invasive portion of its base

Table I. Cases excluded from analysis

Melanoma confined to epidermis	12
Melanoma with an early invasion of dermis (Clarke Level II [5])	16
Disease advanced beyond regional nodes on presentation	9
Treatment inadequate or non-standard	6
Histology of primary lesion not available	20
Follow-up inadequate	7
Died before 5 years from cause other than malignant disease	18
Total	88

appeared to elicit little or no lymphocytic reaction and the basal 'barrier' of round cells appeared to be incomplete.

3. *Slight:* those cases in which a major portion of the tumour and its base had little or no lymphocytic reaction or infiltration.

Results

The results of follow-up at 5 years are summarised in table II.

It can be seen that both the recurrence rate and the mortality rate is three times greater in those cases with only slight lymphocyte response than it is in those with a marked lymphocyte response in the primary tumour. This relationship persists regardless of age, sex, site of the primary lesion, cell type, depth of invasion, rate of growth or ulceration of the primary.

In several cases which were slowly progressive but with multiple recurrences there was a lymphocyte infiltration around or beneath these recurrent lesions. In biopsies taken some weeks after non-excisional treatment of some of these lesions, e.g. by diathermy or regional cytotoxic perfusion, there was an increase in the numbers of round cells composing this infiltration. It could be postulated either that the treatment reduced the total tumour mass to such an extent that the available immune response to the residual malignant cells became more apparent, or that the malignant cells damaged by the treatment stimulated the immune mechanism by release of tumour antigen in more effective quantities. When serial biopsies were available in cases with multiple recurrences, it was noted that the degree of lymphocyte reaction decreased as the disease became advanced, and absence of any lymphocyte reaction usually heralded rapid progression of the disease. However, at this stage it was usually apparent on clinical grounds that the balance had turned in favour of the malignancy.

Table II. Summary of results showing correlation of prognosis with degree of lymphocyte infiltration

Lymphocyte infiltration	Total No. of cases	Deaths from melanoma	% deaths	Had recurrence treated but alive	%
Marked	36	5	14	3	8
Intermediate	43	15	37	6	14
Slight	52	28	54	10	19

Discussion

It seems most probable that the degree of lymphocyte infiltration of the lesion represents a measure of the effectiveness of the hosts cellular immune defence against malignant melanoma, or if BURNET's concept of immunological surveillance [4] is accepted, the absence of lymphocytes is a measure of the degree of failure of this mechanism allowing emergence of a clinical malignancy.

It is clear that there are other factors which influence the prognosis. Some deaths do occur in cases with an apparently vigorous immune response, and few surgeons would take all the credit for saving those cases which survive despite an apparently poor cellular immune response. The overall results of the analysis can be shown in a graph such as figure 1 in which the

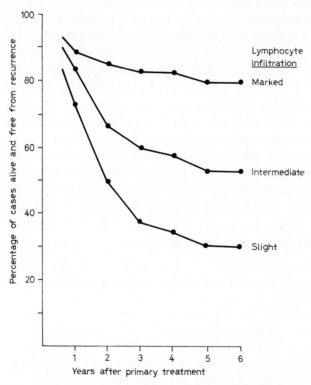

Fig. 1. Graph demonstrating percentage of cases remaining alive and free from recurrence during the 9 years after primary treatment in each classification of lymphocyte response.

percentage of cases alive and without recurrence are plotted each year. If then cases which are known to have a better than average prognosis, i.e., in this series those tumours arising on legs and face, are separated from the remainder and the outcome of this group plotted in a similar fashion (fig. 2) the results in both groups in the categories of slight and intense lymphocyte response remain much the same but the improved prognosis in the good risk group is found to occur in those with moderate lymphocyte infiltration; that is to say, a lesser degree of observed immune response is effective in these good risk cases, whilst in the bad risk cases only the maximum immune response can alter the outcome, indicating perhaps that there are other factors influencing the prognosis.

The reality of a cell-mediated host defence against spontaneous human malignancy is now accepted but the effectiveness of this mechanism and the

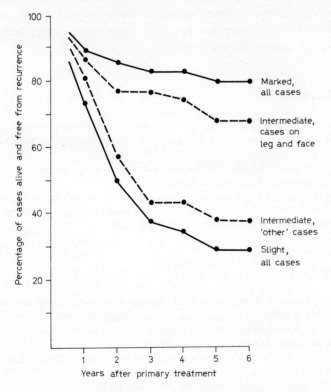

Fig. 2. Graph demonstrating the improved prognosis in the intermediate 'grade' of lymphocyte response occurring in a group of cases known to have a better outcome for other reasons.

precise role it can play once a malignancy has become established is far from clear. Attempts at boosting the immune response by specific or non-specific means have had little therapeutic success in the management of advanced solid tumours and it is possible that it can be effective only when very small numbers of malignant cells are involved. Viewed in this light it would seem that the surgeon's task, or indeed the task of any form of cytoreductive therapy, in the treatment of the patient with primary or recurrent disease, is to reduce the total number of viable malignant cells sufficiently to tip the balance back in favour of the immunological surveillance mechanism.

Conclusion

As in some other tumours prognosis can be correlated with the extent of lymphocytic infiltration in primary cutaneous malignant melanoma, and the degree of this infiltration seems to represent a measure of the effectiveness of the hosts defence against the malignancy.

Acknowledgements

I am grateful to Mr. D. C. BODENHAM for his stimulation and encouragement in this investigation and in allowing use of the Malignant Melanoma Registry at Frenchay Hospital.

Also to Dr. O. C. LLOYD for advice and access to the histological material on file at the Department of Pathology, University of Bristol.

This work was supported by an anonymous benefaction for research in the Department of Plastic Surgery at Frenchay Hospital. The Malignant Melanoma Registry is supported by funds from the British Empire Cancer Campaign.

References

1. BLACK, M. M.; OPLER, S. R. and SPEER, F. D.: Structural representation of tumour – host relationship in gastric carcinoma. Surg. Gynec. Obstet. *102:* 599–603 (1956).
2. BLOOM, H. J. G.; RICHARDSON, W. R. and FIELD, J. R.: Host resistance and survival in carcinoma of breast. Brit. med. J. *iii:* 181–187 (1970).
3. BODENHAM, D. C.: A study of 650 observed malignant melanomas in the South-West region. Ann. roy. Coll. Surg. Engl. *43:* 218–239 (1968).
4. BURNET. McF.: Immunological surveillance (Pergamon Press, 1970).
5. CLARKE, W. H.; FROM, L.; BERNADINO, E. A. and MIHM, M. G.: The histogenesis and biologic behaviour of primary malignant melanomas of skin. Cancer Res. *29:* 705–726 (1969).

6   COUPERUS, M. and RUCKER, R. C.: Histopathological diagnosis of malignant melanoma. Arch. Derm. Syph. *70:* 199–206 (1954).
7   CURRIE, G.A.; LEJEUNE, F. and FAIRLEY, G.H.: Immunisation with irradiated tumour cells and specific lymphocyte cytotoxicity in malignant melanoma. Brit. med. J. *ii:* 305–310 (1971).
8   DIXON, F. J. and MOORE, R.A.: Testicular tumours, a clinico-pathological study. Cancer *6:* 427–438 (1953).
9   ELSTON, C.W.: Cellular reaction to chorioncarcinoma. J. Path. Bact. *97:* 261–268 (1969).
10 HANDLEY,W.S.: Pathology of melanotic growths in relation to their operative treatment. Lancet *i:* 927–933 (1907).
11 HELLSTROM, I.; HELLSTROM, K. E.; SJOGREN, H. O. and WARNER, G.A.: Demonstration of cell-mediated immunity to human neoplasms of various histological types. Int. J. Cancer *7:* 1–16 (1971).
12 LLOYD, O. C.: Regression of malignant melanoma as a manifestation of cellular immunity response. Proc. roy. Soc. Med. *62:* 543–545 (1969).
13 MARTIN, R. F. and BECKSMITH, J. B.: Lymphoid infiltration in neuroblastoma: their occurrence and prognostic significance. J. paediat. Surg. *3/ Suppl.:* 161–163 (1968).

Author's address: Dr. P. G. THOMPSON, Department of Plastic and Jaw Surgery, Frenchay Hospital, *Bristol* (England)

Macromolecular Pathology of Pagetoid Melanoma

Y. MISHIMA and M. MATSUNAKA

Department of Dermatology, Wakayama Medical University, Wakayama-shi, Japan.

Introduction

Malignant melanoma is not a single entity but can be divided into two biologically and clinically distinct entities developing through naevocytic and melanocytic ontogeny [6, 7]. Compared with the malignant melanoma developing from junctional naevi, melanoma developing from Dubreuilh's melanosis is less malignant judged by growth rate, metastasis and invasiveness [8]. Dubreuilh's precancerous melanosis in contrast to the junctional naevus occurs later in life in exposed areas and has a high incidence of malignant transformation. It also differs histologically and cytologically from the junctional naevus. Typical Dubreuilh's precancerous melanosis (fig. 1) exhibits active proliferation of melanocytes having the appearance of atypical clear cells without the formation of theques at the epidermal dermal junction [8]. In contrast, junctional naevi have a tendency to form well-circumscribed cell nests in the lower epidermis [9]. The cytoplasm of the individual naevus cell is homogeneous with an oval or cuboidal, distinctly outlined perikaryon. The dopa reaction of Dubreuilh's melanosis, reveals the melanocytic nature of these cells by their intensely positive dendritic appearance. Further, the cells of Dubreuilh's melanosis invade the dermis only after malignant transformation acquiring intense tyrosinase activity, while junctional naevus cells generally become intradermal naevus cells by the 'dropping off' process, gradually losing their tyrosinase activity and increasing cholinesterase activity [9]. Electron microscopy [10] also reveals the melanocytic rather than naevocytic nature of the cells of Dubreuilh's melanosis. Dubreuilh's melanocytes (fig. 2) proliferate in a disorganized fashion among the keratinocytes without the formation of distinct theques. Another sub-

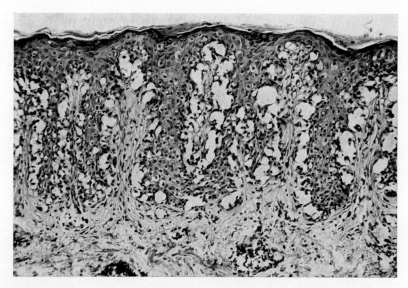

Fig. 1. Typical histological characteristics of Dubreuilh's precancerous melanosis exhibiting the proliferation of atypical clear cells with a honey-combed appearance which are aligned in a palisade fashion at the epidermal-dermal junction. H–E, x 120.

cellular difference between Dubreuilh's melanosis and junctional naevus is found in melanosome polymorphism. Melanosomes synthesized in melanocytes of well developed lesions of Dubreuilh's melanosis appear rod-shaped averaging 500 × 250 mμ in maximum diameter when cut through their short axis. They appear as round bodies or ring-like in partially melanized granules averaging up to 500 mμ when cut parallel to their long axis. Three-dimensionally, these melanosomes can be considered to be shaped like red blood cells. In addition to melanosomes of this type, the earlier stages of Dubreuilh's melanosis contain many elongated melanosomes which possibly represent a carry-over from normal melanosome synthesis. On the other hand, melanosomes synthesized by junctional naevus cells are short football-shaped having a maximum average size of 500 × 150 mμ and they also have a different internal structure. Following our reports [6, 7, 10], CESARINI et al [1], KLUG and GUNTHER [4] and others [3, 11] reported their observation essentially similar to ours. Not only on the basis of melanosome polymorphism but on the basis of our accumulated comprehensive findings seen in biological, clinical, cellular, subcellular as well as outgenic characteristics, I previously proposed [3,4] that malignant melanoma is not a single entity but is composed of two biologically and clinically separate entities arising from naevocytes and

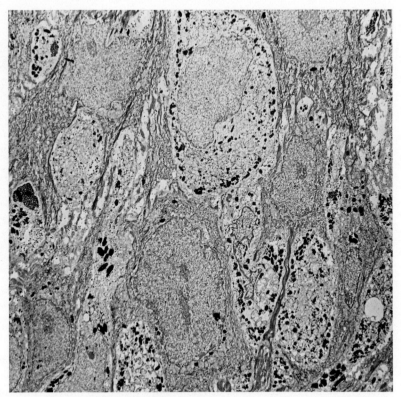

Fig. 2. Electron-microscopic features of Dubreuilh's precancerous melanosis, revealing melanocytic rather than naevocytic nature by the disorganized proliferation of dendritic cells without the formation of theques. Fixed with O_sO_4 and stained with phosphotungstic acid (PTA) and lead citrate. x 3100.

melanocytes and, therefore, may be called malignant naevocytoma and malignant melanocytoma respectively. The differences in malignancy, radiosensitivity, and prognosis, together with differences in biological behavior, may be explained by the separate ontogenesis of these two conditions. The neural crest differentiates to melanoblasts and further into melanocytes which can undergo initial neoplastic changes resulting in lentigo senilis. Lentigo senilis can assume premalignant changes, becoming Dubreuilh's precancerous melanosis, later generally becoming transformed into less invasive malignant melanocytoma. The neural crest can also give rise to the bipotential naevoblast, which can form lentigo simplex leading further to junctional naevus. When junctional naevi undergo malignant change the result is the highly invasive malignant naevocytoma.

Recently, pigmented lesions, clinically resembling Dubreuilh's precancerous melanosis but differing from it by a tendency to occur in both exposed and unexposed areas, and by their worse prognosis, have been reported under the name of pagetoid melanoma by MCGOVERN [5], or superficially spreading malignant melanoma by CLARK and his associates [2]. However, the subcellular and enzymic characterization of pagetoid melanoma in naevocytic or melanocytic differentiation has not yet been established.

In 1960 [8], I first reported the histological criteria to differentiate Dubreuilh's precancerous melanosis from active junctional naevus. Since then, Dr. H. PINKUS and I occasionally have encountered a number of cases which clinically resembled Dubreuilh's precancerous melanosis, but which occurred in unexposed areas and, unlike Dubreuilh's melanosis, histologically resembled junctional naevi by the presence of distinct junctional theques. Furthermore, these puzzling cases usually had a worse prognosis than Dubreuilh's melanosis.

Materials and Methods

Our collected cases of malignant melanoma and of antecedent stages have been reexamined. These include 7 cases generously provided by MCGOVERN, 331 cases from Pinkus, and 94 of our own cases, totalling 432 in all.

Results and Discussion

Our pagetoid melanoma cases have been found to have clinically different characteristics from Dubreuilh's melanosis by close examination. They occurred in both non-exposed and exposed skin and mucous membranes in middle age rather than old age and had a moderately rapid growth. They were usually irregularly elevated, flat plaques of rather small size with brown hyperpigmentation and had shades of violaceous pink or rose colour, often forming arc-shaped outlines with multiple nodules (fig. 3). Pagetoid melanosis and melanoma in our series had a peak incidence in the range of 40 to 50 years of age, in agreement with the figures of MCGOVERN [5] and CLARK [2]. On the other hand, Dubreuilh's precancerous melanosis and its melanocytic melanoma had a peak incidence in the range of 70 to 80 years of age. In contrast to the predominant incidence of Dubreuilh's melanosis on the face, pagetoid melanosis has a wider distribution in both exposed and unexposed areas.

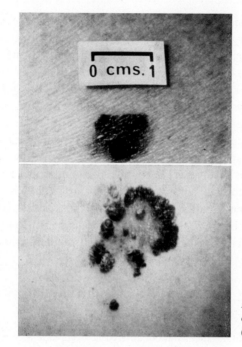

Fig. 3a and b. Clinical appearance of pagetoid premalignant melano- (a) and pagetoid melanoma (b).

We used the term pagetoid premalignant melanosis to indicate the lesion that is still confined to the epidermis. When dermal invasion occurwith the acquisition of the power to metastasize, the term pagetoid melanoma is then applied.

Dubreuilh's precancerous melanosis is characterized by irregular proliferation of atypical melanocytes whereas in pagetoid melanosis there is an intraepidermal pagetoid proliferating pattern of nest-forming large ovoid or fusiform atypical pigment cells which contain varying degrees of fine melanin granules. Compared to Dubreuilh's melanosis, pagetoid melanosis is rather similar in general configuration to junctional naevi but has no tendency to develop tumour cell nests predominantly in the rete ridges; instead, pagetoid cell nests proliferate laterally in all levels of the epidermis (fig. 4). Cells of pagetoid melanoma invading the dermis have similar morphological appearances to those seen within the epidermis. Pagetoid melanosis cells demonstrate not only an intense dopa-positive reaction but also distinctly positive tyrosinase reactions in both epidermal and dermal pagetoid cell nests. As in junctional naevi cholinesterase activity can be observed, in the intraepidermal cell nests and when invasion of the dermis occurs, the invading cells also have this activity.

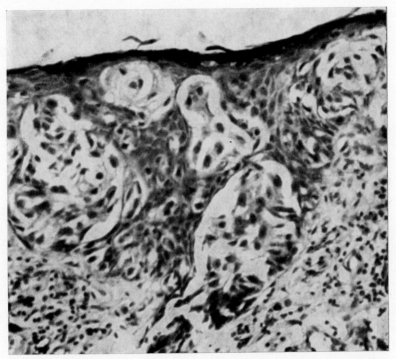

Fig. 4. Histological characteristics of pagetoid premalignant melanosis exhibiting an intraepidermal pagetoid proliferating pattern of nest-forming large ovoid or fusiform atypical pigment cells. H–E. x 530.

Electron-microscopically, Dubreuilh's precancerous melanosis (fig. 2) shows irregular proliferation of atypical dendritic melanocytes within the epidermis; pagetoid melanosis is characterized by the proliferation of closely associated ovoid or cuboidal tumour cells forming intraepidermal pagetoid nests (fig. 5). Unlike the cells of Paget's disease of the breast, these pagetoid cells completely lack desmosomes. These subcellular patterns appear to be quite similar to those of junctional naevi [10]. The cuboidal cells of pagetoid melanosis have been found to synthesize unique spheroid melanosomes containing finely granular, densely distributed melanin particles and, occasionally, a whirlpool-like internal membrane structure. In addition, tumour cells of pagetoid melanosis and melanoma occasionally exhibit round, uniformly electron-dense giant melanosomes compared to the 500 to 700 mμ size usually seen.

To account for the differences between pagetoid melanosis and Dubreuilh's melanosis, a revised concept of differentiation is required.

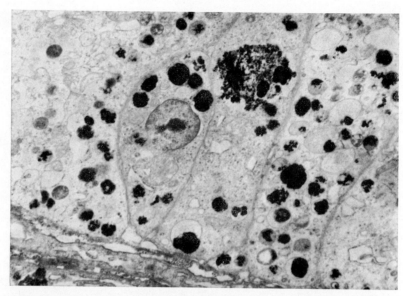

Fig. 5. Closely associated pagetoid cuboidal cells forming spheroid melanosomes and aligned on the basement membrane. Fixed with O_sO_4 and GTA, and stained with uranyl acetate and lead citrate. x 8,000.

Conclusion

Although it is difficult to make a final conclusion at present, as shown in my revised scheme of neural crest differentiation (fig. 6), the above cel-

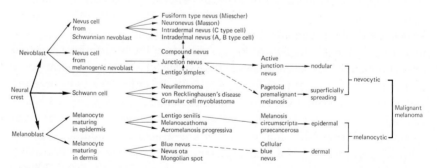

Fig. 6. New proposed ontogenies of melanotic tumors based on current findings, Benign juvenile melanoma is a variant of naevus cell naevus and similarly has junctional compound, and intradermal forms. Congenital giant nevus may develop nevocytic melanoma not only from junctional naevus cells but also from intradermal naevus cells.

lular and subcellular findings as well as clinical characteristics seem to indicate that pagetoid melanosis and melanoma macroscopically resemble Dubreuilh's precancerous melanosis and resulting melanoma at first glance, but belong to a particular type of nevocytic pigment cell tumor.

References

1 CESARINI, J.P., BONNEAU, H., et CALAS, E.: Apports de l'ultrastructure au diagnostic des mélanomes. Bull. Soc. fr. Derm. Syph., 76: 479 (1969).
2 CLARK, W.H. Jr.; FROM, L.; BERNARDINO, E.A. and MIHM, M.C.: The histogenesis and biologic behavior of primary human malignant melanomas of the skin. Cancer Res. 29: 705–726 (1969).
3 HIRONE, T., NAGAI, T. MATSUBARA, T. and FUKUSHIRO, R.: Human malignant melanomas of the skin and their pre-existing conditions. In biology of Normal and Abnormal Melanocytes. KAWAMURA, FITZPATRICK and SEIJI, (Eds.,) pp. 329–349 (University of Tokyo press, Tokyo 1971).
4 KLUG, H. and GÜNTHER, W.: Ultrastructural differences in human malignant melanomata. Brit. J. Derm., 86: 395 (1972).
5 MCGOVERN, V.J.: The classification of melanoma and its relationship with prognosis Pathology 2: 85–98 (1970).
6 MISHIMA, Y.: Melanocytic and nevocytic malignant melanomas: cellular and subcellular differentiation. Cancer 20: 632–649 (1967).
7 MISHIMA, Y.: Changes in the current concept of malignant melanoma; in J.W.H. MALI Current problems in dermatology, pp. 51–81 (Karger, Basel/New York 1970).
8 MISHIMA, Y.: Melanosis circumscripta praecancerosa Dubreuilh, a non-nevoid premelanoma distinct from junction nevus. J. Invest. Derm. 34: 361–375 (1960).
9 MISHIMA, Y.: Macromolecular changes in pigmentary disorders. III. Cellular nevi: subcellular and cytochemical characteristic with reference to their origin. Arch. Derm. 91: 536–557 (1965); ibid. 92: 393 (1965).
10 MISHIMA, Y.: Melanotic tumors; in A.S. ZELICKSON Ultrastructure of normal and abnormal skin, pp. 388–424 (Lea and Febinger, Philadelphia, Pa. 1967).
11 WAYTE, D.M.: Pathology of Nevi and Melanomas. In the Skin. HELWIG and MOSTOFI. (Eds.,) pp. 490–532 (The Williams and Wilkins Co., Baltimore 1971).

Author's address: Prof. YUTAKA MISHIMA, Department of Dermatology, Wakayama Medical University, *Wakayama-shi 640* (Japan)

Local Recurrence and Diffuse Multiple Melanoma Following Chronic Oral Administration of L-Dopa[1]

E. F. Gilbert, R. G. McCord, J. L. Skibba, J. F. Fallon, W. A. Croft and W. F. Jaeschke

University of Wisconsin, School of Medicine, Madison, Wis.

Introduction

Deficiency of dopamine in the corpus striatum of patients with Parkinson's disease has led to therapeutic attempts to restore function by replenishing dopamine [11]. Since dopamine cannot pass the blood brain barrier, its immediate precursor, levodopa (L-3, 4-dihydroxyphenylalanine), which does so readily, has been used in large sustained doses with marked functional improvement [10, 18]. While side effects have been noted [10], dermatologic reactions have not included stimulation or augmentation of the pigmentary system.

L-dopa is an intermediate in the biosynthesis of catecholamines and melanins and is largely dependent upon tyrosinase. L-dopa and other phenol and indole melanogens are excreted in the urine of patients with malignant melanoma [29]. Tyrosinase activity has been found in the serum [25] and urine [23] of melanoma patients.

The many studies of excretory melanoma metabolites offer no clue as to whether the chronic administration of L-dopa might stimulate, augment or perhaps promote the growth of a melanoma. Recently observed was a patient with Parkinson's disease who developed local recurrence of a melanoma four years after primary excision and four months after initiation of L-dopa therapy with the simultaneous appearance of multiple melanomata.

Case History

The patient was a white male who had developed Parkinson's disease in 1964 at the age of 48 years and had been treated with benztropine and procyclidine. He was admitted

[1] This work was supported in part by National Institutes of Health, General Research Support Sub-grant, Medical School, University of Wisconsin; graduate training program in pathology 5-TO1-GM-00130; and the National Science Foundation, GB 24704.

to the University of Wisconsin Hospital in May 1966 at the age of 50 for excision of a pigmented lesion on his back which had changed in character over the 6 months prior to admission. It was 1 cm in diameter, black and raised about 2 mm above the skin surface. There was no evidence of metastatic disease. A wide surgical excision was performed and split thickness skin graft was applied. Pathologic examination disclosed a nodular melanoma with invasion to the level of the subpapillary plexus of vessels (McGovern [19]; level III of Clark et al. [9]). There was epidermal invasion and focal ulceration. After discharge he continued taking benztropine and procyclidine for control of his Parkinsonism.

L-dopa therapy was initiated on September 1, 1970. The maximum dose achieved was 4.5 g daily. Four months later, black raised areas within the skin graft site in the back as well as multiple dark pigmented nodules over the trunk, extremities and scalp were noted. The pigmented nodules in the grafted area and several other lesions were excised and found to be malignant melanomas. L-dopa therapy was discontinued.

During the ensuing two months he developed further recurrence in the center of the skin graft site. He continued to develop multiple raised nodular pigmented lesions over his skin surface which gave a striking clinical appearance. He was given systemic chemotherapy with 4(5)-(3,3-dimethyl-1-triazeno)imidazole-5(4)-carboxamide (NSC 45388, DIC). Chest x-ray, bone survey, and liver function studies at this time were normal. However, he rapidly developed symptoms and signs of widespread metastatic disease and expired 10 months after the initiation of L-dopa therapy. An autopsy was not granted; however, a liver biopsy before death confirmed the presence of metastatic disease.

The multiple lesions excised from the skin grafted area of the back showed the presence of recurrent melanoma. These lesions were confined to the dermis with a clearly defined subepidermal Grenz zone. Intense pigmentation of both the cytoplasm and the nuclei of the tumor cells tended to obscure the cytological details.

Pathology

Sections from the multiple primary lesions which appeared after L-dopa therapy demonstrated evidence of malignant melanoma with dermal invasion to Level III. There was intense pigmentation, loss of cohesiveness and minimal to moderate mitotic activity. Enzyme histochemistry using L-dopa as substrate indicated high dopa oxidase (tyrosinase) activity, indicative of active melanogenesis.

Ultrastructure

Tissue for electron microscopic examination was fixed in 3% glutaraldehyde, post-fixed in 2% osmium tetroxide and embedded in epon. Thin sections were examined with an RCA EMU-3G electron microscope at 50 kV. The melanoma cells (fig. 1) contained large irregular nuclei with peripheral margination of nuclear chromatin with large granular nucleoli. The cyto-

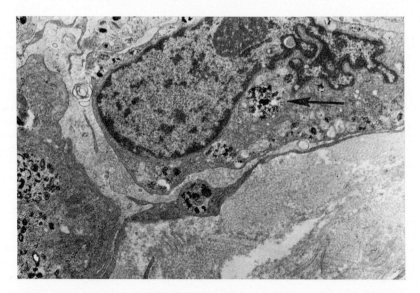

Fig. 1. Low-power electron micrograph of melanoma cell. The nucleus is irregular with peripheral margination of chromatin. Nucleolus is large and granular. Autophagosomes containing melanosomes are abundant (arrow). x 6000.

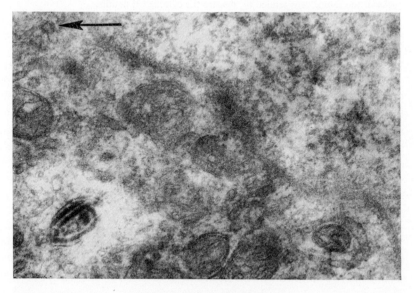

Fig. 2. Medium-power electron micrograph of a melanocyte. A virus-like particle (arrow) appears to be budding from an ER cisterna. x 56,000.

plasm contained many large polyribosomes in rosette formation as well as much fibrillar material. Mitochondria were abundant, large and of irregular shape and the Golgi apparatus was large and complex. The rough endoplasmic reticulum was extensive and dilated. Premelanosomes and melanosomes were numerous and the latter contained electron-dense accumulations of melanin. Many cells contained autophagosomes containing primarily melanosomes. Virus-like particles were observed which were round or slightly oval and measured 80–100 mμ in diameter (fig. 2). The nucleoid was unevenly dense and spherical with fine projections extending toward the limiting membrane resembling C type particles [3] (fig. 3). They appeared within the cisternae of RER and some appeared to be budding from the plasmalemmal membrane. After imidazole therapy, the virus-like particles were more abundant and found within the cytoplasm, usually within the vicinity of melanosomes and free between cells. They were also seen in some of the prickle cells of the epidermis.

Discussion

The recurrence of a malignant melanoma in this patient and the simultaneous appearance of multiple primary melanomas occurring after the initiation of L-dopa therapy has suggested a cause and effect relationship between exacerbation of this patient's disease and L-dopa administration. A distinctive characteristic feature of pigmented melanomas of human origin is the presence of the enzyme tyrosinase in a highly active form, in contrast to its relative inactive state in mammalian melanocytes [8]. An increase in tyrosinase activity is associated with proliferation of melanocytes, and is present in most malignant melanomas [17], although it has been difficult to

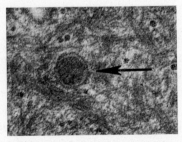

Fig. 3. High-power electron micrograph of a virus-like particle (arrow) in the cytoplasm of a prickle cell. The dense spherical nucleoid and double membrane resemble a C-type particle. x 70,000.

correlate the specificity of tyrosinase activity with type, viability, location, or the biological behavior of the tumor cells [7].

The relationship of this unique metabolic pathway, the tyrosine-tyrosinase system, to the benign and malignant proliferation of the melanocyte remains obscure, although this system appears to be important in the energy metabolism of the melanocytes. Melanin granules from mouse melanomas couple phosphorylation to the oxidation of certain Krebs cycle intermediates and to terminal electron transport [14, 24], thereby demonstrating close functional similarity to liver mitochondria. Electron spin resonance studies have shown that melanins are stable free radicals and are theoretically potent electron acceptors [21]. The facility with which melanin undergoes reversible oxidation and reduction suggests that it could function as a respiratory catalyst [28]. Studies by DEMOPOULOS [13] on pigmented human and mouse melanomas utilizing phenylalanine-tyrosine restricted diets or tyrosinase inhibitors have suggested that a large segment of vital respiration which is coupled to oxidative phosphorylation might be dependent on tyrosinase. A phenylalanine-tyrosine restricted diet has been associated with regressive changes in some cases of metastatic melanoma [27]. In addition, dopa has frequently been excreted in large amounts by patients with melanoma [29]. These studies suggest that administration of L-dopa to a patient who previously had undergone surgical excision of a malignant melanoma may enhance or stimulate growth of any residual melanoma tissue. Levodopa has been shown to inhibit extracerebral dopa decarboxylase necessary for catecholamine synthesis [22]. This may then lead to accumulation of dopa and allow uncompetitive metabolism to melanin. This patient developed a recurrent tumor in the site of previous surgical excision. The multiple primary melanomas perhaps reflect the statement by PACK et al. [22] that 'these patients have a tendency to multicentricity of malignant melanomas' which in this case might have been promoted by the administration of L-dopa.

Although the incidence of a second primary melanoma among the melanoma population is greater than the incidence of primary melanoma in the general population [6, 20], multiple primary melanomas are rare [20] with the exception of hereditary malignant melanoma [1, 2]. This patient's family history excluded him from this group. A patient with a melanoma exhibits a diathesis for the activation of junctional nevi in the skin of other parts of the body [22]; i.e., co-existent nevi of the skin of these patients may show increased pigmentation and many of these nevi show histologic evidence of very active junctional changes. CHANG et al. [7], found that the number of tyrosinase-positive nevi seems increased in patients with melanoma, especially those with disseminated metastases, but no secondary primary melanomas developed from nevi of any of the patients studied.

The patient presented could have developed a second primary coincidental with the L-dopa therapy, however, the appearance of diffuse multiple melanomas is unique in this patient.

There has been no indication that L-dopa is carcinogenic; however, adequate testing for carcinogenic properties has not been studied. The chemical structure of L-dopa lends itself to possible structural changes with conversion to an alkylating agent, this binding to nucleic acids or protein of the target cell. The role of growth hormone in carcinogenesis has been the subject of investigation. In regard to melanomas, the action of somatotrophic hormone appears to be dependent upon the synergistic effect of other hormones [26]. Of particular interest, therefore, is recent evidence demonstrating elevation of plasma growth hormone during therapy with L-dopa [4]. In this particular case the administration of large doses of L-dopa may have stimulated melanoma cells by providing a biological advantage to the growth of dormant melanocytic cells. It is of interest to note that EPSTEIN et al. [15, 16] were able to induce malignant melanomas in hamsters 3–6 weeks after injection of cell-free ultrafiltrates prepared from malignant melanoma; furthermore, particles resembling viral C-forms were seen in the ER cisternae of the melanocytes of the induced tumor. The relationship of virus-like particles to melanoma formation in experimental animals remains unknown. They may transform cells to increase growth rate and account for the paradox of very rapid growth of heavily pigmented melanomas, heavy pigment formation usually being associated with a slowly growing tumor.

Tryptophan metabolism during administration of L-dopa to patients with Parkinsonism have been shown to produce abnormal tryptophan metabolites [12]. Abnormal tryptophan metabolites have been suggested as a possible role in the genesis of human cancer [5].

Possible mechanisms, therefore, of melanoma recurrence in this case include:

a) Conversion of L-dopa to an alkylating agent (carcinogen).

b) Promotion of carcinogenesis by L-dopa or one of its metabolites

c) Induction of abnormal tryptophan metabolism which in turn may form a carcinogen.

d) Production of growth hormone with subsequent induction of malignancy.

e) Activation of a latent oncogenic virus.

All factors may have played a role in the induction or promotion of malignancy. In addition, the possibility of tumor recurrence unrelated to L-dopa administration cannot be entirely excluded.

On the basis of this, albeit single observation, it would seem prudent to use L-dopa with extreme caution, if at all, in a patient with a history of melanoma, since this compound, in a susceptible subject, may stimulate recurrence or development of further malignancy.

Acknowledgement

We are grateful for the assistance of Dr. THOMAS BARBER, Mrs. KRIS MUETZEL and Miss CYNTHIA HARRIS.

References

1 ALLEN, A. C. and SPITZ, S.: Malignant melanoma. Cancer 6: 1–45 (1953).
2 ANDERSON, D. E.; SMITH, J. L., Jr. and McBRIDE, C. M.: Hereditary aspects of malignant melanoma. J. Amer. med. Ass. 200: 741–746 (1967).
3 BERNHARD, W.: The detection and study of tumor viruses with the electron microscope, Cancer Res. 20: 712–726 (1960).
4 BOYD, D. E.; LEBOVITZ, H. E. and PFEIFFER, J. B.: Stimulation of human growth hormone secretion by L-dopa. New Engl. J. Med. 283: 1425–1429 (1970).
5 BRYAN, G. T.: The role of urinary tryptophan metabolites in the etiology of bladder cancer. Amer. J. clin. Med. 24: 841–847 (1971).
6 CARTER, A. P. and LEE, J. A. H.: Death certification of malignant melanoma: a problem in epidemiology. Amer. J. Epidemiol. 93: 77–78 (1971).
7 CHANG, J. P.; RUSSELL, W. O.; STEHLIN, J. S., Jr. and SMITH. J. L., Jr.: Chemical and histochemical analyses of tyrosinase activity in melanoma and related lesions. Ann. N.Y. Acad. Sci. 100: 951–964 (1963).
8 CHEN, Y. M. and CHAVIN, W.: Comparative biochemical aspects of integumental and tumor tyrosinase activity in vertebrate melanogenesis; in W. MONTAGNA and F. HU Advances in biology of the skin, vol. 8: The pigmentary system, pp. 253–268 (Pergamon Press, Oxford 1967).
9 CLARK, W. H., Jr.; FROM, L.; BERNARDINO, E. A. and MIHM, M. C.: The histogenesis and biologic behavior of primary human malignant melanomas of the skin. Cancer Res. 29: 705–727 (1969).
10 COTZIAS, G. C.; PAPAVASILIOUS, P. S. and GELLENE, R.: A critical review of nine years' experience. Canad. med. Ass. J. 101: 791–800 (1969).
11 COTZIAS, G. C.; VAN WOERT, M. H. and SCHIFFER, I. M.: Aromatic amino acids and modification of Parkinsonism. New Engl. J. Med. 276: 374–379 (1967).
12 COZZOLINO, G.; CAMPRIANI, S.; ALUNNO, S.; GIANNINI, R. and GIOMBINI, L.: Tryptophan/nicotinic acid metabolism during levodopa treatment of parkinsonism. Letter to editor. Lancet ii: 1042 (1971).
13 DEMOPOULOS, H. B.: Effects of reducing the phenylalanine-tyrosine intake of patients with advanced malignant melanoma. Cancer 19: 657–664 (1966).

14 DORNER, M. and REICH, E.: Oxidative phosphorylation and some related phenomena in pigment granules of mouse melanomas. Biochem. biophys. Acta 48: 534–546 (1961).
15 EPSTEIN, W. L. and FUKUYAMA, K.: Light and electron microscopic studies of a transplantable melanoma associated with virus-like particles. Cancer Res. 30: 1241–1247 (1970).
16 EPSTEIN, W. L.; FUKUYAMA, K.; BENN, M.; KESTON, A. S. and BRANDT, R. B.: Transmission of a pigmented melanoma in golden hamsters by a cell-free ultrafiltrate. Nature, Lond. 219: 979–980 (1968).
17 FITZPATRICK, T. B. and KUKITA, A.: Tyrosinase activity in vertebrate melanocytes; in M. GORDON Pigment cell biology, pp. 389–524 (Academic Press, New York 1959).
18 McDOWELL, F.; LEE, J. E.; SWIFT, T.; SWEET, R. D.; OGSBURY, J. S. and KESSLER, J. T.: Treatment of Parkinson's syndrome with L-dihydroxyphenylalanine (levodopa). Ann. intern. Med. 72: 29–35 (1970).
19 McGOVERN, V. J.: The classification of melanoma and its relationship with prognosis. Pathology 2: 85–98 (1970).
20 McLEOD, R.; DAVIS, N. C.; HERRON, J. J.; CALDWELL, R. A.; LITTLE, J. H. and QUINN, R. L.: A retrospective survey of 498 patients with malignant melanoma. Surg. Gynec. Obstet. 126: 99–108 (1968).
21 MASON, H. S.; INGRAM, D. J. E. and ALLEN, G.: The free radical property of melanins. Arch. Biochem. Biophys. 86: 225–230 (1960).
22 PACK, G. T.; SCHARNAGEL, I. M. and HILLYER, R. A.: Multiple primary melanoma. Cancer 5: 1110–1115 (1952).
23 RICHTERICH, R.; CANTZ, B. and DAUWALDER, H.: Increased urinary dihydroxyphenylalanine oxidase activity in patients suffering from malignant melanoma. Clin. chim. Acta 29: 295–301 (1970).
24 RILEY, V.; HOBBY, G. and BURK, D.: Oxidizing enzymes of mouse melanomas: their inhibition, enhancement, and chromatographic separation; in M. GORDON Pigment cell growth. Proc. 3rd Conf. on Biology of Normal and Atypical Pigment Cell Growth, pp. 231–266 (Academic Press, New York 1953).
25 SOHN, N.; GANG, H.; GUMPORT, S. L.; GOLDSTEIN, M. and DEPPISCH, L. M.: Generalized melanosis secondary to malignant melanoma: report of a case with serum and tissue tyrosinase studies. Cancer 24: 897–903 (1969).
26 STARR, K. W.: in I.M. ARIEL Progress in clinical cancer, vol. 4, p. 6 (Grune and Stratton, New York 1969).
27 Tate, S. S.: SWEET, R.; McDOWELL, F. H. and MEISTER, A.: Decrease of the 4,4-dopa decarboxylase activities in human erythrocytes and mouse tissue after administration of dopa. Proc. nat. Acad. Sci. 68: 2121–2123 (1971).
28 VAN WOERT, M. H.: Reduced nicotinamide-adenine dinucleotide oxidation by melanin: inhibition by phenothiazines. Proc. Soc. expt. Biol. Med. 129: 165–171 (1968).
29 VOORHESS, M. L.: Urinary excretion of dopa and metabolites by patients with melanoma. Cancer 26: 146–149 (1970).

Author's address: Dr. ENID F. GILBERT, University of Wisconsin, School of Medicine, *Madison, Wis.* (USA)

Biochemistry of Melanoma

Metabolism of L-Dopa-3-^{14}C (3,4-Dihydroxyphenylalanine) in Human Subjects[1]

McC. GOODALL

University of Texas Medical Branch, Galveston, Tex.

Introduction

DL-dopa was first synthesized by FUNK in 1911 [3]. However, it was not until 1950 that L-dopa, along with dopamine (3,4-dihydroxyphenylethylamine), was shown to occur naturally in mammalian tissue [4]. It is currently being used in the treatment of parkinsonism [1]. L-dopa and tyrosine are the principle precursors of melanin; however, dopamine, noradrenaline and adrenaline are also precursors to melanin but to a lesser degree [2, 7–9].

Methods

General

Three normal healthy males were infused with 100 µc of 3,4-dihydroxyphenylalanine-3-^{14}C (L-dopa-3-^{14}C). The radioactive L-dopa was dissolved in 1000 ml of physiological saline and infused at a constant rate for 4 h. The urine was collected at the end of the infusion, at 2 h, 4 h, 8 h, and 24 h and thereafter daily for the next 4 days. The urine samples were stored at −20 °C until assayed.

Isolation and Quantitation of Metabolic Products of 3,4-Dihydroxyphenylalanine (L-Dopa)

The details of this procedure have been previously described [5]. In brief, the method used to separate the acidic metabolites is as follows: an aliquot of urine containing 100,000 dpm along with carrier compounds were placed on a 1 × 35 cm column of Dowex-1-X2 acetate anion exchange resin. The column was attached to a specially

---

[1] This research was supported by the Hoffmann-La Roche Foundation.

designed flow monitoring system and eluted with water followed by a variable gradient elution consisting of ammonium acetate buffers of varying molarity and acidity. In this flow monitoring system, the column eluate first passed through a UV spectrophotometer where the optical density was measured and recorded on one channel of a dual recorder. From the UV spectrophotometer the eluate then passed through a scintillation counter where the radioactivity was monitored and recorded on the second channel of the dual recorder. The scintillation counter was instructed by a volume measuring device to digitally record the total counts accumulated in each preset volume collection and punch tape the same information for computer use. In this manner a continuous integration was performed during the course of the column elution. The punch tape was fed into a computer which was programmed to calculate and print out the percentage of radioactivity of each metabolite in terms of the amount of radioactivity infused and in terms of the collection period.

To separate the basic metabolites, an aliquot of urine containing 500,000 dpm was placed on a 30 × 0.9 cm column of Dowex-50-X4 resin along with appropriate carrier compounds. The procedure used for separation of the basic metabolites was quite similar to that of the acidic metabolites except that the resin and elution patterns were different.

Results

Of the infused L-dopa-3-$^{-14}$C, $71.6 \pm 3.1\%$ of the radioactivity was recovered in 24 h and 80.6 ± 3.4 in 120 h. Approximately 11% of the radioactivity was recovered as metabolic products of dopa, 64% as dopamine or metabolic products of dopamine, and 5% as noradrenaline or metabolic products of noradrenaline. There were no detectable amounts of radioactive adrenaline or direct metabolic products of adrenaline. The remaining 20% of the infused radioactivity was unaccounted for (see fig. 1).

A total of 35 radioactive products were separated from the urine and of these 16 were identified, 2 tentatively identified and the rest unidentified. The identified products represented 65.7% of the total radioactivity recovered in 120 h. Most of the identified products were either related to noradrenaline or were metabolic products of dopamine [5].

The principle metabolic products of dopamine were 3,4-dihydroxyphenylacetic acid (DOPAC) and 3-methoxy-4-hydroxyphenylacetic acid (homovanillic acid). Other dopamine metabolic products were 3-methoxytyramine, 3,4-dihydroxyphenylethanol, 3-methoxy-4-hydroxyphenylethanol and conjugates of dopamine, 3-methoxy-4-hydroxyphenylethanol, homovanillic acid and DOPAC plus several unidentified products.

The largest single direct metabolic product of L-dopa was 3-methoxytyrosine. There were several unidentified radioactive products of L-dopa [5].

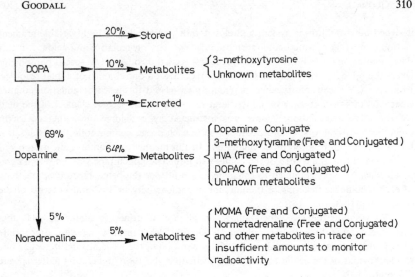

Fig. 1. Distribution of the intravenously infused L-dopa-3-^{14}C and its metabolites at 120 h post infusion.

Discussion

In that 69% of the infused L-dopa was converted to dopamine (64%) and noradrenaline (5%), this would mean that the circulating L-dopa was taken up by various tissues including the dopaminergic neurons, sympathetic system, kidney, liver, heart, etc. and rapidly decarboxylated to dopamine. Why such a large percentage of the infused L-dopa-3-^{14}C was converted to dopamine can best be explained on the basis of the ubiquity and activity of the enzyme dopa decarboxylase [6, 10]. Since dopa decarboxylase is found in most tissues and since it rapidly decarboxylates dopa to dopamine it would then seem that the infused L-dopa was systemically exposed to decarboxylation; consequently, dopamine was rapidly formed.

It is generally agreed that vertebrate melanin is a polyquinone and that the precursors of these quinones are specific phenols or catechols. The most commonly cited precursors of melanin are tyrosine and L-dopa and to a lesser degree dopamine, adrenaline and noradrenaline [2, 7–9]. Further, the enzymes tyrosinase and dopa-oxidase are explicably involved in the formation of melanin. From the L-dopa-3-^{14}C infusion experiments, 20% of the infused radioactivity was unaccounted for and presumably was distributed through the dopaminergic-sympathetic nervous system, in the epidermis as melanin and in other tissues. Therefore, it is reasonable to conclude that of this 20% of unaccounted-for radioactivity, at least a part was synthesized into melanin. Certainly, the previous work of HEMPEL using labeled dopa in mice with melanomas [7] and that of DUCHON and others in the identification of 5-hydroxy-6-methoxyindole-2-carboxylic acid and 5-methoxy-6-hydroxyindole-2-carboxylic acid in the urine of melanoma patients supports the tyrosine-L-dopa pathway to the formation of melanin [2, 9, 11].

References

1 BARBEAU, A. and MCDOWELL, F. H. (eds.): L-dopa and parkinsonism (F. A. Davis, Philadelphia, Pa. 1970).
2 DUCHON, J.: Identification of 2 new metabolites in melanoma urine: 5-hydroxy-6-methoxyindole-2-carboxylic and 5-methoxy-6-hydroxyindole-2-carboxylic acids. Clin. chim. Acta *16:* 397–402 (1967).
3 FUNK, C.: Synthesis of dl-3:4:dihydroxyphenylalanine. J. chem. Soc. *99:* 554 (1911).
4 GOODALL, McC.: Dihydroxyphenylalanine and hydroxytyramine in mammalian suprarenals. Acta chem. scand. *4:* 550 (1950).
5 GOODALL, McC. and ALTON, H.: Metabolism of L-dopa (3,4-dihydroxyphenylalanine) in human subjects. Biochem. Pharmacol. ([In press] 1972).
6 HAKANSON, R. and OWMAN, C.: Pineal dopa decarboxylase and monoamine oxidase activities as related to the monoamine stores. J. Neurochem. *13:* 597–605 (1966).
7 HEMPEL, K.: Investigation on the structure of melanin in malignant melanoma with H^3 and C^{14}-dopa labeled at different positions; in Structures and control of the melanocyte, pp. 162–175 (Springer, Berlin 1966).
8 KAWAMURA, T.; FITZPATRICK, T. B. and SEIJI, M. (eds.): Biology of normal and abnormal melanocytes (University Park Press, Baltimore, Md. 1971).
9 KOPF, A. W. and ANDRADE, R. (eds.): 1967–1968 Yearbook of Dermatology (Year Book Medical Publishers, Chicago, Ill. 1968).
10 SOURKES, T. L.: Dopa decarboxylase: substrates, coenzyme, inhibitors. Pharmacol. Rev. *18:* 53–60 (1966).
11 SWAN, G. A.: Current knowledge of melanin structure. Proc. VIIIth Int. Pigment Cell Conf., Sydney, Australia 1972, pp. 151–157 (Karger, Basel/New York 1972).

Author's address: Dr. MCCHESNEY GOODALL, University of Texas Medical Branch, *Galveston, Tex.* (USA)

Precursors of Melanin in the Urine and 3,4-Dihydroxyphenylalanine in the Blood of Patients with Malignant Melanoma

H. HINTERBERGER, A. FREEDMAN and R. J. BARTHOLOMEW

Division of Clinical Chemistry, Department of Pathology, Prince Henry Hospital and the New South Wales State Cancer Council's Special Unit for Investigation, Prince of Wales Hospital, Sydney

Introduction

Malignant melanoma, like the pigment cell from which it arises, may synthesise melanin. Although the amount of pigment elaborated by the tumour tissue varies from case to case, colourless precursors, melanogens, are frequently found in the urine of patients harbouring such tumours. The melanocyte and the tumour cell derived from it are capable of converting phenylalanine and tyrosine received through the circulation, into 3,4-dihydroxyphenylalanine (dopa) [3]. The pigment cell shares this biosynthetic pathway with sympathetic neural and adrenal medullary tissue so that dopa is the precursor of melanin as well as of the catecholamines.

In the pigment cell, dopa is converted to a polymer by concerted oxidation, cyclisation and coupling, then this complex molecule is incorporated into the melanoprotein by a series of reactions yet to be elucidated. Only minute quantities of precursors of melanin and their metabolites normally appear in the urine, but considerable quantities of melanogens may be excreted by patients in advanced stages of malignant melanoma. Demonstration of increased urinary levels of melanogens has been used as a confirmatory test for malignant melanoma for many years.

The Thormählen test as adapted for quantitative use by PECHAN [8] and applied to assessment of melanogenuria by DUCHON and PECHAN [1] measures indoles with an unsubstituted pyrrole moiety (fig. 1). DUCHON and MATOUS [2] further demonstrated the presence in urine of derivatives of indole-2-carboxylic acids and their methyl ethers. Increased urinary excretion of dopa was first reported by SCOTT [9], TAKAHASHI and FITZPATRICK

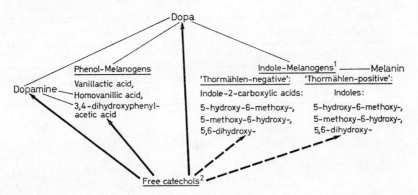

Fig. 1. The relation of 'free catechols' to other urinary melanogens. Includes the free compounds, their glucuronides and sulphate esters; solid arrows refer to components which have been identified, broken arrows to possible components.

[10], and more recently by VOORHESS [11]. For the past 9 years we have studied a group of urinary metabolites related to melanin precursors, which we called 'free catechols' to distinguish them from nitrogen-free phenol-melanogens [1, 8], and have attempted to relate urinary levels of these melanogens to tumour mass in patients in different stages of malignant melanoma.

Materials and Methods

Free catechols were isolated from urine by passage through alumina at pH 8.5 and were eluted with dilute acid, together with the noradrenalin and adrenalin [4]. This separation procedure ensures adsorption of compounds containing a catechol, i.e., phenol moiety with 2 vicinal hydroxy groups. Separation of free catechols on Dowex-50 ion exchange resin showed that dopa, DA and DOPAC are present regularly, while minor, as yet unknown, components occur infrequently, and then only in the urine of patients in advanced stages of the disease. These components are believed to be 5,6-dihydroxyindole-2-carboxylic acid and/or 5,6-dihydroxyindole. Dopa and DA were estimated fluorimetrically after oxidation with sodium metaperiodate and rearrangement to the dihydroxyindole [5]. DOPAC was extracted with ether and estimated fluorimetrically as described by MELLINGER and HVIDBERG [7]. In figure 1 arrows point to those components of 'free catechols' which occur regularly, broken arrows to possible components the chemical nature of which is still to be elucidated.

Table I. Urinary free catechols, dopa, DA and DOPAC in various clinical stages of malignant melanoma

Stage	Mean ± S.E.M.			
	Free catechols µg per 24 h	Dopa	DA	DOPAC
0	122±17 n = 19	–	–	–
I	146±18 n = 19	41±17 n = 4	127±15 n = 4	1213±367 n = 4
II	238±26 n = 25	170±65 n = 4	78±19 n = 4	304±117 n = 3
III	316±40 n = 23	393±120 n = 8	63±18 n = 8	900±197 n = 8
IV	1347±434 n = 27	1076±576 n = 10	275±102 n = 10	4889±2837 n = 9
Controls[1]	87±47 n = 100	6 0–24 n = 6	78 42–130 n = 6	600 0–780 n = 6

1 Controls for free catechols were hypertensive patients, but excluding those with phaeochromocytoma and related tumours. Controls for dopa, DA and DOPAC from 6 healthy laboratory workers.

Results

Table I lists urinary free catechols of 114 patients grouped according to tumour mass (for criteria of grouping see HINTERBERGER *et al.* [4]). Minor increases were observed in patients in stage I and II and significant increases in stage III and IV when compared with the excretion values in patients after removal of a primary malignant melanoma (stage 0). The dependence of urinary free catechols on tumour mass was confirmed by observation of patients over prolonged periods: the amount of free catechols in urine increased in parallel with the spread of the disease.

Separate estimation of dopa, DA and DOPAC (table I) showed significant increases in the urinary levels of the latter 2 metabolites only in stage IV, defined as disseminated melanoma with hepatic metastases. It could be speculated that dopa formed in hepatic tumour deposits could be metabolised along the catecholamine pathway by adjoining liver tissue. This assumption is supported by recent experience in the treatment of patients with Parkinson's disease with high oral doses of L-dopa: a large proportion of the

Table II. Concentration of dopa in plasma and blood cells of patients with various stages of malignant melanoma

Stage		Dopa	
		Plasma ng/ml	Cells ng/ml
I	Mean	1.37	2.93
n = 3	Range	0–3.3	0–7.5
II	Mean	8.3	13.83
n = 3	Range	0–17.9	0–40.70
III	Mean	20.7	25.33
n = 6	Range	0.9–61.0	1.7–95.0
IV	Mean	80.76	52.18
n = 9	Range	1.6–499.0	0–248.0
Controls	Mean	0.0	0.1
n = 6	Range	–	0–0.6

aminoacid is converted to DA and the latter further metabolised to DOPAC and other metabolites during passage of the drug through the alimentary canal and portal system [5, 6].

Levels of dopa in the blood of 21 patients with various stages of malignant melanoma are summarised in table II. Separate measurements were made in plasma and blood cells because equilibration between these two pools occurs relatively slowly [5]. Although the levels varied widely, even among patients at the same clinical stage, a general trend towards higher blood concentrations was recognizable in the more advanced cases of melanoma, in which they reached levels comparable to therapeutic levels observed in patients with Parkinson's disease treated with L-dopa [5]. Not unexpectedly, there appeared to be no obvious correlation between blood levels of dopa and urinary excretion of free catechols since the former test measures the momentary state of dopa secretion while the latter gives insight into the overall output during 24 h. Further investigations into factors which may influence the levels of dopa in blood are now in progress.

We concluded from these observations that estimation of free catechols in urine provides a useful confirmatory test for melanotic malignant melanoma at stages II to IV of the disease. Similarly, assay of dopa in blood may provide evidence for increased production of this amino acid. Both tests are useful in following the course of the disease and in assessing the efficacy of treatment.

References

1 DUCHON, J. and PECHAN, Z.: The biochemical and clinical significance of melanogenuria. II. In RILEY and FORTNER (Ed.), The Pigment Cell, Ann. N.Y. Acad. Sci. *100:* 1048–1068 (1963).
2 DUCHON, J. and MATOUS, B.: Identification of two new metabolites in melanoma urine: 5-hydroxy-6-methoxyindole-2-carboxylic and 5-methoxy-6-hydroxyindole-2-carboxylic acids. Clin. chim. Acta *16:* 397–402 (1967).
3 FITZPATRICK, T. B.; MIYAMOTO, M. and ISHIKAWA, K.: The evolution of concepts of melanin biology. Arch. Derm. *96:* 305–323 (1967).
4 HINTERBERGER, H.; FREEDMAN, A. and BARTHOLOMEW, R. J.: Precursors of melanin in the urine in malignant melanoma. Clin. chim. Acta *18:* 377–382 (1967).
5 HINTERBERGER, H.: The distribution of 3,4-dihydroxyphenylalanine (dopa) and of some of its metabolites in the blood of patients during oral therapy with L-dopa. Biochem. Med. *5:* 412–424 (1971).
6 HINTERBERGER, H. and ANDREWS, C. J.: Catecholamine metabolism in Parkinson's disease during oral medication with 2,3-dihydroxyphenylalanine (L-dopa). Arch. Neurol. *26:* 245–252 (1972).
7 MELLINGER, T. J. and HVIDBERG, E. F.: Spectrofluorimetry of dihydroxyphenylacetic acid in urine. Amer. J. clin. Path. *51:* 559–565 (1969).
8 PECHAN, Z.: Studien über Melanine und Melanogenese. III. Auswertung der Thormählen'schen Reaktion für die Bestimmung des Indol-Melanogens. Neoplasma Sav. *6:* 397–403 (1959).
9 Scott, J. A.: 3,4-dihydroxyphenylalanine (dopa) excretion in patients with malignant melanoma. Lancet *ii:* 861–862 (1962).
10 TAKAHASHI, H. and FITZPATRICK, T. B.: Quantitative determination of dopa: its application to measurement of dopa in the urine and in the assay of tyrosinase in serum. J. Invest. Derm. *42:* 161–166 (1964).
11 VOORHESS, M. L. Urinary excretion of dopa and metabolites by patients with melanoma. Cancer *26:* 146–149 (1970).

Author's address: Dr. H. HINTERBERGER, Senior Research Officer, Division of Clinical Chemistry, Prince Henry Hospital, *Little Bay, NSW 2036* (Australia)

Dopa and Its Metabolites in Melanoma Urine

J. Duchŏn and B. Matouš

Department of Biochemistry, Faculty of Medicine, Charles University, Prague

Dopa (3,4-dihydroxyphenylalanine) and its metabolites in melanoma urine are phenolic and indolic derivatives which are often called 'urinary melanogens' since most of them are unstable and can be converted to the dark, secondarily formed, 'urinary melanin' by oxidation. Until now, at least 11 compounds have been identified in melanoma urine. They can be divided into two principal groups: Thormählen-positive and Thormählen-negative. Another criterion of their classification is their solubility: i.e., conjugates are soluble in water and free acids in organic solvents (e.g., ethylacetate or ether) [8] (fig. 1).

It is generally believed that urinary melanogens are precursors (or metabolites of these precursors) on the tyrosine-to-melanin biosynthetic pathway. To date, however, the detailed mechanism of their formation has not been explained accurately. There is a possibility that they might be also metabolites, i.e., degradation products of the melanin already formed. However, no evidence for such degradation *in vivo* (from the chemical point of view) has as yet been found.

Since it is impossible to give here a detailed review of contemporary methods of urinary melanogens detection and determination, it can be stated only that there are, at present, three types of methods used for their quantitative determination:

1. Colorimetric methods, such as the colorimetric determination of the total sum of Thormählen-positive melanogens [5, 11, 13, 17, 18].

2. Fluorimetric methods, such as the fluorimetric determination of dopa [14, 21, 23] or the determination of so-called free catechols [13].

3. Chromatographic methods, such as the determination of homovanillic [6, 11, 12, 17, 19] and vanillactic [9, 16] acids, the separation of Thor-

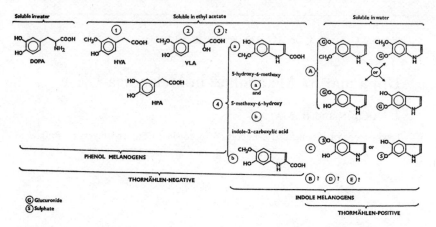

Fig. 1. Urinary melanogens. *Thormählen-negative melanogens:* dopa = 3,4-dihydroxyphenylalanine [14,20,21,23]; HPA = homoprotocatechuic acid (3,4-dihydroxyphenylacetic acid) [22]; compound 1 = HVA = homovanillic acid (3-methoxy-4-hydroxyphenylacetic acid [6,10–14,23]; compound 2 = VLA = vanillactic acid (3-methoxy-4-hydroxyphenyllactic acid) [4,9,16]; compound 3 = structure hitherto unidentified; compound 4: a) = 5-hydroxy-6-methoxy-indole-2-carboxylic acid, b) = 5-methoxy-6-hydroxyindole-2-carboxylic acid [5. 7, 8].

Thormählen-positive melanogens [15]: compound A = 5-hydroxy-6-methoxyindole-5-monoglucuronide or 5-methoxy-6-hydroxyindole-6-monoglucuronide according to [2], or 5,6-dihydroxyindole-5-monoglucuronide or 5,6-dihydroxyindole-6-monoglucuronide according to [11]; compound C = 5,6-dihydroxyindole-5-monosulphate or 5,6-dihydroxyindole-6-monosulphate [1, 11]; compounds B, D and E = structure hitherto unidentified. Indolic melanogens react also with *p*-dimethylaminobenzaldehyde (Ehrlich's reagent) but this reaction is non-specific. The Thormählen test, on the other hand, has a considerable specificity: the condition of Thormählen-positivity is the unsubstituted pyrrole ring of the indole nucleus [15]. (The positive result of the Thormählen test: after the addition of an aqueous solution of nitroferricyanide – $Na_2Fe(CN)_5NO$ – and of potassium hydroxide the solution turns violet and by the subsequent acidification with acetic acid blue or blue-green.)

mählen-positive melanogens [1, 2, 5, 10, 11, 15], or the separation of isomeric 5-hydroxy-6-methoxy- and 5-methoxy-6-hydroxy-indole-2-carboxylic acids [3, 7, 8].

Some of the urinary melanogens occur in the urine of melanoma patients only (indolic melanogens), others occur also in some other pathological circumstances, as well as in healthy persons, although usually in much lower concentrations (the majority of phenolic melanogens).

The clinical significance of urinary melanogens depends on: a) their specificity; b) the frequency of their occurrence, and c) the mutual relationship between the stage and type of disease and the corresponding analytical data. It seems that Thormählen-positive melanogens and the isomeric 5-hydroxy-6-methoxy- and 5-methoxy-6-hydroxy-indole-2-carboxylic acids are highly specific for melanoma [10, 11, 16]. On the other hand, dopa, homoprotocatechuic, vanillactic and homovanillic acids are less specific. They occur in large amounts also in neuroblastomas, pheochromocytomas and some other pathological conditions. It seems that dopa occurs most frequently in melanoma urine: on the average in 65 % of melanoma patients, whereas other types of melanogens occur on the average in about 20 % only (table I). The clinical significance of urinary melanogens also depends on the mutual relationship between the stage and type of disease and the corresponding analytical data. No complete evaluation in this respect is known from the literature, only some partial results are available. For example, from our 263 melanoma patients [11], only in 56 cases (21 %) were the results of Thormählen-positive melanogens determinations positive (i.e., the level of Thormählen-positive melanogens in the urine was higher than 10 $\mu g/ml$).

When we divided our patients into groups according to the stage of the disease, we found that in stage one (i.e., in the primary tumour) the results were positive in only 5 %. In stage two (i.e., with swelling of regional lymph nodes) the number of positive results increased to 20 %. However, in stage three (i.e., with the metastases and generalized tumour) it increased to 80 %. An important conclusion following from this is that the quantitative determination of Thormählen-positive melanogens has practically no value for very early diagnosis. In stage one there are only 5 % positive results, i.e., in the stage when the physician needs the results of laboratory examinations most urgently for further treatment. This is in agreement with results of other authors, e.g., CRAWHALL et al. [5], HINTERBERGER et al. [13] and others [18], and also in other types of melanogens, e.g., dopa [13, 20, 21, 23]. Nevertheless, the determination of urinary melanogens may be useful from the prognostic point of view [10, 11].

Unfortunately, it has not been possible to give a full survey of contemporary problems of the biochemical and clinical significance of dopa and its metabolites in melanoma urine, i.e., of urinary melanogens. There are numerous other problems in this field which may be solved by greater international cooperation in this field.

Table I. The occurence of urinary melanogens in the urine of melanoma patients

Melanogen	No. of melanoma patients	No. of positive results	Positivity %	Author, year and reference
Dopa	28	22	78	Scott 1962 [20]
	9	7	77	Takahashi and Fitzpatrick 1964 [21]
	16	7	44	Voorhess 1970 [23]
'Free	11	10	91	Käser et al. 1970 [14]
catechols'	34	18	53	Hinterberger et al. 1967 [13]
Homoprotocatechuic acid	2	2	100	Thannhauser and Weiss 1922 [22]
Total	100	66	66	
Homovanillic acid	5	5	100	Duchŏn and Gregora 1962 [6]
	15	7	47	Duchŏn and Pechan 1964 [11]
	7	1	14	Schwartze and Conradi 1966 [19]
	5	3	60	Coward et al. 1967 [4]
	34	4	11	Hinterberger et al. 1967 [13]
	2	2	100	Duke and Demopolos 1968 [12]
	16	4	25	Voorhess 1970 [23]
	74	8	11	Matous et al. 1971 [16]
Total	158	34	21	
Vanillactic acid	5	5	100	Duchŏn and Gregora 1962 [6]
	15	7	47	Duchŏn and Pechan 1964 [11]
	5	3	60	Coward et al. 1967 [4]
	74	8	11	Matous et al. 1971 [16]
Total	99	23	23	
5H-6M-I-2C	5	5	100	Duchŏn and Gregora 1962 [6]
and	15	7	47	Duchŏn and Pechan 1964 [11]
5M-6H-I-2C[1]	1	1	100	Badinand et al. 1969 [3]
	74	8	11	Matouš et al. 1971 [16]
Total	95	21	22	
Sum of	263	56	21	Duchŏn and Pechan 1964 [11]
Thormählen	32	7	22	Crawhall et al. 1966 [5]
positive	54	20	37	Schwartze and Grüneberg 1966 [18]
melanogens	34	8	23	Hinterberger et al. 1967 [13]
	74	19	26	Matouš et al. 1971 [16]
Total	457	110	24	

1 5H-6-M-I-2C and 5M6H-I-2C = 5-hydroxy-6-methoxy- and 5-methoxy-6-hydroxy-indole-2-carboxylic acids.

References

1 ANDERSON, A. B.: Urinary melanogens. Biochem. J. *83:* 10 (1962).
2 ATKINSON, M. R.: Isomeric methoxyindolyl glucosiduronic acids in melanotic urine. Biochim. biophys. Acta *74:* 154–155 (1963).
3 BADINAND, A.; PASQUIER, J.; VALLON, J. J.; GUILLUY, R. et MILON, H.: Identification chromatographique en couche mince de deux mélanogènes indoliques urinaires. Clin. chim. Acta *25:* 357–364 (1969).
4 COWARD, R. F.; SMITH, P. and MIDDLETON, J.E.: Urinary phenols in melanoma. Nature, Lond. *213:* 520–521 (1967).
5 CRAWHALL, J. C.; HAYWARD, B. J. and LEWIS, C. A.: Incidence and significance of melanogenuria. Brit. med. J. 1: 1455–1457 (1966).
6 DUCHŇ, J. and GREGORA, V.: Homovanillic acid and its relation to tyrosine metabolism in melanoma. Clin. chim. Acta *7:* 443–446 (1962).
7 DUCHŇ, J. and MATOUŠ, B.: Identification of two new metabolites in melanoma urine: 5-hydroxy-6-methoxyindole-2-carboxylic and 5-methoxy-6-hydroxyindole-2-carboxylic acids. Clin. chim. Acta *16:* 397–402 (1967).
8 DUCHŇ, J.; MATOUŠ, B. and PECHAN, Z.: On the chemical nature of urinary melanogens; in G. DELLA PORTA and O. MÜHLBOCK, Structure and control of the melanocyte, pp. 175–184 (Springer, Berlin 1966).
9 DUCHŇ, J.; MATOUŠ, B. and PROCHÁZKOVÁ, B.: The vanillactic acid in the urine of melanoma patients. Clin. chim. Acta *18:* 318–319 (1967).
10 DUCHŇ, J. and PECHAN, Z.: The biochemical and clinical significance of melanogenuria. Ann. N. Y. Acad. Sci. *100:* 1048–1068 (1963)
11 DUCHŇ, J. and PECHAN, Z.: Biochemie melaninů a melanogenese. (In Czech). (Státní zdravotnické nakladatelství Praha 1964).
12 DUKE, P. S. and DEMOPOULOS, H. B.: One-dimensional paper chromatographic method for determination of urinary homovanillic acid. Clin. Chem. *14:* 212–221 (1968).
13 HINTERBERGER, H.; FREEDMAN, A. and BARTHOLOMEW, R. J.: Precursors of melanin in the urine in malignant melanoma. Clin. chim. Acta *18:* 377–382 (1967).
14 KÄSER, H.; TÜRLER, K. und OTT, F.: Biochemische Untersuchungen beim malignen Melanom. Schweiz. med. Wschr. *100:* 960–963 (1970).
15 LEONHARDI, G.: Papierchromatographische Trennung von Harnmelanogenen. Naturwissenschaften *41:* 141 (1954).
16 MATOUŠ, B.; PAVEL, S. and DUCHŇ, J.: Naše dosavadní zkušenosti s hodnocením nálezů melanogenurie u nemocných melanoblastomem. (In Czech). (Celostát, sjezd čs. dermatologů Hradec Králové 1971).
17 PECHAN, Z. und DUCHŇ, J.: Melanogenbestimmung im Urin: Hydroxyindolkonjugate und Homovanillinsäure. Med. Lab. *18:* 73–80 (1965).
18 SCHWARTZE, G. und GRÜNEBERG, T.: Die Ausscheidung von Indolmelanogenen beim malignen Melanom. Arch. klin. exp. Derm. *225:* 207–217 (1966).
19 SCHWARTZE, G. und CONRADI, G.: Über die Ausscheidung von Homovanillinsäure im Urin beim malignen Melanom. Hautarzt *17:* 348–350 (1966).
20 SCOTT, J. A.: 3,4-dihydroxyphenylalanine (dopa) excretion in patients with malignant melanoma. Lancet pp. 861–862 (1962).

21 TAKAHASHI, H. and FITZPATRICK, T. B.: Quantitative determination of dopa; its application to measurement of dopa in urine and in the assay of tyrosinase in serum. J. Invest. Derm. *42:* 161–165 (1964).
22 THANNHAUSER, S. J. und WEISS, S.: Über das Melanogen bei melanotischen Tumoren und seinen Zusammenhang mit der normalen Pigment-Bildung. Verh. dscht. Ges. inn. Med. *34:* 156–160 (1922).
23 VOORHESS, M. L.: Urinary excretion of dopa and metabolites by patients with melanoma. Cancer *26:* 146–149 (1970).
24 A more complete review of the literature up to 1961 may be found in [10] and up to 1963 in [8, 11].

Author's address: Dr. J. DUCHON, Department of Biochemistry, Faculty of Medicine, Charles University, *Prague* (Czechoslovakia)

Chemical Studies on Urinary Melanogens[1]

L. TASKOVICH, P. W. BANDA and M. S. BLOIS

Department of Dermatology, University of California, San Francisco, Calif.

Introduction

A previous report [1] has pointed out the usefulness of α, α-diphenyl-β-picrylhydrazyl (DPPH) as a reagent for assaying antioxidants, particularly in biological fluids. Many other reports are in the literature on the reaction of DPPH with various phenolic derivatives and higher aromatic amines in organic solvents [4, 11, 12, 13, 15]. A method for determining α-tocopherol with DPPH has been reported [1, 3], and the breakdown of indole-3-acetic acid has been investigated using DPPH as a free radical trapping agent [14]. However, little attention has been given to the reaction of DPPH with a wider range of biological metabolites. We wish to report here on the use of DPPH as a clinical reagent for measuring phenolic and indolic derivatives in the urine of patients with melanoma.

The spontaneous darkening of urine from patients with advanced melanoma has long been taken as evidence for the excretion of metabolites characteristic of this tumor. The metabolites released to the urine are comprised of substituted phenolic and indolic compounds, and as a group are termed 'melanogens'. These include dihydroxyphenylalanine (dopa), 3-methoxy-4-hydroxyphenylacetic acid (homovanillic acid), 3,4-dihydroxyphenylacetic acid, 5-hydroxy-6-methoxy-indole-2-carboxylic acid, and related compounds [6]. It is quite likely, however, that not all the melanogens have thus far been identified, despite the notable work of DUCHON and collaborators [7]. Neither is it known whether the melanogens leak from

---

[1] This work was supported in part by the U.S. Public Health Service Grants CA-08064 and CA-12043.

tumor cells during pigment synthesis or result from the breakdown of completely formed melanin pigment

Many investigators have examined the urine of melanoma patients using a variety of lengthy techniques designed to measure either the phenolic melanogens, such as dopa [10, 16, 18, 19] or the indolic melanogens as a group by the Thormählen reaction [5, 8, 10]. Attempts at correlating the level of phenolic or indolic melanogens in the urine with the clinical state of the patient have been variable, and in the reports thus far, dopa appears to be the single melanogen most frequently excreted in large amounts [19]. We have found that DPPH reacts with all the phenolic and indolic melanogens so that total melanogens in a urine sample can be estimated by a single test. DPPH reacts with dopa in a 10:1 ratio, and thus provides a high sensitivity for the detection of dopa. High melanogen levels measured by the DPPH reaction generally correlate with dissemination of malignant melanoma, although normal individuals may occasionally give elevated values due to dietary influences. Attempts to normalize urinary constituents by the use of a synthetic diet has been reported [20], and the use of a standardized diet will also be necessary for precise melanogen measurements.

Materials and Methods

a) Urine, Collection

All urines used in this work were 24-h samples, refrigerated during the collection period, but with no preservatives added. All urines should be analyzed within a few hours after completion of the collection. The use of thoroughly rinsed, brown glass containers is recommended in order to avoid contamination by low molecular weight organic compounds which may be leached from some plastic containers.

b) The DPPH Reaction

A DPPH stock solution is freshly mixed as follows: 15 mg α,α-diphenyl-β-picrylhydrazyl (Eastman Organic Chemicals) are dissolved in 1–2 ml chloroform and 100% ethanol from glass pint bottles is added to 100 ml. The solution should be stored in a dark glass bottle at 4 °C. DPPH reagent is prepared as needed by diluting the stock solution 1:5 with ethanol that is .05M in acetic acid. The O.D. at 520 nm of such a solution is approximately 0.75. The exact starting O.D. is not critical, since the calibration curves are linear.

A 0.1 ml aliquot of urine is transferred to a 50 ml glass stoppered flask and diluted to 1.0 ml with distilled water. 50 ml of DPPH reagent is added, the solution mixed, placed in the dark, and the O.D. at 520 nm is read after 24 h. The O.D. readings are taken relative to a blank prepared by using water in place of the urine. Standard calibration curves of O.D. vs μM of reactant may be prepared according to this procedure. All operations are carried out at room temperature.

DPPH is a stable free radical and forms deeply violet solutions in ethanol with a broad absorption maximum centered at 520 nm. As DPPH is reduced, the solution loses color stoichiometrically with the number of electrons or hydrogen radicals taken up. Cysteine is known to react rapidly with DPPH in a 1:1 ratio [1], and for this reason, cysteine is chosen as the standard for recording the DPPH reactivity of urine samples, even though cysteine is not normally found in significant quantities in the urine. An O.D. reading of the reacting solution is converted to the equivalent μM of cysteine by using a standard calibration curve of O.D. vs μM cysteine. The μM cysteine equivalent is then multiplied by 121.5, the molecular weight of cysteine, to give mg cysteine equivalents per 100 ml (mg % cys. eq.). For example, if the O.D. reading corresponds to 1.0 μM cysteine according to the standard curve, then

$$\frac{1.0 \ \mu M \ \text{cys. eq.}}{.1 \ \text{ml urine}} \times \frac{121.5 \ \text{mg}}{1000 \ \mu M} = \frac{121.5 \ \text{mg cys.eq.}}{100 \ \text{ml urine}}$$

Patients are evaluated on the basis of their urine concentration of melanogens expressed in mg % cysteine equivalents and of their total 24-h excretion of cysteine equivalents (mg cys. eq./24 h) by taking into account the total excretion volume. The results are thus expressed as though a single monovalent constituent were responsible, even though a number of constituents are known to be present, each with a complex stoichiometry. Although this procedure is arbitrary, it has proven useful.

c) The Thormählen Reaction

We include here our procedure for carrying out the Thormählen reaction. The method is a slight modification of that given by DUCHON and PECHAN [7].

To 0.5 ml of urine, add 0.5 ml of 0.2% sodium nitroferricyanide. 2.0 ml of 2 N sodium hydroxide are then added, and after 60 sec, add 1.0 ml 30% acetic acid. The O.D. at 620 nm is read after 10 min. A blank is prepared by using water in place of urine. A linear calibration curve of O.D. vs concentration is prepared using indole as a standard. The curve converts O.D. readings to μg indole per 0.5 ml. The values are doubled to the standard form of μg indole/ml. In our work, we designate the readings of the Thormählen reaction 'indole equivalents per ml'. In addition, we have found it useful to multiply by the urine volume and calculate the total indole equivalents excreted per 24 h.

Results

a) Reactivity of DPPH

DPPH reacts rapidly with 3,4-dihydroxy phenols, containing a primary or secondary amino group, and 5,6-dihydroxy-substituted indoles. For example, DPPH reacts with dopa and dopamine, but not with tyrosine or tyramine. DPPH reacts with 5,6-dihydroxyindole acetic acid, as well as with the 5-hydroxy analog, but will not react with indole or tryptophan. The reaction with compounds containing one hydroxy and one methoxy group on the ring, such as HVA, is slow. As stated above, DPPH reacts readily with cysteine, presumably via its exchangeable sulfhydryl hydrogen, but will not react with exchangeable hydrogens in general; i.e., there is no reaction with aliphatic-amino acids ($-NH_2$, $-COOH$) or sugars ($-OH$). Reduced metallic ions, such as ferrous iron, will react with DPPH. The reaction of DPPH with the foregoing types of compounds and its lack of reaction with other urinary constituents, such as simple amino acids and reducing sugars, make DPPH a valuable reagent for clinical melanogen studies.

Although it has been primarily employed in clinical investigations on whole urine samples, in order to determine total urinary melanogens, DPPH is equally useful in assaying urine fractions which have been separated using column chromatography, and as a developing reagent for use in the detection of reducing compounds separated by paper chromatography. A method for the use of DPPH on thin layer chromatograms has been reported [9].

b) Stoichiometry of the DPPH Reaction

DPPH is found to be de-colorized (reduced to the hydrazine form) on a mole-for-mole basis by cysteine, which we have accordingly chosen as a standard and in terms of which we express the reducing power of the urines. It might be expected that a reactive dihydroxy compound would reduce DPPH in a ratio of 2:1, with the hydroxyl groups acting independently. Ascorbic acid, for example, reacts in the predicted 2:1 ratio. Such is not always the case, however, and the stoichiometry of DPPH with certain hydroxy phenols, is found to be higher; e.g., hydroquinone and catechol – 3:1, 3,4-dihydroxyphenylacetic acid – 6:1, dopa – 10:1. The reaction of catechol with up to four molecules of DPPH has been reported [15]. The higher than expected 4:1 stoichiometry is attributed to further radical reactions and appeared to be dependent on the presence of water.

The stoichiometry of the DPPH reagent is very dependent on the polar nature of the solvent system, its acid concentration, and when working with buffered systems, on the final molarity of the reaction mixture. At low pH's the color intensity of the solution decreases, and at molarities above .05, the reaction ratio decreases.

The DPPH reaction, as reported here, intrinsically affords ten times more sensitivity in detecting dopa than conventional colorimetric reactions. The reaction works equally well in blood chemistries, and has been used to monitor dopa levels in the plasma of patients with Parkinson's disease being treated with dopa [2]. We believe the DPPH reaction to be of value in a wide range of clinical applications.

Discussion

Laboratory techniques which can provide evidence of dissemination of melanoma or which would provide a measure of tumor burden are urgently needed if the clinical management of the melanoma patient is to be an informed one.

The standard procedures including chest x-rays, liver function tests, and more recently, liver scans, leave much to be desired. It is our conclusion that at the present time, urinary melanogen determinations (DPPH and Thormählen tests) are better early indicators of hepatic metastases than are liver function tests in the vast majority of cases. In our experience, the urinary melanogens have, more often than not, become elevated before hepatic metastases can be demonstrated by isotopic scanning techniques.

On the other hand, urinary melanogens do not apparently become elevated at an early stage when tumor spread is primarily local and cutaneous. We have had several patients with recurrent localized cutaneous or subcutaneous tumors, some quite large, who have had normal melanogen levels until late in their disease when the melanogens rose precipitiously, shortly before their death, and presumably in coincidence with systemic spread. It is thus tempting to infer that hepatic metastases are responsible for the rise in urinary melanogens, but the explanation may not be this simple.

The greatest limitation in the interpretation of urinary melanogen data is the overlap between the distribution of levels in normals, and those of early melanoma patients. We believe the origin of this difficulty is primarily an artifact due to uncontrolled dietary intake, and only a brief glance at other quantitative urine chemistry data obtained on patients who are dietarily not controlled, reinforces this conviction. The reliability of urinary melanogens is, of course, increased by multiple repetition of the test procedures, and sustained elevations repeatedly obtained over a period of time has, in our experience, always been associated with disseminated melanoma.

The Thormählen test seems not to have achieved a wide acceptance; partly it would seem, because as a laboratory procedure, it lacks sensitivity and is poorly reproducible, involving as it does, unstable colored intermediates. Detractors of its use cite its less than complete correlation with the stage of disease and the frequency of false negative test results, and continue instead to obtain chest films, liver function tests and liver scans, thus generating false negatives of different origins.

At the present stage of development, it is not clear that the DPPH test is more *clinically* sensitive than the Thormählen test; it is, however, a technically more satisfactory procedure for reasons of reproducibility and simplicity, and has greater chemical sensitivity. It has the additional advantage of detecting both phenolic and indolic melanogens while the Thormählen test measures only the latter. Since the two tests measure different metabolites, we routinely perform both and find that in better than 80 % of the cases, the correlation is good. There have been patients, however, who have had elevations by one test, but not by the other, with disseminated melanoma being later demonstrated (see table I). Because of the generally poor sensitivity of the standard laboratory procedures in detecting disseminated melanoma, and the simplicity of the melanogen assays, we believe the latter should be a routine procedure in the management of all patients with melanoma.

Finally, there is a great need, in the clinical investigation of melanoma, for quantitative and objective indicators of the response of an individual patient to experimental therapy. In table II, the serial melanogen values obtained on a patient undergoing clinical remission while under treatment with transfer factor, are shown. In this patient, the remission of visible,

Table I. Urinary melanogens in patients with disseminated melanoma

Patient	Thormählen (μg indole eq.)		DPPH	
	Concentration per ml upper limit or normal (8)	Excretion per 24 h Upper limit of normal (10,000)	Concentration mg % cys. eq. Upper limit of normal (130)	Excretion mg cys. eq./24 h Upper limit of normal (2,100)
EB	8.8	–	166	–
FB	21.0	–	131	–
KC	8.8	24,000	148	2,610
BF	15.4	–	316	–
KF	13.2	–	181	–
TG	16.5	7,440[1]	212	990[1]
TL	18.8	–	181	–
LO	11.0	13,200	121	1,820
DS	8.8	–	163	–
GS	5.5	–	264	–
WT	13.2	–	213	–
NV	53.0	–	231	–
PY	13.2	–	217	–

[1] Patient is a 90 lb. female with daily urine volumes ranging from 400-1000 ml.

Table II. Sequential urinary melanogens in a patient with disseminated melanoma undergoing clinical remission during treatment with transfer factor

Date	DPPH mg % cys. eq.	Thormählen mg indole eq./ml
June 10	200	–
June 11	186	11.0
July 21	171	13.2
October 6	75	6.0
December 1	101	7.5
January 10	106	4.4

cutaneous metastatic spread correlates with reduction in urinary melanogen levels, and improved cellular immunity as measured by migration inhibitory factors and lymphocyte stimulation [17].

References

1 BLOIS, M.S.: Antioxidant determinations by the use of a stable free radical. Nature, Lond. *181:* 1199–1200 (1958).
2 BLOIS, M. S. and TASKOVICH, L.: (unpublished).
3 BOGUTH, W. and REPGES, R.: Photometric determination of α-tocopherol with 1,1-diphenyl-2-picrylhydrazyl. Int. Z. Vitaminforsch. *39:* 289–295 (1969).
4 BRAUDE, E. A.; BROOK, A. G. and LINSTEAD, R. P.: Dehydrogenation reactions with diphenylpicrylhydrazyl. J. chem. Soc. 3574–3578 (1954).
5 CRAWHALL, J. C.; HAYWARD, B. J. and LEWIS, C. A.: Incidence and significance of melanogenuria. Brit. med. J. *i:* 1455–1457 (1966).
6 DUCHON, J. and MATOUS, B.: Identification of two new metabolites in melanoma urine; 5-hydroxy-6-methoxyindole-2-carboxylic and 5-methoxy-6-hydroxyindole-2-carboxylic acids. Clin. chim. Acta *16:* 397–402 (1967).
7 DUCHON, J. and PECHAN, Z.: Biochemie malaninu a melanogenese (Statni zdravotnicke nakladatelstvi, Prague 1964).
8 DUCHON, J. and PECHAN, Z.: The biochemical and clinical significance of melanogenuria. II. Ann. N.Y. Acad. Sci. *100:* 1048–1068 (1963).
9 GLAVIND, J. and HOLMES, G.: Thin-layer chromatographic determination of antioxidants by the stable free radical α,α-diphenyl-β-picrylhydrazyl. J. Amer. Oil Chem. Soc. *44:* 539–542 (1967).
10 HINTERBERGER, H.; FREEDMAN, A. and BARTHOLOMEW, R. J.: Precursors of melanin in the urine of malignant melanoma. Clin. chim. Acta *18:* 377–382 (1967).
11 MCGOWAN, J. C.; POWELL, T. and RAW, R.: The rates of reaction of α,α-diphenyl-β-picrylhydrazyl with certain amines and phenols. J. chem. Soc. 3103–3110 (1959).
12 PAPARIELLO, G. J. and JANISH, M. A. M.: Diphenylpicrylhydrazyl as an organic analytic reagent. Anal. Chem. *37:* 899–902 (1965).

13 PAPARIELLO, G. J. and JANISH, M. A. M.: Diphenylpicrylhydrazyl as an organic analytic reagent in the spectrophotometric analysis of phenols. Anal. Chem. *38:* 211–214 (1966).
14 PARUPS, E. V.: Involvement of free radicals in the oxidative degradation of indole-3-acetic acid. Canad. J. Biochem. *47:* 220–224 (1969).
15 SCHENK, G. H. and BROWN, D. J.: Free radical oxidation of dihydric phenols with diphenylpicrylhydrazyl. Talanta *14:* 257–261 (1967).
16 SCOTT, J. A.: 3,4-dihydroxyphenylalanine excretion in patients with malignant melanoma. Lancet *ii:* 861–862 (1962).
17 SPITLER, L. E.; LEVIN, A. S.; BLOIS, M. S.; EPSTEIN, W. L.; FUDENBERG, H.; HELLSTROM, I. and HELLSTROM, K. E.: Lymphocyte responses to tumor specific antigens in patients with malignant melanoma and results of transfer factor therapy. J. clin. Invest. (to be published).
18 TAKAHASHI, H. and FITZPATRICK, T. B.: Quantitative determination of dopa; its application to measurement of dopa in urine and in the assay of tyrosinase in serum. J. Invest. Derm. *42:* 161–165 (1964).
19 VOORHESS, M. L.: Urinary excretion of dopa and metabolites by patients with melanoma. Cancer *26:* 146–149 (1970).
20 YOUNG, D. S.; EPLEY, J. A. and GOLDMAN, P.: Influence of a chemically defined diet on the composition of serum and urine. Clin. Chem. *17:* 765–773 (1971).

Author's address: Dr. LINA TASKOVICH, Alza Corporation, *Palo Alto, Calif.* (USA)

Plasma and Urine Amino Acid Changes Associated with Melanoma

V. RILEY, D. SPACKMAN and M. A. FITZMAURICE

Department of Microbiology, Pacific Northwest Research Foundation, and the Fred Hutchinson Cancer Research Center, Seattle, Wash.

Introduction

Although numerous reports are available on melanuria associated with melanoma, relatively little work has been reported concerning the presence of abnormal substances appearing in the blood of either patients or of experimental animals bearing pigmented tumors. Since the metabolism of the pigmented cell, whether normal or malignant, is intimately involved with the amino acids phenylalanine, tyrosine, dopa, and their metabolic products, we have examined the blood plasma of several melanoma patients, as well as both the plasma and urine of mice bearing transplanted B-16 melanoma, for indications of changes in these and other plasma amino acids or related metabolites.

As a preamble to a description of these studies, it is appropriate to discuss the curious reaction which takes place between various melanoma-associated materials and a rather simple chemical, paraphenylenediamine, commonly known as PPDA by most biochemists who use it routinely as a cytochrome-c reductant in studying cellular metabolism. When this compound is added to either the plasma or the urine of a melanoma patient there is a striking oxygen consumption as compared with analogous materials from a normal individual. The time-course of this reaction is followed by standard Warburg manometric procedures which measure the uptake of molecular oxygen [2, 5, 6, 7, 9, 10, 20].

It is pertinent to note that this reaction is not restricted to the urine, since when paraphenylenediamine (PPDA) is added to the plasma of established melanoma patients a similar oxygen-consumption occurs. Plasma by itself, either from a normal individual, or from a melanoma patient, exhibits no significant activity in the absence of PPDA. Paraphenylenediamine by itself, under these circumstances also causes no appreciable oxygen consumption [7, 9]. It is thus obvious that an oxygen-consuming reaction occurs between PPDA and some unknown substance, or substances, in the plasma and urine of melanoma patients.

It may be of interest to briefly mention the curious history of the discovery of this reaction. It had been reported many years ago by a distinguished biochemist that there was a 'powerful cyanide-insensitive oxidative enzyme system' present in melanoma tissue, which was absent in all other normal and malignant tissues tested. The data which led to this erroneous conclusion was obtained by standard studies of cellular extracts from melanoma patients or experimental animals, and compared with comparable preparations derived from other cancer patients [3, 4].

It was confirmed that oxygen consumption does indeed occur in the presence of KCN [5, 7, 9], which is unusual for oxidative enzymes, since they are inactivated by cyanide. However, it was also established that an identical oxygen consumption may be obtained also following heating of the tissue extract in such a manner as to inactivate most mammalian enzymes, and that the 'powerful cyanide-insensitive enzyme' was also heat-insensitive [7]. It was thus concluded that the reaction was non-enzymic, and probably indicated the presence of one or more heat-stable substances capable of a non-enzymic reaction with the added paraphenylenediamine. It is gratifying to note that a distinguished, Hillebrand Prize-winning biochemist anticipated such experimental results by intuition and rationale [2].

Efforts were thus initiated to determine the nature of such reactants. Since it seemed metabolically reasonable, as well as biochemically logical, that dopa (dihydroxyphenylalanine) or dopa-like intermediate pigmentation metabolites should be present in melanoma tissues, and thus probably in the plasma and urine of melanoma-bearing animals or patients, we proceeded to combine experimentally *in vitro*, in the absence of cells or tissue, the two simple compounds, PPDA and dopa [6, 7, 9, 10]. No oxygen uptake was observed when the two substances were tested individually; whereas, a vigorous oxygen consumption resulted when the two compounds were combined *in vitro*.

This finding appeared to provide a reasonable clue as to the possible nature of the PPDA-reacting substances present in the plasma and urine of melanoma patients. The question naturally arose as to the relative specificity of this reaction. Was it restricted to dopa or would other compounds with comparable molecular configuration also react?

Subsequent studies provided basic information in this regard [5, 7, 12]. For example, it was demonstrated that a partial reaction could be obtained with catechol which is, of course, identical to dopa with the exception that it lacks the alanine residue. It thus appears that the alanine side chain in dopa provides additional oxygen-consuming activity, possibly by feeding electrons in a more effective way, or by providing a more favorable combination geometry. However, incubating PPDA with phenylalanine, or even with tyrosine, induced no appreciable reaction beyond the small autoxidation of PPDA by itself. Such data indicated that the two ortho hydroxyl groups are of greater importance than the other molecular features of the molecules tested. A variety of related compounds were tested by this method; although despite certain useful generalizations, it was not always possible to predict which compounds will react with PPDA in this oxygen-consuming reaction, since some closely related compounds had variable capabilities. However, it was established that there are a variety of known dopa-like or dopa-related, dihydroxy compounds capable of reacting *in vitro* with PPDA [7, 9].

Experimental data also indicated that there was a reasonable PPDA reaction specificity insofar as melanoma patients are concerned. This was in contrast to patients bearing non-melanotic tumors, such as carcinoma of the cervix, and non-pigmented malignancies of other tissue origins which did not exhibit the PPDA reaction [5, 7, 9).

Other experiments indicated that the extent of the melanoma disease may be reflected by the quantitative response of the PPDA reaction. In a particular study [11, 12], where the PPDA reaction was applied to the urine of a patient bearing a rare malignant blue nevus, it was found that there was a significantly greater oxygen uptake during the terminal stages, as compared with the reaction observed in the earlier stages of the disease.

These findings, and other related observations, made it quite clear that unique metabolic substances were being produced as a result of the neoplastic process involving the malignant pigmented cell, or its relationship to the host. Pertinent questions would seem to be: what is the biochemical nature of these substances, and can they provide us with instructive information on the special differential, metabolic properties of this malignancy which might lend themselves to possible therapeutic exploitation?

At this point it is pertinent to recall that among the clinical manifestations observed with some melanoma patients, especially with those in the terminal stages, have been indications of neurological involvement. In view of the known tendency of melanomas to undergo widespread metastasis rather early, it has been generally assumed that such central nervous system aberrations represented the physical effects of brain metastasis. However, at Memorial Hospital in New York, it was frequently noted that melanoma patients who had exhibited bizarre personality changes showed no evidence of brain metastasis upon autopsy. In the face of this information, together with an appreciation of the close pharmacological relationships between dopa and related compounds such as adrenaline, noradrenaline, and other substances known to have central nervous system activities, it appears possible that unknown metabolites resulting from melanoma growth may exert pharmacological effects upon the central nervous system, in place of, or in addition to the consequences of metastasis. It thus becomes more imperative that the nature of these substances be established and characterized.

As a consequence, experiments were initiated to determine what alterations of significance, if any, occur in the free amino acids of either the plasma or the urine of melanoma patients, or of experimental animals bearing transplanted melanoma.

Materials and Methods

Mice
Hybrid (C57BL/6 × A) F_1 female mice between 6 and 10 weeks of age, weighing 18 to 22 g were employed.

Tumors
The B-16 melanotic melanoma employed was obtained from K. ADACHI of the Oregon Regional Primate Center, and the EARAD-1, asparaginase-sensitive leukemia was obtained from L. J. OLD of the Sloan-Kettering Institute. Tissue culture passage was utilized to free the tumors of the LDH-virus, since this inconspicuous entity is capable of compromising host response under a variety of experimental conditions. Mice were implanted subcutaneously in the hip with tumor cell suspensions in physiological saline containing between 10^5 and 10^6 cells in a 0.1 ml volume. Standard three-dimensional tumor volume measurements, reflecting tumor growth, were made with calipers.

Mouse Blood Samples
Careful sampling procedures were employed to minimize amino acid alterations during bleeding and processing of plasma samples for amino acid analysis [15]. Blood samples of about 0.2 ml per mouse were removed by the orbital bleeding technique [8] using disposable bleeding tubes containing sodium heparinate. Following the bleeding procedure, which requires only a few seconds, the tubes were quickly stoppered and plunged into an ice bath, chilling the blood immediately to 0 °C. The red cells and plasma were separated in a refrigerated centrifuge and the cold plasmas were transferred to preweighed, disposable tubes containing sulfosalicylic acid (SSA), a protein precipitating agent which inactivated all enzymes and stabilized the amino acid pattern. Further sample

preparations, or storage, were carried out in such a way as to preserve as nearly as possible the original concentrations of free amino acids present at the time the samples were drawn.

Mouse Urine Samples

Samples of urine were obtained from each group of mice as needed. Each mouse was held by hand over a collection beaker; gentle pressure on the bladder area would usually result in the voiding of 0.1 to 0.3 ml of urine, although in the latter stages of tumor growth, urine sample volumes were smaller and more difficult to obtain. The specific gravity was determined on each sample and a measured volume was pipetted into sulfosalicylic acid. The SSA served two purposes, namely, precipitating any proteins present in the urine, and inactivating viruses, specifically the LDH-elevating virus, which was present in both the blood and urine of some intentionally infected mice. The urine of healthy, normal mice was virtually protein-free, while mice with advanced melanotic tumors exhibited significant quantities of protein in the urine. Such urine protein, of course, was precipitated by sulfosalicylic acid.

Patient Blood Samples

Samples of venous blood were drawn from three melanoma patients of 18, 30, and 49 years of age (one male, two females) who had moderate to advanced stages of melanoma. In two of the patients the original tumor had metastasized, however, blood samples were drawn before the initiation of treatment. In contrast to the use of mouse plasma, the patient blood samples were allowed to clot and were then centrifuged to provide serum samples which were placed in frozen storage to await further processing. At the time of analysis, de-proteinization and final sample preparations were carried out as described above for the mouse plasma samples.

Amino Acid Determinations

Analysis of free plasma and urine amino acids were carried out on a Beckman Model 120 B amino acid analyzer modified to provide accelerated, semiautomatic runs of high sensitivity [16]. Recently revised and improved methods of analysis were employed. An improved sodium citrate buffer system for basic amino acids [19] allows the routine analysis of E-N-monomethyllysine [18] along with all other basic amino acids. Lithium citrate buffers were used for the acidic and neutral amino acids [1, 17]. The latter buffer system in its modified form is essential for the satisfactory separation of asparagine and glutamine which are of critical importance in amino acid studies of tumor-bearing subjects [14, 15]. This report concerns itself only with the free amino acids appearing in the plasma and urine, as opposed to non-protein bound amino acids.

Results and Discussion

Is is appropriate to recall that there are few physiological parameters, either in the human or in experimental animals, more stable than the composition of the circulating free amino acid pool. This has been demonstrated by the use of hemodialysis, both experimentally with laboratory models and with patients. In these experiments we have removed relatively large quantities of amino acids from the blood, with only minimal short-lived influences upon the normal levels of the circulating amino acid pool. Because of the effective homeostatic capabilities of the organism, it is thus difficult to significantly alter the quantitative relationship of the various free amino acids simply by their physical removal. Thus, any systematic changes in such entities observed during a pathological process assume a special significance.

Table I demonstrates that in melanoma patients, at least in the limited number studied, there are significant alterations induced in the circulating plasma amino acid pool during the course of the disease.

This table lists 10 plasma amino acids which were elevated in the three melanoma patients studied. It may be noted that the greatest increase was found with threonine, which was increased by 600 to 700 %, followed by taurine, with an increase of 300 to 400 %; ornithine, by 100 to 200 %; alpha aminobutyric acid, 90 to 120 %; and serine, 70 to 90 %. Phenylalanine was increased by 50 to 70 %, whereas alanine was increased 40 to 50 %. Glycine, leucine, and tyrosine all showed somewhat smaller increases of 10 to 60 %, depending upon the individual patient.

Moving from the human to the mouse situation, it is pertinent to examine the decreases, as well as the increases, in the plasma amino acids of melanoma-bearing mice. Table II cites the alterations which occurred in the plasma of mice bearing the B-16 transplanted melanoma and indicates that taurine, proline, and alanine increased over 100 % above normal; whereas, in contrast, aspartic acid and glutamine were decreased 40 to 50 % below their normal levels. Amino acids showing somewhat lesser alterations, but

Table I. Amino acids elevated in the plasma of melanoma patients

Amino acid	% elevation
Threonine	600–700
Taurine	300–400
Ornithine	100–200
Alpha aminobutyric acid	90–120
Serine	70–90
Phenylalanine	50–70
Alanine	40–50
Glycine	20–60
Leucine	20–50
Tyrosine	10–60

nevertheless significant changes, were phosphoethanolamine, threonine, and phenylalanine, which were all increased, but only of the order of 50% above normal. In contrast, tryptophan, serine, asparagine, and tyrosine were decreased more than 25% below normal.

Table III represents the relative alterations observed in the free amino acids in the urine of the same animals during the rapid growth period of the

Table II. Changes in plasma amino acids during rapid melanoma growth in the mouse

Increases	Decreases
Over 100% above normal	More than 50% below normal
Taurine Proline Alanine	Aspartic acid Glutamine
Over 50% above normal	More than 25% below normal
Phosphoethanolamine Threonine Phenylalanine	Tryptophan Serine Asparagine Tyrosine
Over 25% above normal	
Histidine Hydroxyproline Valine	

Table III. Changes in urine amino acids during rapid melanoma growth in the mouse

Increases	Decreases
Over 200% above normal	More than 50% below normal
Proline Phenylalanine	Ornithine Arginine
Over 50% above normal	More than 25% below normal
Glutamine Alanine Isoleucine	Histidine Creatinine Glycine
Over 25% above normal	
Phosphoserine Aspartic acid Cystine	

melanoma. It may be noted that the alterations in the urine were greater in some instances than those observed in the plasma. Proline and phenylalanine were increased approximately 50% above normal. Phosphoserine, aspartic acid, and cystine increased over 25% above normal. The changes which occurred in the various free amino acids were not all in the same direction inasmuch as some exhibited decreases, and others increases. Amino acids showing decreases include ornithine and arginine, which were depleted by about 50% below normal, while histidine, creatinine, and glycine were reduced more than 25% below normal.

The absolute values of the plasma amino acids observed in the experimental mice are listed in table IV; while the amino acid data obtained from analysis of mouse urines during the course of melanoma growth, in comparison with normal control values, are listed in table V.

So far, we have described only alterations in known amino acids. However, it is of interest to note that in the urine of experimental animals growing the B-16 mouse melanoma, there was a substantial increase in two unidentified compounds that made an appearance on our amino acid chromatograms (see table V). One of these unknown substances is illustrated in figure 1. The upper section of the figure illustrates a relatively small segment of a chromatogram of the acidic and neutral amino acids found in the urine of normal mice. The chromatogram at the bottom of figure 1 shows the same region, but was obtained with the urine from B-16 melanoma-bearing mice in advanced stages of the disease. It may be noted that phenylalanine increased substantially, whereas the tyrosine level remained approximately the same. However, a point of special interest concerning the melanoma urine was the substantial increase in an unknown peak, labelled AN 358, in figure 1. Studies are under way to establish the identity of this compound. It is appropriate to inquire whether this, or a related substance, may be responsible for the PPDA reaction occurring in both melanoma-bearing experimental animals and in melanoma patients. This peak is not dihydroxyphenylalanine.

To return again to the remarkable stability of the plasma amino acid pool, it is instructive to contrast this normal homeostatic tendency with the data shown in figure 2 which shows a systematic decrease in plasma glutamine during the growth of an experimental tumor. These amino acid data were obtained from studies with the transplanted EARAD-1 asparagine-dependent mouse leukemia. The metabolic capabilities of this tumor growth for altering the plasma concentration of a key amino acid are intriguing, inasmuch as the malignancy accomplished alterations in the circulating

Table IV. Plasma amino acids in mice during B-16 melanoma growth[1]

	Day following tumor impantation						
	'Day 0' (Controls)[2] [3]nm/ml	Day 8 nm/ml	Day 11 nm/ml	Day 15 nm/ml	Day 18 nm/ml	Day 22 nm/ml	Day 25 nm/ml
Basic amino acids	(S.D.)						
Ornithine	41±3	36	33	33	32	35	46
Ethanolamine	13±1	12	10	12	13	9	13
Ammonia	77±3	73	105	82	99	101	168
1-methylhistidine	3±1	3	3	2	3	4	5
Lysine	209±36	214	217	194	216	271	262
Methyllysine	9±1	8	8	9	12	12	15
Histidine	48±1	43	44	41	47	54	63
3-Methylhistidine	6±1	5	5	7	6	11	14
Tryptophan	84±9	75	71	57	51	67	61
Arginine	114±14	117	110	93	84	86	64
Acidic and neutral amino acids							
Taurine	264±26	234	212	271	173	137	591
Phosphoethanolamine	14±3	11	10	14	14	13	25
Urea	6,882±150	7,104	6,254	5,835	7,076	6,814	17,022
Aspartic acid	17±2	18	17	13	14	14	12
Hydroxyproline	19±2	23	26	25	28	26	31
Threonine	104±7	104	105	92	89	119	145
Serine	115±8	114	103	102	89	103	88
Asparagine	34±3	34	31	32	29	29	33
Glutamic acid	23±2	25	23	29	26	30	35
Glutamine	644±17	604	546	595	533	510	409
Proline	67±8	60	66	94	95	142	189
Glycine	239±14	237	238	269	280	309	309
Alanine	278±16	285	259	381	338	506	489
Citrulline	85±9	80	79	58	59	57	78
α-aminobutyric acid	3±1	4	4	6	4	9	7
Valine	139±5	128	121	83	104	154	174
Cystine	36±5	35	29	43	39	35	42
Homocitrulline	1±1	1	1	1	2	3	4
Methionine	37±2	39	39	40	35	38	38
Isoleucine	61±7	59	58	39	54	53	75
Leucine	100±6	89	85	68	180	17	118
Tyrosine	43±6	44	47	37	39	52	40
Phenylalanine	51±6	48	48	46	55	66	82

1 Free plasma amino acids.
2 Day zero controls are normal plasma amino acid concentrations determined by averaging 4 separate analyses; each pool sample was obtained from 10 normal mice at various times during the experiment. Other values are from single analyses of pooled plasmas collected from the surviving mice whose number varied from 20 at the beginning to 10 on Day 25.
3 Nanomoles per ml of plasma with their standard deviation (S.D.).

Table V. Mouse urine amino acids during B-16 melanoma growth[1]

	Days following tumor implantation			
	'Day 0' (Controls)[2] [3]nm/ml	Day 11 nm/ml	Day 18 nm/ml	Day 25 nm/ml
Basic amino acids	(S.D.)			
γ-aminobutyric acid	65±8	79	63	46
Ornithine	102±19	131	93	53
Ethanolamine	117±28	544	153	91
1-merhylhistidine	111±17	132	152	121
Lysine	417±50	514	495	332
Histidine	56±14	51	57	31
3-methylhistidine	248±38	295	265	182
Creatinine	7,949±1115	11,734	13,740	4,780
Carnosine	172±29	257	325	172
Arginine	82±18	117	86	41
Acidic and neutral amino acids				
Phosphoserine	496±51	660	934	592
Taurine	7,833±935	3,503	1,763	
Phosphoethanolamine	560±77	714	894	647
Urea	1,123,000±117,000	1,313,500	1,074,500	740,300
Aspartic acid	48±9	44	56	55
Threonine	168±21	215	260	204
Serine	94±10	119	144	117
Asparagine	151±28	189	218	166
Glutamic acid	95±11	114	136	96
Glutamine	281±26	359	536	476
Glycine	321±47	376	535	262
Alanine	173±24	219	333	292
Citrulline	56±17	41	51	103
α-aminobutyric acid	72±16	67	100	100
Valine	82±16	99	122	124
Cystine	134±20	168	258	222
Homocitrulline	16±5	13	15	16
Methionine	51±10	49	56	84
Isoleucine	47±13	52	71	108
Leucine	138±41	245	191	140
Tyrosine	55±7	70	76	71
Phenylalanine	85±12	149	332	303
β-alanine	153±20	179	152	165
β-aminoisobutyric acid	110±20	148	155	144
AN-358	87±26	252	526	471
AN-438	1,270±307	1,785	1,760	1,695

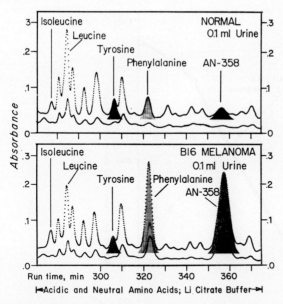

Fig. 1. Chromatograms showing some changes observed in the free amino acids n the urine of mice. The upper curve is from normal mice and the lower, from the urine of mice with advanced B-16 melanoma. The peak designated AN-358 is an unknown amino acid or peptide (or closely related compound). Dopa, when present, appears just before and incompletely separated from isoleucine.

plasma amino acid pool that could not be readily accomplished by hemodialysis wherein large quantities of glutamine and other amino acids were physically removed. We therefore posed the questions: is this a general phenomenon, and does it also occur in the case of melanomas?

Figure 3 demonstrates that experimental melanomas produce a similar effect, with a systematic reduction of plasma glutamine occurring during the logarithmic growth phase of the tumor. At present, the mechanism is obscure,

---

1 Free urine amino acids.

2 Day zero controls are normal urine amino acid concentrations determined by averaging 6 separate analyses; each pool sample was obtained from 10 normal mice at various times during the experiment. Other values are from single analyses of pooled urines collected from the surviving mice whose number varied from 20 at the beginning of the experiment to 10 on Day 25.

3 Nanomoles per ml of urine with their standard deviations (s.d.).

The specific gravity of the urine samples varied from 1.043 to 1.081 with the mean being 1.067 for 18 samples tested.

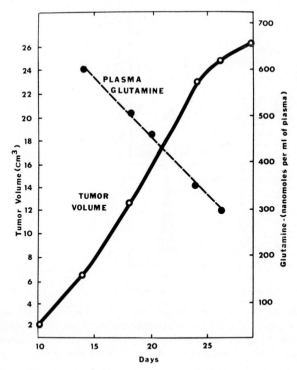

Fig. 2. Decrease in plasma glutamine during the growth of the EARAD-1 asparagine-dependent mouse lymphoma.

since we are reluctant to assume that this glutamine depletion simply represents an unusually high utilization of glutamine by the tumor. Considering the ability of the host to rapidly replace glutamine which is physically removed, it seems more likely that some interference by the tumor or its products with the metabolic processes may be occurring; conceivably in the synthesis, influx, or clearance of this or of critical precursor amino acids, or by the metabolic consequences of tumor growth upon the liver or other organs involved in maintaining physiological equilibrium.

It is appropriate to ask the following question: if protein synthesis is increased as a consequence of tumor growth, should not all plasma amino acids be partially depleted? Figure 3 illustrates that this is not the case. Upon examining the alterations of free alanine in the plasma during tumor growth, we find the contrary trend, with alanine being increased while glutamine was simultaneously depleted. There undoubtedly are simple biochemical and physiological explanations for these and the other observed amino acid altera-

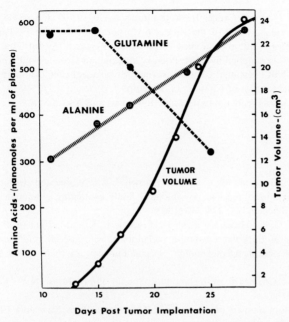

Fig. 3. Decrease in plasma glutamine and concomitant increase in plasma alanine during the growth of the B-16 melanotic mouse melanoma.

tions during the malignant process; however, at this stage of our studies we can only speculate, and attempt to design experiments to resolve these intriguing questions.

Conclusions

Changes occur in the concentration of a variety of both urinary and circulating plasma amino acids during the course of melanoma growth. Some amino acids in serum from patients were increased by more than 600%, whereas others were significantly decreased. In experimental animals, plasma glutamine was systematically depleted while alanine simultaneously increased during tumor growth.

Acknowledgements

These studies were supported in part by research grants from the American Cancer Society (T-522) and the ACS Washington State Division; the National Institutes of Health

(NIH Grant FR-5520); National Cancer Institute Grants CA 03192-09, CA 08748 and CA 12188; and Brown-Hazen Fund of the Research Corporation; the Elsa U. Pardee Foundation; the Ledbetter Fund; and the New York Cancer Research Institute, Inc.

Research animals were supplied by S. M. POILEY, Mammalian Genetics and Animal Production Section, CCNSC, National Cancer Institute. Miscellaneous NCI support was possible through the kind interest of CHARLES G. ZUBROD.

Technical assistance was rendered by LUCILLE BREDBERG and ANN SEVER.

References

1 BENSON, J. V., Jr.; GORDON, M. J. and PATTERSON, J. A.: Accelerated chromatographic analysis of amino acids in physiological fluids containing glutamine and asparagine. Anal. Biochem. 18: 228–240 (1967).
2 BURK, D.; ALGIRE, G. H.; HESSELBACH, M. L.; FISCHER, C. E. and LEGALLAIS, F. Y.: Tissue metabolism of transplanted mouse melanomas, with special reference to characterization of paraphenylenediamine. Special Publications, N.Y. Acad. Sci. IV: 437–446 (1948).
3 GREENSTEIN, J. P.: Chemistry of melanomas. Special Publications, N.Y. Acad. Sci. IV: 433–436 (1948).
4 GREENSTEIN, J. P.: Biochemistry of cancer; 2nd ed. (Academic Press, New York 1954).
5 RILEY, V.: Synergistic reaction between p-phenylenediamine and melanoma components. Proc. Soc. exp. Biol. Med. 93: 57–61 (1958).
6 RILEY, V.: Demonstration of differences between normal and tumor-bearing animals. Proc. Soc. exp. Biol. Med. 97: 169–175 (1958).
7 RILEY, V.: The melanoma as a model in a rational chemotherapy study; in M. GORDON Pigment cell biology, pp. 389–433 (Academic Press, New York 1959).
8 RILEY, V.: Adaptation of orbital bleeding technique to rapid serial blood studies. Proc. Soc. exp. Biol. 104: 751–754 (1960).
9 RILEY, V.: In vitro and in vivo reactions between tumor products and anti-tumor compounds. Acta Un. int. Cancer 16: 735–751 (1960).
10 RILEY, V.: Tumors as unique biological tools in basic biochemical studies. Trans. N.Y. Acad. Sci. 22: 248–364 (1960).
11 RILEY, V. and PACK, G. T.: Enzymic, metabolic, electron microscopic, and clinical characteristics of a human malignant blue nevus. Symp. Structure and Control of the Melanocyte, pp. 184–199 (Springer, Berlin 1966).
12 RILEY, V.: Some enzymatic and metabolic characteristics of malignant pigmented tissues; in W. MONTAGNA and F. HU Advances in biology of skin, vol. 8, pp. 581–619 (Pergamon Press, New York 1967).
13 RILEY, V.: Role of the LDH-elevating virus in leukaemia therapy by asparaginase. Nature, Lond. 220: 1245–1246 (1969).
14 RILEY, V.; SPACKMAN, D. H. and FITZMAURICE, M. A.: Critical influence of an enzyme-elevating virus upon long-term remissions of mouse leukemia following asparaginase therapy; in E. GRUNDMAN and H. F. OETTGEN Experimental and clinical effects of L-asparaginase, pp. 81–101 (Springer, New York 1970).
15 RILEY, V.; SPACKMAN, D. H. and FITZMAURICE, M. A.: Influence of asparaginase

M. Boiron La L-asparaginase, pp. 139–158 (Centre National de la Recherche Scientifique, Paris 1971).
16 Spackman, D. H.; Stein, W. H. and Moore, S.: Automatic recording apparatus for use in the chromatography of amino acids. Anal. Chem. *30:* 1190–1206 (1958).
17 Spackman, D. H.: Improved resolution in amino acid analysis in cancer therapy studies. Fed. Proc. *28:* 898 (1969).
18 Spackman, D. H. and Riley, V.: E-N-monomethyllysine and other plasma amino acids in leukemic mice, and effects of asparaginase. Fed. Proc. *30:* 1067 (1971).
19 Spackman, D. H.: Unpublished data; manuscript in preparation.
20 Umbreit, W. W.; Burris, R. H. and Stauffer, J. F.: Manometric techniques and tissue metabolism (Burgess Publ., Minneapolis 1949).
and glutaminase upon free amino acids in normal and tumor-bearing mice; in

Author's address: Dr. Vernon Riley, Chairman, Department of Microbiology, Pacific Northwest Research Foundation, 1102 Columbia Street, *Seattle, WA 98104* (USA)

Serum Haemopexin in Human Malignant Melanomas

J. J. Bourgoin, Y. Manuel, J. M. Sonneck, K. Gronneberg and M. C. Defontaine

Department of Immunology and Experimental Oncology, Centre Léon Bérard, Lyon, and Department of Research on Protein and Macromolecules Institute Pasteur, Lyon

Introduction

In a previous paper [2], we reported that in some cases of melanoma-bearing patients, the level of serum haemopexin was often higher than in normal serum. A correlation seemed apparent between the evolution of the tumor or the effectiveness of treatment, and this level.

Material and Methods

a) Technique

All our measurements were performed by radial immunodiffusion with an anti-haemopexin serum whose monospecificity had been carefully tested. For the test, we used 30 μl of antiserum in 220 μl of 1.5% agarose, and 4 μl of the patient's serum, diluted to ¼.

b) Sera

We tested 3 groups of subjects: group I, normal population (70 cases); group 2, melanoma-bearing population (100 cases); group 3, diverse tumors-bearing population (20 cases).

The levels were compared with a pool of sera from normal donors.

Results

In the first and third groups, the mean level was 80 units. The physiological level was, therefore, between 60 and 100 units, and the level in diverse non-melanoma cancer patients was about the same.

In the second group, the results were very different, with a variation from 40 to 400 units and a mean of about 124 units. In view of this spread of results, we tried to see if two populations of patients could be distinguished. We found no modification with respect to the stage of the illness: stages 1, 2 and 3 showed the same variations of the results.

By contrast, there existed two groups of subjects if we considered the evolution of the tumor. We call 'slowly evolving melanomas' those cases in which no modification of the clinical status was observed three months before and three months after the measurement. If, however, any new metastases or cutaneous resurgences were observed during this period, then it was considered a 'fast evolving melanoma'.

In this way, we obtained two populations whose haemopexin levels are given in figure 1. For the 'slowly evolving melanomas', they varied in most cases between 40 and 100 units, with a mean value of 77.4, whereas for the 'fast evolving melanomas', in most cases, levels ranged between 100 and 260 units, with a mean value of 170.5. In table I our results are statistically analyzed: the difference between the two groups of melanoma-bearing patients is significant at >0.001 using Student's T-test.

In addition, we tested extracts from normal and involved lymph nodes, and from normal and involved livers, to determine whether there is any evidence of an increase in haemopexin in these organs, in association with the presence of malignant melanoma cells. The extracts, obtained from the histologically controlled lymph nodes and liver of the same patient, were freeze-dried and concentrated ten times. Figures 2–5 show that the involved organs contain considerably more haemopexin than the normal ones.

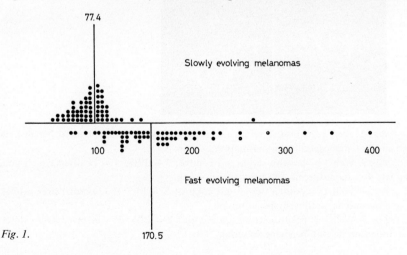

Fig. 1.

Table I

	'Slowly evolving melanomas'	'Fast evolving melanomas'	Normal population
Number of cases	120	60	70
Mean	77.4	170.55	80
σ	17.8	71.8	10

Student's t = 13.5 significance > 0.0001.

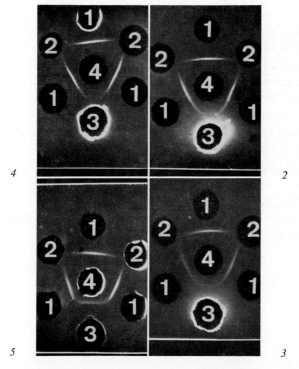

Fig. 2. Well 1 = positive serum test control; 2 = PBS; 3 = normal lymph node extract; 4 = monospecific anti-haemopexin serum.

Fig. 3. Well 1 = positive serum test control; 2 = PBS; 3 = involved lymph node extract; 4 = monospecific anti-haemopexin serum.

Fig. 4. Well 1 = positive serum test control; 2 = PBS; 3 = normal liver extract; 4 = monospecific anti-haemopexin serum.

Fig. 5. Well 1 = positive serum test control; 2 = PBS; 3 = involved liver extract; 4 = monospecific anti-haemopexin serum.

Discussion

To date, very few papers have been published dealing with variations of haemopexin in pathological processes. MÜLLER-EBERHARDT found that the haemopexin level was very low in haemolytic anemias. By comparison, we found that patients in the final stages of 'fast evolving melanomas', with low haemopexin levels in their serum, always had anemia.

Haemopexin is a heme- and cytochrome-binding protein and, as we view it, may be of some importance in cellular metabolism. In fact, the melanoma cell utilizes such a metabolism for the synthesis of melanin. Moreover, a relationship may exist between this metabolism and the results of thermography, evidenced by the fact that the cutaneous malignant melanoma appears as a hot area in thermoscans.

Using indirect immunofluorescence, we actually tried to determine whether haemopexin is synthesized by melanoma cells, and, if so, we would conclude from this, that there is a direct relationship between the melanoma cell and the serum haemopexin level [1].

Conclusion

Within our total group of melanoma patients, our results allowed us to distinguish two sub-groups:

a) the 'fast evolving melanomas', where the haemopexin levels varied between 100 and 250 units, with a mean value of 170.5. In these cases, cutaneous resurgences and metastatic lesions were observed within three months following measurement.

b) The 'slowly evolving melanomas', where the haemopexin level varied less – between 40 and 100 –, with a mean value of 77.4. In these cases, the patients were free of recurrence three months later.

References

1 BOURGOIN, J.J.; MANUEL, Y. and SONNECK, J. M.: Haemopexin levels. Variations in human malignant melanoma. Rev. Inst. Pasteur, Lyon (to be publ.) (1972).
2 MANUEL, Y.; DEFONTAINE, M. C.; BOURGOIN, J. J.; DARGENT, M. and SONNECK, J. M.: Serum haemopexin levels in patients with malignant melanoma. Clin. chim. Acta *31:* 485–486 (1971).

Author's address: Dr. J. J. BOURGOIN, Department of Immunology and Experimental Oncology, Centre Léon Bérard, 28, rue Laénnec, *F-69 Lyon 8ᵉ* (France)

Cellular Immunity in Malignant Melanoma[1]

D. A. CLARK and L. NATHANSON

Department of Medicine, Tufts University School of Medicine Boston, Mass.

Introduction

Circumstantial evidence suggests that immunologic reactions influence the course of human malignant melanoma. For example, primary lesions often contain areas of flattening and depigmentation over sites of dense lymphoid infiltration [13], BCG injection into skin metastases may induce regression of uninjected lesions [17], and biopsies of spontaneously regressing skin metastases demonstrate features of delayed hypersensitivity [2]. In the majority of patients, unfortunately, such spontaneous regressions do not occur. This may not be due to deficient cellular immunity, but to the presence of 'blocking factors' [7, 8, 9]. When present in sufficient concentration, these neutralize and prevent immune cells from reaching the tumor and eliminating it. In spite of blocking factors, the patient may express sufficient cellular immunity *in vivo* to retard tumor growth and dissemination. a) Suppression of this small but significant immune force with cytotoxic drug therapy may account for the transient nature of any response. Certain sequences of drug and antigen exposure may induce tolerance or augmentation of the immune response [21]. Such rebound augmentation correlates with a better prognosis [5]. Schedules of anti-tumor drug therapy which produce the desired effect of increasing the immune response must, therefore, be defined. b) *in vitro* tests can now provide a quantitative measure of specific immune defense. Thus, simplifying the search for immunostimulatory drug schedules.

1   Supported in part by grants: American Cancer Society, No. T-550; Damon Runyon Memorial Fund, No. 1067A, and National Institutes of Health, No. CA-12924-01.
2   Ontario Department of Health Fellowship.

Previous work from this laboratory demonstrated cellular hypersensitivity to a cross-reacting melanoma antigen which did not appear to be related to melanoma pigment [10]. The following is a report of our preliminary data from an *in vitro* assay for cell-mediated immunity and serum blocking utilizing the cross-reacting antigens of melanoma cell lines.

Materials and Methods

Patients

Patients receiving chemotherapy and BCG immunotherapy [16] at the Tufts-NEMCH were tested. Most had progressing disease, but a few had either sustained remissions or were regressing after a 4 day course of therapy with 4 (or 5) (3,3-dimethyl-1-triazeno)-imidazole-5 (or 4)-carboxamide (DTIC). An attempt was made to test patients prior to any therapy, but when this was not possible, at least 4 weeks since last drug therapy was allowed.

Sera and Lymphocytes

Patient and control sera were inactivated at 56 °C for 30 min, filtered through a 0.22 μ Millipore filter, and stored at -20 °C until use. A 1:6 dilution with 0.01 M phosphate buffered saline pH 6.8 was made. Several methods of lymphocyte preparation were tried including hypaqueficoll flotation [19], gelatin sedimentation and glass absorbtion of the remaining phagocytes, plasmagel (Laboratoire Roger Bellon, Neuilly, France) sedimentation and glass absorbtion [8], and the Woods' method [29]. In the latter method, lightly heparinized blood is sedimented with plasmagel and the supernatent deformated with 66 mesh nylon powder. This was used for most assays since it achieved a lymphocyte purity approaching 99%, and fibrin, platelets and lymphocytes which stick to the bottom of the assay wells were eliminated. Viability was greater than 90% by trypan blue exclusion.

Tumor Cells

Metastatic nodules were excised, minced, and established in monolayer culture in Ham's F-10, 20% fetal calf serum (Microbiologic Associates), with penicillin 50 U/ml and streptomycin 50 U/ml. In most cases, target cells were not used until 6 to 8 weeks and several subcultures had elapsed. The cultures were taken off antibiotics, and two weeks later bacterial cultures and cultures for PPLO done. The FOGH FL amnion assay for PPLO has now replaced standard PPLO cultures [4]. Where mycoplasma have been found, intensive antibiotic therapy has been used to suppress them before the cells were used. Tumor line donors and all patients and controls were ABO typed and HLA typed; wherever possible, ABO compatable tumor cells were used as targets. Dopa-oxidase stains identified the cells as melanoma.

Assay

Tumor cells were harvested by trypsinization (0.25–0.025%, Grand Island Biologicals), suspended in media with 20% heat inactivated fetal calf serum, and seeded 50–400 per well in 0.2–0.5 ml media in Linbro Disposotrays. The following day, media was decanted and 0.2–0.5 ml of dilute serum added and incubated 45 min 37°, 8% CO_2. This was also decanted and 0.5×10^6 lymphocytes in 0.5 ml media added. The trays were rocked 45 min on a Bellco rocker and an additional 0.5 ml 40% fetal calf serum in media added. The rocking continued for 48 h. The trays were then washed with saline, fixed, stained, and the number of tumor cells per well counted. A randomized layout of sera and lymphocyte seeding was developed in later assays to eliminate any bias due to non-specific growth and plating effects.

Controls

Lymphocytes were treated with mitomycin-C to prove reactions with tumor cells were lymphocyte mediated and not due to allogeneic inhibition or agglutination cytotoxicity [15]. When available, other types of cancer cell lines were used as specificity controls. Lymphocytes from other types of cancer patients were included.

Analysis

Predesigned comparisons of two samples were analysed using Student's T-test. Comparability was established by F-ratio of the variances, T-test results were confirmed by one-way analysis of variance. General comparison of means was accomplished with the Duncan range test [24].

Results

A series of assays using Falcon Microtest II (Falcon Plastico, Oxnaid, California) plates with 6 mm diameter wells [6, 8] is outlined in table I. The only positive results were found in controls. 4 out of 5 negative patients retested in 16 mm diameter Linbro wells gave positive results: in 2 out of 4, the same target cell line was tested. We observed that rocking small wells on the Bellco rocker was ineffective due to surface tension effects. A parallel comparison of the two well sizes is outlined in table II. Neither cell-mediated effects nor blocking occurred in the 6 mm wells in contrast to the 16 mm wells where striking effects were noted. Cell-mediated immunity and blocking with allogeneic serum from patient BR were highly significant. Although significant cell-mediated immunity was not detected with autochthonous BR lymphocytes, inclusion of BR serum induced a feeder effect and a significant increase in the number of BR cells. Blocking effect implies the existence of cell-mediated immunity which was not detectable due to insensitivity of the assay. This insensitivity was overcome by increasing the number of replicate wells in later assays.

A summary of results of tests for autochthonous and allogeneic cell-mediated immunity and serum blocking is given in table III. 8 out of 10 patients and 1 out of 6 controls showed significant cell-mediated immunity.

Table I. Microcytotoxicity test done in 6 mm diameter wells

Target cell	Lymphocyte method	Lymphocyte donor	LC per well	No. fg tumor cells surviving ±SEM	Significance
ME	Hypaque-ficoll	Melanoma K	0.3×10^6	1.0±0.72	NS
		Control C	0.3×10^6	2.4±0.72	–
		Melanoma K	0.1×10^6	9.0±2.2	NS
		Control C	0.1×10^6	7.3±2.1	–
BR	Hypaque-ficoll	Melanoma BR	0.1×10^6	114±6.04	NS
		Control B[1]	0.09×10^6	127±13.81	–
		Melanoma BR	0.1×10^5	127±5.81	NS
		Control B[1]	0.09×10^5	97±6.32	–
BR	Hypaque-ficoll glass absorbed	Melanoma C	0.5×10^6	76±2.05	NS
		Control N[2]	0.5×10^6	36±4.67	0.01
BR	3% gelatin plastic absorbtion	Melanoma K	0.5×10^6	357±44.3	NS
		Melanoma KPL	0.5×10^6	274±24.8	NS
		Control N[2]	0.5×10^6	185±5.7	0.005
ME	3% gelatin glass absorbed	Melanoma ME	0.5×10^6	27±2.0	NS
		Melanoma M	0.5×10^6	28±7.3	NS
		Melanoma KPL	0.5×10^6	50±4.3	0.05
		Control Y[1]	0.5×10^6	33±5.3	–
		Control B[1]	0.5×10^6	8±1.4	0.05
BR	Woods	Melanoma BR	0.5×10^6	125±19.7	NS
		Melanoma AP	0.5×10^6	154±27.4	NS
		Control JC	0.5×10^6	132±17.9	–

1 Patient with cancer of a different type.
2 Positive control in subsequent assays.

Table II. Comparison of 6 mm to 16 mm well diameters

Lymphocyte donor	Control serum treated			
	6 mm well diameter		16 mm well diameter	P value
Progressor BR	125±19.7	NS	332±26.8	NS
Regressor AP	154±27.4	NS	181± 9.9	0.0025
Control JC	132±19.9	–	412±50.0	–

	Progressor serum treated			
	6 mm well diameter	16 mm well diameter	P value	P control vs. progressor serum
Progressor BR	145±15.6	572±32.2	NS	0.05
Regressor AP	150±27.4	657±28.5	0.01	0.001
Control JC	122± 4.1	530±32.2	–	NS

Table III. 16 mm wells: summary of CMI and blocking antibody test

	Lymphocyte donor	Cell count Control serum	(Mean±SEM) Melanoma serum	Significance of diff. CMI	Serum effect
Autochthonous					
ME	Melanoma ME	149± 9.8	264±16.0	0.005	0.01
	Control C	193±10.6	140± 9.0	–	0.01
BR	Melanoma BR	332±26.8	572±32.2	NS	0.05
	Control JC	412±50.0	530±28.5	–	NS
KPL	Melanoma KPL	107±17.6	133± 8.9	0.05	NS
	Control AS	155±17.4	124±19.2	–	NS
	Control JC	163±18.7	124± 8.8	–	NS
Allogeneic					
BR	Melanoma AP	181± 9.9	657±28.5	0.0025	0.001
	Control JC	412±50.0	530±32.2	–	NS
KPL	Melanoma H Positive	218±43.5	[1]305±47.9	0.01	NS[1]
	Control N	178±38.5	337±40.5	0.005	0.05
	Control L	434±45.6	[1]261±31.6	–	0.05[1]
KPL	Melanoma TH	71± 8.0	–	0.0025	
	Control JC	163±18.7	–	–	
MG	Melanoma SH Positive	24± 3.7	–	0.025	
	Control N	23± 4.3	–	–	
	Control JC	36± 6.0	–	–	
BR	Melanoma M Positive	163±10.6	165±15.8	0.05	NS
	Control N	142±15.6	116±11.3	0.01	NS
	Control JC	254± 9.25	204±15.4	NS	0.05
	Control AF[2]	216±22.1[2]	85±11.3[2]	–	0.001
BR	Melanoma C	67±10.7	–	NS	
	Control JC	50± 2.7	–	NS	
	Control AF	62± 7.1	–		
LC	Melanoma G[2]	128± 7.9	–	.001	
	Melanoma AF[2]	222±14.3	–		

1 Serum ABO incompatable with target: serum blocks positive control. Blocking serum ABO compatable blocks patient.
2 Significant contamination with polymorphonuclear leukocytes.

Correcting for differing sensitivities between target cells from patients BR, KPL, MG, patients AP and T, who were drug responders, showed 56 and 42% cytotoxicity. The positive disease-free control had 44% cytotoxicity, and progressor patients all had less than 36%. The low value of 13.5% for

AP against line MG was obtained when she had recurrent disease. Corrected values were not available for all cell lines.

Serum effects are included in table III. Patients ME, BR, H, KPL, and M had progressive disease. ME, BR, KPL, and H showed blocking with their diluted serum. M did not show blocking, but in contrast to the other patients, his disease was progressing only in the CNS. In early studies, no blocking serum showed cytotoxicity with control lymphocytes. Two sera (ME, M) have now been identified which produce this effect, and a similar trend has been noted with KPL serum. H serum was ABO incompatable with target cells. The effect of the two regressor patients' sera have not been fully evaluated. The serum from 'positive control' N, who has no known disease but has a family history of melanoma, blocked only with KPL target cells.

Cell control data is shown in table IV. Table V documents the effect of mitomycin-C blockade. Only with melanoma lymphocytes was cytotoxicity inhibited. Where the concentration of mitomycin-C had been inadequate to abrogate cell-mediated immunity, the addition of blocking serum led to complete inhibition, and the number of tumor cells became the same as with untreated control lymphocytes.

Table IV. Target cells

Lymphocyte donor	(Mean ± SE		
	MG Melanoma	Carcinoma bowel	Carcinoma parotid
Melanoma SH	24 ± 3.7[1]	322 ± 27.8	60 ± 4.0
Positive Control N	23 ± 4.3[1]	310 ± 20.9	63 ± 5.6
Control JC	36 ± 6.0	285 ± 24.3	60 ± 3.7
	MG melanoma	Carcinoma colon	
Colon CA B[2]	514 ± 29.5	15 ± 1.4	
Melanoma AP	448 ± 17.7[1]	11 ± 1.1	
Control JC	513 ± 29.4	11 ± 1.0	
	LC melanoma	Carcinoma ovary	
Melanoma G[2]	128 ± 7.9[3]	78 ± 7.9[1]	
Control AF[2]	222 ± 14.3	98 ± 4.8	

1 Significantly lower than Control P less than 0.05.
2 Significant contamination with polymorphonuclear leukocytes.
3 Significantly lower than Control P less than 0.001. Correction for non-specific cytotoxicity leaves 21.9 % cytotoxicity against melanoma cells alone.

Table V. Mitomycin-C blockade

Treatment of lymphocytes	Target cell Melanoma Lymphocytes		Survival (Mean±SE) Control lymphocytes
	218±45.5		434±45.6
Mitomycin-C 8.3 mcg/ml[1]	290±26.8		–
Blocking serum[2]	345±17.5	0.05	310±38.3
Mitomycin-C + blocking serum	411±35.6		–
	178±38.5		434±45.6
Mitomycin-C 8.3 mcg/ml[1]	337±57.2		–
Blocking serum[2]	354±42.9	0.05	310±38.3
Mitomycin-C + blocking serum	490±63.0		–
	107±17.6		163±18.7
Mitomycin-C 2.5 mcg/ml[3]	126±17.4		109±19.0
Blocking serum[2]	133± 8.9	0.05	124± 8.8
Mitomycin-C + blocking serum	166± 9.2		51±25.6
			155±17.4
Mitomycin-C 4.1 mcg/ml[3]			164±18.0
Blocking serum[2]			124±19.2
Mitomycin-C + blocking serum			123±25.1

1 25 mcg/10^7 cells, 37°C for 30 min then 2 washes.
2 ABO compatible with target cells.
3 25 mcg/10^7 cells, 37°C for 30 min then 1 wash.

Discussion

Despite the emphasis on blocking factors as the critical difference between progressor and regressor patients [7, 9], we are impressed with the differences in levels of cell-mediated immunity detected by our assay. That this was not due to reduced antigenicity of our lines was confirmed by Dr. Heppner (Rodger Williams Hospital, Providence, Rhode Island) who tested one of our lines with the Hellstrom microcytotoxicity test [8] and found it as sensitive as primary cell cultures. Three possible explanations for our failure to obtain evidence of cell-mediated immunity in 6 mm diameter wells are: 1. non-specific cytotoxicity due to 'sticky cells' in the lymphocyte preparation, and loss of the potentiating effect of such cells where the Woods' method of lymphocyte preparation was used; 2. type and concentration of fetal calf serum; 3. immunosuppressive therapy, and advanced disease. Lymphocytes prepared by the Woods' procedure contain numerous red cells which may interfere with lymphocyte-target cell interaction. Rocking overcomes this problem, but only in larger wells where surface tension does not inhibit motion of the media-air interface. In a recent experiment, extreme tilting and rocking of 6 mm diameter wells enabled detection of lymphocytotoxicity in these wells. The concentration of fetal calf serum we use is double that recommended by the HELLSTROMS, but relating this to well-diameter is difficult.

Striking levels of cell-mediated immunity were detected only in postchemotherapy regressors and the positive control. Where patients with disease of different stages were compared in the same assay, cell-mediated immunity was always greater in the patient with less disease. Mitomycin-C blockade indicates that such lymphocytotoxicity is due to activation of presensitized lymphocytes. Use of other tumor cells as control cells indicates such activation is specific for melanoma cells. More extensive data [8] supporting the specificity of the reaction have been published by the HELLSTROMS. This reaction is an *in vitro* correlate of *in vivo* immunity [12]. Our data suggests a deficiency in cell-mediated immunity in patients with progressing melanoma. Corroborative evidence for such a deficiency has been published for animal models [11, 12], and skin-testing with autologous melanoma extracts has shown specific loss of delayed hypersensitivity with advancing disease [3]. This deficiency of cellular immunity may explain the ineffectiveness of immunotherapy for metastatic disease [16].

The increase in cell-mediated immunity in drug-induced regressors is poorly understood. Others [5] have observed a rebound following chemotherapy which correlated with a better prognosis. Since we have no pretherapy assay values for our two regressors we cannot determine if the high levels of immunity represent such a rebound. Such a phenomenon could result from increased availability of tumor antigen from tumor lysis leading to stimulation of the afferent limb of the immune reflex arc [22, 27]. On the other hand, we have shown DTIC therapy may inhibit tumor antigenemia [18]. Such antigenemia may be immunosuppressive [1, 20, 25, 26, 28] by sequestration of immune lymphocytes in lymph nodes and spleen, and by a direct effect of antigen-antibody complexes on circulating immune cells, both lymphocytes [23] and macrophages [14]. These mechanisms suggest the increase in circulating cell-mediated immunity is a consequence of tumor regression rather than the cause. If cellular immunity levels in patients with untreated pathologically Stage I primary melanomas correlated with prognosis this would suggest that immunity exerted an important anti-tumor effect. More convincing would be an increase in immunity following BCG which correlated with subsequent tumor regression and increased survival. Both studies are currently underway in our laboratory.

Acknowledgements

Grateful acknowledgment is made to Miss ARLENE FLOWERS and Mrs. PATRICIA DOHERTY for technical assistance and to Dr. ROBERT SCHWARTZ for encouragement and criticism.

References

1 BERNSTEIN, I. D.; THOR, D. E. and RAPP, H. J.: Impairment of passive systemic tumor immunity in tumor-bearing hosts. Proc. Amer. Ass. Cancer Res. *12:* 40 (1971).
2 EVERSON, T. C. and COLE, W. H.: The spontaneous regression of cancer (Saunders, Philadelphia, Pa. 1966).
3 FASS, L.; HERBERMAN, R. B.; ZIEGLER, J. L. and KIRYABWIRE, J. W. M.: Cutaneous hypersensitivity reactions to autologous extracts of malignant melanoma cells. Lancet *i:* 116–118 (1970).
4 FOGH, J. and FOGH, H.: Procedures for control of mycoplasma contamination of tissue cultures. Ann. N.Y. Acad. Sci. *172:* 15–30 (1969).
5 HALTERMAN, R. H. and LEVENTHAL, B. G.: Enhanced immune response to leukemia. Lancet *ii:* 704–705 (1971).
6 HELLSTROM, I.; HELLSTROM, K. E. and ALLISON, A. C.: Neonatally induced allograft tolerance may be mediated by serum-borne factors. Nature, Lond. *230:* 49–50 (1970).
7 HELLSTROM, I.; HELLSTROM, K. E. and SJOGREN, H. O.: Serum mediated inhibition of cellular immunity to methylcholanthrene-induced murine sarcomas. Cell. Immunol. *1:* 18–30 (1970).
8 HELLSTROM, I.; HELLSTROM, K. E.; SJOGREN, H. O. and WARNER, G. A.: Demonstration of cell-mediated immunity to human neoplasms of various histologic types. Int. J. Cancer *7:* 1–16 (1971).
9 HELLSTROM, I.; SJOGREN, H. O.; WARNER, G. and HELLSTROM, K. E.: Blocking of cell-mediated tumor immunity by sera from patients with growing neoplasms. Int. J. Cancer *7:* 226–237 (1971).
10 JEHN, U.; NATHANSON, L. and SCHWARTZ, R. S.: In vitro lymphocyte stimulation by a soluble antigen from malignant melanoma. New Engl. J. Med. *283:* 329–333 (1970).
11 KLEIN, W. J.: Lymphocyte-mediated cytotoxicity in vitro. J. exp. Med. *134:* 1238–1252 (1971).
12 LEFRANCOIS, D.; YOUN, J. K.; BELEHRADEK, J. and BARSKI, G.: Evolution of cell-mediated immunity in mice bearing tumors produced by a mammary carcinoma cell line. Influence of tumor growth, surgical removal, and treatment with irradiated tumor cells. J. nat. Cancer Inst. *46:* 981–987 (1971).
13 LLOYD, O. C.: Regression of malignant melanoma as a manifestation of a cellular immune response. Proc. roy. Soc. Med. *62:* 543–545 (1962).
14 MITCHELL, M. S.; MOKYR, M. B. and GOLDWATER, D. J.: Specific inhibition of receptors for cytophilic antibodies on macrophages by isoantibody. Proc. Amer. Ass. Cancer Res. (1972).

15 MOLLER, G. and LUNDGREN, G.: Aggressive lymphocytes and sensitive target cells: two pathways for cytotoxicity, in R. T. SMITH and R. A. GOOD Cellular recognition, (Appleton-Century-Crofts, New York 1969).
16 MORTON, D. L.; EILBER, F. R.; MALMGREN, R. A. and WOOD, W. C.: Immunologic factors which influence response to immunotherapy in malignant melanoma. Surgery 68: 158–164 (1970).
17 NATHANSON, L.: Experience with BCG in malignant melanoma. Proc. Amer. Ass. Cancer Res. 12: 99 (1971).
18 NATHANSON, L.; JEHN, U. and SCHWARTZ, R. S.: Disappearance of a tumor-associated antigen in malignant melanoma after imidazole carboxamide therapy. Cancer 27: 411–416 (1971).
19 PEPPER, R. J.; ZEE, T. W. and MICKELSON, M. M.: Purification of lymphocytes and platelets by gradient centrifugation. J. Lab. clin. Med. 72: 842–848 (1968).
20 SCHLOSSMAN, S. F.; LEVIN, J. A.; ROCKLIN, R. E. and DAVID, J. R.: The compartmentalization of antigen-reactive lymphocytes in desensitized guinea pigs. J. exp. Med. 134: 741-750 (1971).
21 SCHWARTZ, R. S.: Alteration of immunity by antimetabolites, in Immunity, cancer, and chemotherapy (Academic Press, New York 1967).
22 SIMMONS, R. L. and RIOS, A.: Immunotherapy of cancer: immunospecific rejection of tumors in recipients of neuramonidase-treated tumor cells plus BCG. Science 174: 591–593 (1971).
23 SJOGREN, H. O.; HELLSTROM, I.; BANSAL S. C. and HELLSTROM, K. E.: Suggestive evidence that the 'blocking antibodies' of tumor bearing individuals may be antigen-antibody complexes. Proc. nat. Acad. Sci., Wash. 68: 1372–1375 (1971).
24 SNEDECOR, G. W. and COCHRANE, W. G.: Statistical methods (Iowa State University Press, Ames, Iowa 1971).
25 STUTMAN, O.: Growth of antigenic tumors in the presence of specific cellular immunity. Proc. Amer. Ass. Cancer Res. 12: 14 (1971).
26 VAAGE, J.: Specific depression of immunity by residual or injected syngeneic tumor tissue. Proc. Amer. Ass. Cancer Res. 12: 33 (1971).
27 WATKINS, E.; OGATA, Y.; ANDERSON, L. L.: WATKINS, E. III and WATERS, M. F.: Activation of host lymphocytes cultures with cancer cells treated with neuraminidase. Nature (new Biol.) 231: 83–85 (1971).
28. WEPSIC, H. T.: Abrogation of passively transferred tumor immunity in vitro by antigenically related tumor cells. Proc. Amer. Ass. Cancer Res. 12: 80 (1971).
29. WOODS, A. H.: A closed system for large-scale lymphocyte purification. Blood 35: 39–43 (1970).

Author's address: Dr. DAVID A. CLARK, Department of Medicine, Tufts University School of Medicine, *Boston, Mass,* (USA)

Cell-Mediated Immunity to Malignant Melanoma

A. J. Cochran, U. W. Jehn and B. P. Gothoskar

University Department of Pathology, The Western Infirmary, Glasgow and Department of Tumour Biology, Karolinska Institute, Stockholm

Introduction

A proportion of patients with malignant melanoma appear to make an immune response to antigens present on their tumour cells. More is known of the humoral aspects of this response [7, 9, 12, 14] but some authors have reported tumour-directed cell-mediated immunity [5, 6, 13]. Jehn et al. [8] found that melanoma extracts stimulated autologous lymphocytes and one preparation, fluid from a cystic melanoma, also stimulated lymphocytes from other melanoma patients. We have examined the cyst fluid for activity in the leucocyte migration test and attempted to isolate the active component.

Materials and Methods

a) Fluid From Cystic Melanoma
This material has been described and characterised in part previously [8].

b) The Leucocytes
The clinical status of the cell donors was unknown at the time of testing. The leucocytes from 30 cc of heparinised blood were washed three times in Eagles minimum essential medium (MEM), counted and resuspended in MEM to a concentration of 3×10^6 leucocytes per 20µl.

c) 'Antigens'
The protein concentration of the cyst fluid and its fractions was assessed by spectrophotometry. These materials were then diluted with MEM to provide a range of protein concentration from 12.5 to 100 µg/ml.

d) Leucocyte Migration Test

20 µl samples of the cell suspension containing 3×10^6 cells were drawn into capillary tubes. The migration technique employed has been described in detail elsewhere [3] and is modified from that of BENDIXEN and SØBORG [2].

e) Fractionation Procedures

Cyst fluid was passed through a Sephadex G200 column (column size 2.5×100 cm) using sodium phosphate buffer pH 7.4. A Sephadex G100 column was employed in an attempt to detect substances with a molecular weight less than 100,000. Characterisation of the fractions was attempted with specific antisera and the Ouchterlony technique.

f) Statistical Analysis

Groups of test and control samples from each individual were compared using the WILCOXON-MANN-WHITNEY U test of ranking with double ranking where appropriate. Groups of results from each diagnostic category were compared by the chi^2 test. Significance is assessed at the 5% level.

g) Mixed Haemadsorption

A modification of the test described by FAGRAEUS et al. [4] was used. Five µl samples of the cyst fluid and its fractions were smeared on to slides and acetone fixed. The slides were flooded with a 1 in 2 dilution of sera known to react with melanoma cell antigens in membrane immunofluorescence or with negative control sera and incubated at room temperature (22°C) for 60 min in a moist chamber. The smears were then rinsed twice with phosphate buffered saline (PBS) and covered with a thin layer of indicator cell suspension, sheep erythrocytes (SE) coated with human antiserum to SE and then with sheep antiserum to human gammaglobulin G. After a further incubation for 1 h at room temperature the smears were rinsed twice with PBS and examined microscopically. Slides retaining a large number of indicator erythrocytes were scored as positive.

Results

1. Whole Cyst Fluid (Table I)

Migration tests were performed on the peripheral blood leucocytes of 22 melanoma patients using unfractionated cyst fluid as the antigen and significant migration inhibition occurred in 11 tests. Analysis of the patients by clinical stage shows that the leucocytes of 10 out of 16 patients with disease localised to the primary site were inhibited while the leucocytes of only 1 out of 6 patients with recurrence or metastasis were inhibited. Leucocytes from 21 control individuals were examined and migration inhibition occurred in 5 out of 21. Five of these individuals had skin carcinoma and inhibition was observed in 3 of these 5. Inhibition occurred in only 2 out of 16 tests employing leucocytes from normal healthy individuals.

Table I. The frequency of significant leucocyte migration inhibition by contact with melanoma derived cyst fluid and the 'tumour antigen' fraction

Category of cell donor	Whole cyst fluid	P value[1]	Cyst derived 'tumour antigens'	P value[1]
Malignant melanoma Non-recurrent	11/22 (50%)	.0001	5/10 (50%)	<.005
malignant melanoma Recurrent	10/16 (63%)	<.0001	4/6 (67%)	<.01
malignant melanoma	1/6 (17%)	<.05	1/4 (25%)	<.05
Skin carcinoma	3/5 (60%)	–	2/5 (40%)	–
Normal	2/16 (13%)	–	1/9 (11%)	–

1 P value obtained by comparing frequency of positive tests in group with frequency of positive tests in normal control group by chi^2 technique.

2. Fractionation Results

Passage of the cyst fluid through Sephadex G200 produced identifiable IgM, IgG and albumin peaks. Sephadex G100 fractionation provided peaks similar to those found with G200 and an additional late fourth peak which did not react with antisera to normal human serum components. This was regarded as possibly corresponding to the 'tumour antigen' fraction obtained by starch block electrophoresis in the previous study [8]. This view was supported by the activity of this late fraction in the migration inhibition test and by the results of mixed haemadsorption tests.

Samples of the cyst fluid and all fractions derived from it were exposed in the mixed haemadsorption test to sera which had given a positive immunofluorescent reaction with melanoma tissue culture lines [GOTHOSKAR, unpublished observations] and to negative control sera. Strong positive mixed haemadsorption occurred only with the Sephadex G100 separated 'tumour antigen' fraction. A very weak reaction was noted with the IgG fraction but all other fractions were negative.

3. Studies With the 'Tumour Antigen' Fraction

This material produced a pattern of reactivity very similar to that of the whole cyst fluid. Leucocytes from 5 out of 10 melanoma patients were inhibited by contact with this material. Analysis of the patients by clinical stage showed that the leucocytes of 4 out of 6 patients with melanoma localised to the primary site were inhibited while the leucocytes of only 1 out of 4 patients with recurrence or metastases were inhibited.

Inhibition was seen in 3 out of 14 tests with control leucocytes, 2 out of 5 tests of skin carcinoma patients and 1 out of 9 tests of cells from normal controls.

4. Other Fractions

Migration inhibition did occur with most of the fractions at some concentrations in a minority of tests. There were, however, no significant differences between the occurrence of inhibition with these materials in the tumour-bearing group and the control group. Information available from this part of the study does not allow any conclusions to be drawn at present.

Discussion

These results confirm the findings of JEHN *et al.* [8] that there are materials present in this melanoma-derived fluid which react in an apparently specific manner with the leucocytes of some patients with malignant melanoma. Leucocyte inhibitory activity was associated with a low molecular weight (less than 100,000) fraction appearing after Sephadex G100 fractionation. JEHN *et al.* found their active fraction to have a molecular weight of approximately 40,000.

All tests involved allogeneic leucocytes and the results support the view that lymphocytes of melanoma patients are sensitised to a common antigen or closely similar group of antigens. Cross reactivity of melanoma antigens with allogeneic lymphocytes from melanoma patients has been reported by previous workers [5, 6, 8, 15].

Inhibition of leucocyte migration by contact with antigen(s) was relatively uncommon in patients with recurrent or metastatic tumour. It seems likely that these patients have a deficit of circulating lymphoid cells reactive with the melanoma antigens. This could be due either to an absolute deficiency of specifically reactive cells, to a failure of responsiveness of such cells, or to a redistribution of specifically active cells which might no longer be available in the peripheral blood.

HELLSTRÖM *et al.* [6] found no correlation of the lymphocyte donor's clinical stage with lymphocyte cytotoxicity or cell-mediated inhibition of melanoma cell colony formation. VAN 'T HOOG *et al.* [15], however, found specific cellular reactivity to melanoma cells most common in patients 'in an early stage of the disease or ... showing spontaneous tumour regression'.

It should be noted, however, that the 6 patients whose lymphocytes were stimulated by contact with the cyst fluid used in this study and in the study of JEHN et al. [8] all had metastatic disease.

Inhibition with the cyst fluid derived fractions other than the 'tumour antigen' fraction occurred with similar frequency in all groups indicating that the effect was not specifically related to the possession of a malignant melanoma. In the case of the Ig fractions we considered the possibility that naturally occurring antibodies might be involved but could demonstrate no relationship to blood groups.

The high frequency of leucocyte migration inhibition by the melanoma antigens using lymphoid cells from skin carcinoma patients is interesting. The numbers are, however, very small and this interesting cross-reactivity will require further study to allow any conclusions to be drawn.

Using material of known antigenic activity we have found the leucocyte migration technique relatively simple and apparently discriminatory. This confirms the experience of ANDERSEN et al. [1]. Our findings that inhibition occurred in some cases with fractions of the cyst fluid other than those expected to contain tumour-derived antigenic material gives cause for concern. It is certainly clear that care must be taken to employ well categorised homogeneous antigen sources in the test.

References

1 ANDERSEN, V.; BJERRUM, O.; BENDIXEN, G.; SCHIØDT, T. and DISSING, I.: Effect of autologous mammary tumour extracts on human leucocyte migration *in vitro*. Int. J. Cancer *5:* 357–363 (1970).
2 BENDIXEN, G. and SØBORG, M.: A leucocyte migration technique for *in vitro* detection of cellular (delayed type) hypersensitivity in man. Dan. med. Bull. *16:* 1–6 (1969).
3 COCHRAN, A.J.: Tumour cell migration. Europ. J. clin. biol. Res. *16:* 44–47 (1971).
4 FAGRAEUS, A. and EPSMARK, Å.: Use of a mixed haemadsorption method in virus infected tissue cultures. Nature, Lond. *190:* 370–371 (1961).
5 FOSSATTI, G.; COLNAGHI, M. E.; DELLA PORTA, G. and CASCINELLI, N.: Cellular and humoral immunity against human malignant melanoma. Veronesi, U. Int. J. Cancer *8:* 344–349 (1971).
6 HELLSTRÖM, I.; HELLSTRÖM, K. E.; SJÖGREN, H. O. and WARNER, G. A.: Demonstration of cell-mediated immunity to human neoplasms of various histological types. Int. J. Cancer *7:* 1–16 (1971).
7 IKONOPISOV, R. L.; LEWIS, M. G.; HUNTER-CRAIG, I. D.; BODENHAM, D. C.; PHILLIPS, T. M.; COOLING, C. I.; PROCTOR, J.; HAMILTON FAIRLEY, G. and ALEXANDER, P.: Autoimmunisation with irradiated tumour cells in human malignant melanoma. Brit. med. J. *ii:* 752–754 (1970).

8 JEHN, U.W.; NATHANSON, L.; SCHWARZ, R. S. and SKINNER, M.: *In vitro* lymphocyte stimulation by a soluble antigen from malignant melanoma. New Engl. J. Med. *283:* 329–333 (1970).
9 LEWIS, M. G.: Possible immunological factors in human malignant melanoma in Uganda. Lancet *ii:* 921–922 (1967).
10 LEWIS, M. G.; IKONOPISOV, R. L.; NAIRN, R. C.; PHILLIPS, T. M.; HAMILTON FAIRLEY, G.; BODENHAM, D. C. and ALEXANDER, P.: Tumour-specific antibodies in human malignant melanoma and their relationship to the extent of the disease. Brit. med. J. *iii:* 547–552 (1969).
11 MORTON, D. L.; MALMGREN, R. A.; HOLMES, E. C. and KETCHAM, A. S.: Demonstration of antibodies against human malignant melanoma by immunofluorescence. Surgery *64:* 233–240 (1968).
12 MUNA, N. M.; MARCUS, S. and SMART, C.: Detection by immunofluorescence of antibodies specific for human malignant melanoma cells. Cancer *23:* 88–93 (1969).
13 NAGEL, G. A.; PIESSENS, W. F.; STILMANT, M. M. and LEJEUNE, F.: Evidence for tumor-specific immunity in human malignant melanoma. Europ. J. Cancer *7:* 41–47 (1971).
14 OETTGEN, H. F.; AOKI, T.; OLD, L. J.; BOYSE, E. A.; DE HARVEN, E. and MILLS, G. M.: Suspension culture of a pigment producing cell line derived from a human malignant melanoma. J. nat. Cancer. Inst. *41:* 827–832 (1968).
15 VAN'T HOOG, A. G.; DE VRIES, J. E. and RÜMKE, P.: Tumour specific humoral and cellular immunity in patients with malignant melanoma. Proc. 1st Congr. Europ. Ass. Cancer Research Brussels 1970, p. 94.

Author's address: Dr. A. J. COCHRAN, University Department of Pathology, The Western Infirmary, *Glasgow* (Scotland)

Cytoplasmic Antigens in Human Malignant Melanoma Cells

J. J. BOURGOIN and A. BOURGOIN

Department of Immunology and Experimental Oncology, Centre Léon Bérard, Lyon

Introduction

Since an increasing number of researchers have reported on the presence of cytoplasmic antigens in human malignant melanoma cells, it was decided to undertake a study verifying cross reactivity against these antigens. We used the indirect immunofluorescence technique, supplemented, in some cases, by the complement fixation test for comparative purposes.

Material and Methods

a) Sera

Sixty melanoma-bearing patients were tested and, if possible, the test was performed at different stages of the illness. All sera were filtered with Millipore filters, with a final filtration of 0.22 μ, freeze-dried, and stored at $-70\,°C$.

b) Cells

Two types of cells have been tested: 1. cells obtained by Lewis' method from tumors subsequent to surgery; 2. cells maintained in long-term suspension culture in our laboratory: M CLB 1 strain in Waymouth suspension.

c) Immunofluorescence Technique

We used the indirect technique for cold fixed cells (isopentane and liquid nitrogen), as previously described [6].

For nearly all of our cases, deamination was performed between the contact of the serum and of the FITC conjugate, using Burtin's technique.

For microscopical observation, we used a Zeiss microscope, with BG 12 excitation filter and 50–53 barrier-filter.

d) Complement Fixation Test

The separation of the cytoplasm gave us an extract of 100,000 g (granule and ribosome free) and the supernatant (cytoplasmic antigen) was stored at $-70\,°C$. The test is the same as used by Sohier et al. [7] for the detection of antibodies in viral processes and we used a dilution of ¼ for the sera.

Results

a) Cross Reactivity by the Indirect Immunofluorescence Technique

Sera which were found to be positive against the autologous melanoma cells, were tested against other tumors. 23 sera were positive within the autologous system. In 22 cases, cross reactivity was detected with at least two other tumors.

A different reaction pattern was observed in the case of a malignant melanoma-bearing young woman, whose newborn had antibodies to his mother's cells. We tested the mother's and the child's serum against the tumor cells of another young woman, but found no cross reaction. The second patient's serum, however, cross-reacted with the tumor cells of the mother. These observations led us to think that multiple factors may be involved in the antigenicity of the melanoma cell.

This case, and a test in another newborn, moreover, show that the class of antibodies involved in melanoma processes, is mainly IgG. When we tested the serum of newborns against their mother's tumor cells, the reaction was found to be positive. Three months later, however, the test was negative: the presence of antibodies in the newborn relies on a transplacental transfer of the mother's own antibodies (table I).

The immunofluorescence technique revealed good cross reactivity of the positive sera. All negative sera in the autologous system (49) were found to be negative against other cells.

Table I.

Cells	Sera	Test
	Mother R (autologuous) :	+
R	Newborn R :	+
	3-month-old child :	0
	Mother G (autologuous) :	+
G	Newborn G :	+
	3-month-old child :	0

b) Cross Reactivity by the Complement Fixation Test

Using this method, 10 sera were tested. The major problem encountered was the anti-complement activity of these sera at the dilution used. We found only 8 positive cases in the autologous system and 7 cross reactions. The last case was not interpretable. The results of this technique too, seem to suggest that there exists a good cross reactivity against cytoplasmic antigens.

c) Cross Reactivity Against the Strain of Melanoma Cells in Suspension Culture (M CLB 1)

All tests were performed by indirect immunofluorescence. From 28 sera, 20 were positive, 15 of which had also been positive with the technique using fresh tumor cells (the remaining 5 positive cases were not tested in this way).

d) Relationship Between Stages and Positivity

Most of our positive cases (21 out of 27) were in stage I or II, i.e., with local tumor and regional lymph node involvement. Most of the negative cases (31 out of 33) were in stage III.

Within the group of patients in stage I or II, 77% were positive. When the tumor had metastasized to any important degree, a positive reaction was not regularly found: 2 out of 15 cases in stage III were positive (table II).

e) Number of Positive Cells

It was thought that a numerical assessment of positive cells by immunofluorescence might serve a useful purpose. It is quite impossible to arrive at such an assessment when fresh tumor cells are used, because among the positive cells, several may not be melanoma cells. An example of a positive, specific reaction with this type of cells is given in figure 1.

This method has the disadvantage of giving false positive reactions. Figure 3 represents one such case of false positive, indirect reaction. In

Table II.

	Number of cases	Positive fluorescence test	Negative fluorescence test
Stage I and II	27	21	6
Stage III	33	2	31
Total of cases	60	23	37

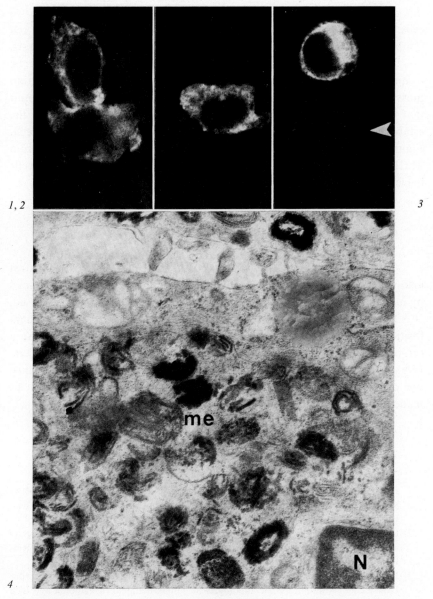

Fig. 1–3. Indirect cytoplasmic immuno-fluorescence. – *Fig. 1.* Fresh tumor cells: specific positive cell. – *Fig. 2.* Cell line M CLB 1 (culture): specific positive cell. – *Fig. 3.* Fresh tumor cells: 'false positive', with PBS (plasma cell) among negative cells (arrow).

Fig. 4. Electron microscopy of the cell line M CLB 1 (culture). N = nucleus; me = melanosomes.

this case, we used PBS instead of the patient's serum. Among many negative cells, we found a very positive one which we believe to be a plasma cell synthesizing immunoglobulins. Its morphology supported this hypothesis.

These various problems seem to suggest that the best method is the technique using cells in culture. Naturally, the ultrastructure of the strain must be verified from time to time to ensure its purity, taking into account that the antigenic modulation may undergo modifications and that the established line may no longer have the same antigens as the first cell. Figure 2 shows an indirect immunofluorescent reaction with our strain M CLB 1, and figure 4 shows the structure of the melanosomes of these cells in electron microscopy.

Conclusion

Our findings suggest the presence of a good cross reactivity against cytoplasmic antigens of malignant melanoma cells. The reactions obtained by indirect immunofluorescence are unreliable when fresh tumors are used. It is recommended to use in these cases a micro-technique of complement fixation. The most suitable approach for immunofluorescence is to use a cell-culture line, but the diversity of antigens in the cells remains a problem. If a serum is negative against one cell type, then it has to be tested against one or two other types of melanoma cells before an interpretation can be made.

References

1 BOURGOIN, J.J.: Cytoplasmic antigen in malignant melanoma cells detected by immunofluorescence. 2nd Symp. Biology of Malignant Melanoma, Lyon 1971. Rev. Inst. Pasteur, Lyon (to be publ.).
2 LEWIS, M.G.: The relationships between serum antibody and dissemination of malignant melanoma. 2nd Symp. Biology of Malignant Melanoma, Lyon 1971. Rev. Inst. Pasteur, Lyon (to be publ.).
3 LEWIS, M.G.; IKONOPISOV, R.L.; NAIRRN, R.C.; PHILLIPS, T.M. and HAMILTON-FAIRLEY, G.: Tumour specific antibodies in human malignant melanoma and their relationship to the extent of the disease. Brit. med. J. 547–552 (1969).
4 MORTON, D.L.; MALMGREM, R.A.; HOLMES, E.C. and KETCHAM, A.S.: Demonstration of antibodies against human malignant melanoma by immunofluorescence. Surgery 64: 233–244 (1968).
5 PHILLIPS, T.M.: Immunofluorescence techniques in the study of malignant melanoma.

2nd Symp. Biology of Malignant Melanoma, Lyon 1971. Rev. Inst. Pasteur, Lyon (to be publ.).
6 PHILLIPS, T. M. and LEWIS, M. G.: A system of immunofluorescence in the study of tumour cells. Europ. J. clin. Biol. *15:* 1016–1020 (1970).
7 SOHIER, R.; PEILLARD, M. M.; GINESTE, J. et FREYDIER, J.: Microméthode en tubes pour la réaction de fixation du complément appliquée au diagnostic des infections à virus. Ann. Biol. clin. *14:* 281–290 (1956).

Author's address: DR. J. J. BOURGOIN, Department of Immunology and Experimental Oncology, Centre Léon Bérard, 28, rue Laénnec, *F-69 Lyon* (France)

Immunoglobulins Associated with Human Malignant Melanoma Tumors

I. S. Cox and M. M. Romsdahl

M.D. Anderson Hospital and Tumor Institute, Houston, Tex.

Introduction

Clinical evidence has for a long time indicated the existence of a host reaction against primary and metastatic malignant melanoma. Early attempts to demonstrate immunological mechanisms by detecting gamma globulin on the surface of live melanoma cells failed [8]. Extension of their studies by treatment of tumor cells with patient serum and then staining with fluorescent anti-serum also met with failure. By the anti-globulin consumption test, Anthony and Parsons [2] showed that 95% of human solid tumor brei contained globulin. Eluates from the tumor brei lowered the titer of antiglobulin serum while eluted brei did not. Refinement and improvement of immunological techniques have resulted in a surge of new information on the immunologic relationship between the patient and his tumor. The existence of both surface and intracellular antigens on and in melanoma tumor cells is now well documented [9, 11, 12, 14]. The Hellströms [5] have shown that patients' lymphocytes are sensitized to their own tumors and other tumors of the same histologic type with a variety of human neoplasms. They have also demonstrated that cell-mediated immunity can in some types of cancers be abrogated by the patients' own serum [4, 6]. Further investigation by Sjögren has resulted in the isolation from an eluate of the tumor cells of two fractions, a high molecular weight fraction containing the immunoglobulins and a low molecular weight fraction presumably containing the antigen(s) [15]. Neither fraction alone is reactive when placed on target cells. However, mixed 1:1 there is inhibition of lymphocyte reactivity on the target cells. However, prolonged incubation of the target cells with the low molecular weight fraction alone did inhibit the reactivity of sensitized lymphocytes.

Following our immunofluorescent study of live and fixed melanoma cells prepared from fresh tumor material, attempts were made to study the reactivity of these tumors and sera by immune precipitation (Ouchterlony method). The immune mechanisms of this tumor are still poorly understood; it was hoped study by other methods would further illuminate the mechanisms underlying the immune response mounted against melanoma tumors.

Materials and Methods

Serum Collection
Bloods were obtained by venipuncture in sterile tubes, allowed to clot at room temperature and the serum decanted after centrifugation. Sera for control purposes were obtained from healthy hospital employees, and adult relatives of patients who had no history of transfusions.

Tumor Collection and Processing
Malignant melanoma tissue was obtained as fresh surgical specimens immediately after excision. Any contiguous adipose and stromal tissue were dissected away and the tumor was prepared for homogenization as a 20% suspension W/V in distilled water and subsequent extraction with Genetron 113 as described by McKenna et al. [10].

Extracts were also prepared by the freeze-thaw method using a dry-ice-acetone bath for freezing and water at 37°C for thawing. After the initial homogenization in the VirTis, the tumor homogenates were frozen and thawed for a total of four times. The cell debris was discarded after centrifugation at 20,000 rpm for 30 min.

Immunodiffusion Tests
The tumor extracts prepared by fluorocarbon extraction and freeze-thaw cleavage were tested in Ouchterlony plates against a variety of melanoma patient sera and appropriate controls. In addition, the tumor extracts were tested against caprine anti-human IgG, IgA and IgM antisera. The immunodiffusion plates were incubated in a humid atmosphere at 37°C for 7 to 10 days and for 20 days at 4°C. The gels were then removed from the plates and pre-developed in 0.0125% cadmium sulfate for 30' [3]. After the appropriate washing and dehydration the gels were stained in 0.1% amido-schwarz and then mounted on plain glass slides with a thin coating of Elvanol [13]. For preservation the gel mounts were sprayed with high gloss polyurethane.

Results

1. Genetron Extracted Melanoma Tumor Extracts
Of the 11 melanoma extracts tested 5 developed a precipitin band against one or more of 77 patient sera tested (table I). In addition the tumor

Table I. Immunodiffusion studies of genetron extracts of human melanoma tumor

Extracts	Patient	Melanoma sera No. positive/ No. tested	%	Normal sera No. positive/ No. tested	%	Antiglobulin sera		
						IgG	IgM	IgA
Melanomas	L.J.[1]	0/77	0	0/11	0	–	–	–
	L.J.[2]	0/77	0	0/11	0	–	–	–
	A.B.	N.D.[1]	–	N.D.	–	+	–	+
	D.M.	0/77	0	0/11	0	–	–	–
	O.M.	4/77	5	0/11	0	+	–	+
	G.P.	0/77	0	0/11	0	+	–	–
	O.P.	17/77	22	0/11	0	+	–	+
	T.H.[1]	N.S.[2]	–	N.S.	–	+	–	+
	T.H.[2]	1/77	1	0/11	0	+	–	+
	T.W.	1/77	1	0/11	0	+	–	–
	A.G.	41/77	53	1/11	0	–	–	–
	J.C.	0/77	0	0/11	0	N.D.	N.D.	N.D.
	M.D.	N.D.	–	N.D.	–	+	–	+
	C.H.	N.D.	–	N.D.	–	+	–	+
	D.D.	N.D.	–	N.D.	–	–	–	N.D.
	J.T.	N.D.	–	N.D.	–	+	–	+
Sarcomas	K.Y.	0/77	0	0/11	0	N.D.	N.D.	N.D.
	W.P.	0/77	0	0/11	0	N.D.	N.D.	N.D.
	P.R.	6/77[3]	8	0/11	0	N.D.	D.N.	N.D.
Controls	Normal liver	0/77	0	0/11	0	–	+	–
	Normal brain	0/77	0	0/11	0	–	–	–
	Normal lung	N.D.	–	N.D.	–	+	–	+
	Normal muscle	N.D.	–	N.D.	–	–	–	–

1 N.D. – not done; 2 N.S. – non-specific precipitation
3 Very dense precipitate around all antigen wells. Six had specific precipitin bands as well.

extract of one patient (T.H.[1], first excision) developed precipitin lines against control as well as patient sera. One of the sarcoma tumor extracts reacted specifically with 6 different melanoma sera. Extracts of normal tissue controls did not react. One melanoma tumor extract reacted with 1 of 11 normal control serum tested.

Ten of 15 soluble melanoma extracts produced precipitin bands when reacted against undiluted anti-human IgG (table I). As many as 3 to 4 lines were detectable with some extract preparations suggesting the presence of IgG with different specificities and/or products of catabolized IgG (fig. 1).

Five of 14 extracts developed precipitin lines when reacted against IgA (table I, fig. 2). IgM was not detectable in any of the extracts, except for a diffuse light staining immediately outside the antiserum reservoir with some of the extract preparations. Extracts of normal muscle and brain did not react with any of the immunoglobulins. Normal lung extract reacted against

Fig. 1. Left: The center well contains undiluted anti-human IgG. The peripheral wells all contain different melanoma tumor extracts. Precipitin bands are absent in one preparation (Po) only. *Center:* Center well contains anti-human IgG; the peripheral wells contain melanoma tumor extracts. Bry = Genetron-extracted melanoma; Bry vac = a 1×10^8/ml cell suspension of the same tumor irradiated for 10 min. with 10,000 rads and finally extracted by freeze-thawing. More immunoglobulin was apparently freed by the fluorocarbon treatment as evidenced by the larger number of precipitin bands obtained. Pu = melanoma tumor extract diluted 1:8; W = melanoma extract; and Gra = melanoma extract. *Right:* The center well contains undiluted anti-human IgG. Br = brain extract; NL = normal lung extract; Ha[1] = melanoma extract; MF = normal muscle extract; Pu = melanoma extract. Precipitin lines were formed against the tumor preparations and possibly the 'normal' lung but not against the other normal tissue extracts. The precipitin lines are very weak. These tumor extracts were prepared a long time previously and had been thawed and refrozen several times prior to this experiment.

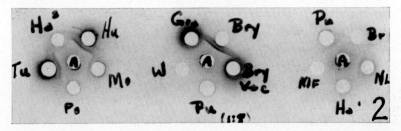

Fig. 2. This plate is a dupulicate in arrangement of figure 1, except that the center wells in all three patterns contain undiluted goat anti-human IgA. *Left:* The precipitin lines are much weaker but quite distinct. IgA is present in all but one extract, Po. This extract also lacked in IgG. *Center:* Of the five preparations tested IgA is detected in only one (Bry). *Right:* Faint precipitin lines are discernible from both tumor extracts (Pu and Ha[1]) and from normal lung. IgA is normally found in secretions of the lung.

antisera IgG and IgA did produce weak bands. Normal liver extract reacted against IgM antiserum.

2. Frozen-and-Thawed Tumor Extracts

The results obtained by incubation at the two temperatures are depicted in table II. Incubation at 4°C was performed with the hope of increasing the intensity of the reaction arcs. The reactions obtained between patient serum and the tumor extract are generally weak, irrespective of incubation temperature and method of extraction.

Extracts of melanoma and patient serum when placed in neighboring reservoirs and reacted against IgG antiserum in the central reservoir showed the precipitin line obtained with the tumor extracts was in identity with the one found in the serum (fig. 3).

Two of the tumor extracts prepared by this method, N.M. and N.L., formed dense zones of precipitated protein immediately outside the reservoir in which they were placed regardless of the reactant run against them. These extracts did not react by distinct band formation when reacted against immunoglobulin antisera. The precipitin zones may represent precipitated antigen-antibody complexes still bound to insoluble microsomal proteins which inhibit their diffusion. In general incubation at 4°C for an extended period does appear to result in more distinct precipitin lines. However,

Table II. Results obtained by immunodiffusion of frozen and thawed extracts of malignant melanoma tissues incubated at 37°C and 4°C

Freeze-thawed extracts	Melanoma sera				Control sera			
	No. positive/No. tested		%		No. positive/No. tested		%	
	37°C	4°C	37°C	4°C	37°C	4°C	37°C	4°C
J. T.	1/13	0/13	8	0	0/6	0/6	0	0
H. M.	8/13	6/13	62	46	2/6	0/6	33	0
L. L.	0/13	0/13	0	0	0/6	0/6	0	0
B. D.	4/13	13/13	31	100	2/6	0/6	33	0
M. F.	1/13	4/13	8	31	0/6	2/6	0	33
C. M.	0/13	10/13	0	77	0/6	3/6	0	50
R. S.	0/13	5/13	0	38	0/6	0/6	0	0
N. M.	13/13[1]	13/13[1]	–	–	6/6[1]	6/6[1]	–	–
B. S.	0/13	1/13	0	8	0/6	0/6	0	0
T. C.	0/13	11/13	0	85	0/6	/06	0	0
N. L.	13/13[1]	13/13[1]	–	–	6/6[1]	6/6[1]	–	–
A. C.	12/13	8/13	92	62	1/6	0/6	17	0

1 Dense precipitation occurred but was non-specific.

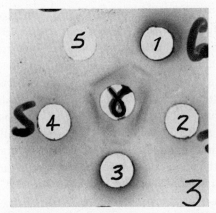

Fig. 3. The center reservoir contains anti-human IgG. Reservoirs 1 and 3 contain a melanoma tumor extract. Wells 2 and 4 contain two different melanoma patient sera. Well 5 contains a normal human serum. The precipitin lines formed against the IgG from all sources are in identity with no spurring.

IgA does not migrate well at 4 °C and was found to be absent at this temperature in some tumor extracts (table III). The reverse was true of IgM which at this reduced temperature tended to form very distinct bands (fig. 4c). In figure 5 is shown a reaction between melanoma tumor extract and an

Table III. Results obtained by immunodiffusion after incubation at 37 °C and 4 °C of frozen and thawed melanoma extracts reacted against anti-human immunoglobulin antisera

Melanoma extract	Immunoglobulin class tested					
	37 °C			4 °C		
	IgG	IgA	IgM	IgG	IgA	IgM
J.T.	+	+	−	+	−	+
H.M.	+	+	−	+	−	−
L.L.	+	+	−	+	+	+
B.D.	+	+	−	+	+	+
M.F.	+	−	−	+	−	+
C.M.	+	−	+	+	−	+
R.S.	+	+	−	+	−	−
N.M.	+	+	+	+	+	+
B.S.	+	+	−	+	+	+
T.C.	+	+	−	+	−	−
N.L.	+	+	+	1	1	1
A.C.	+	+	−	+	−	+

1 There was a diffuse precipitation but no distinct bands.

homologous serum. Figures 6 and 7 depict the reactions obtained from freeze-thawed extracts reacted against IgG, IgA and IgM antisera, respectively.

Of the two methods of tumor extract preparation, fluorocarbon extraction is the superior method resulting in more consistent results. Although immunoglobulins are detected in a larger number of freeze-thawed extracts there is a more rapid deterioration than in fluorocarbon extracts. Also, there are more indeterminate results obtained probably due to incomplete cleavage and solubilization. In parallel experiments, it was observed that there was no greater loss of immunoglobulins by fluorocarbon extraction. Incubation at 4°C does increase the number of positive reactions and more clearly defines indefinite precipitin lines. However, incubation at 37°C is necessary for detection of all classes of immunoglobulins.

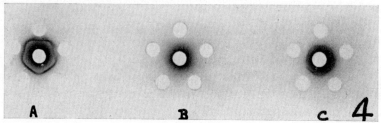

Fig. 4. The center wells in each pattern contain a melanoma extract prepared by freeze-thawing. This gel was incubated at 4°C for 20 days. The peripheral wells in pattern A all contain antihuman IgG; in pattern B, anti-human IgA; and in pattern C, anti-human IgM. The distinct bands developed against IgM are obtained only at 4°C. However, precipitin lines against IgA did not form at this temperature, but were formed at 37°C.

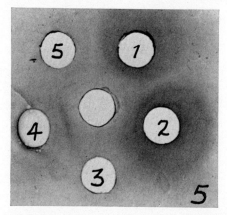

Fig. 5. Melanoma tumor extract is located in the center well. Five different melanoma patient sera are in the outside wells. A prectpitin line developed between the patient serum in reservoir 4 and the tumor extract in the center.

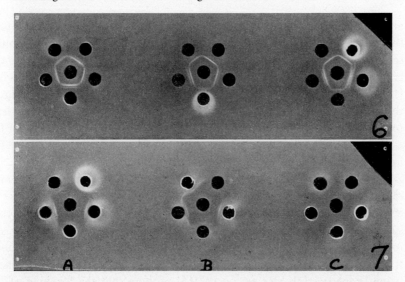

Fig. 6. This gel was an unstained wet mount photographed after predevelopment in cadmium sulfate. The center wells in each pattern contain anti-human IgG. The peripheral wells contain 15 different melanoma tumor extracts. IgG was detected in all 15 extracts. With some extracts two or more bands developed.

Fig. 7. The center wells in patterns A and B contain anti-human IgA. The center well in pattern C contains anti-human IgM. In the patterns A and B are depicted the immuno-precipitin lines developed by 10 different melanoma tumor extracts against anti-IgA. No bands were developed when the first 5 extracts were reacted against anti-IgM. This gel is also a wet mount photographed after pre-development in cadmium sulfate. The gels in all other figures were stained in 0.1 % amidoschwarz.

There were no attempts at quantitation of the immunoglobulins here. However, they are found in lower concentration, if at all, in normal tissues.

Discussion

The demonstration of tumor specific antibody in the malignant melanoma system requires the use of sensitive techniques such as immunofluorescence. Weak serum antibody has been detected by double gel diffusion. Of great interest is the demonstration of three classes of immunoglobulins localized within the tumor mass. The low serum antibody found in melanoma patient sera may be due to the absorption of the serum antibody by the tumor.

The finding of immunoglobulins associated with the tumor by immunodiffusion raises the disturbing question as to why the immunoglobulins are not detected by immunofluorescence. As one of the controls in the test, the target cells are reacted with the fluorocein-conjugated anti-globulin serum alone. Particular consideration may be given in this connection to the type of antigenic modulation described by TAKAHASHI [17] and STACKPOLE [16] where the attachment site of specific antibody and labeled anti-Ig to the cell membrane is vacuolated and consequently ingested. This would lead to a diminution of antigenic expression on the cell surface and account for a weak antibody response. Cleavage of the cell membrane would release the immunoglobulins detectable by immunodiffusion.

The number of precipitin lines developed against IgG and IgA may reflect the number of antigens involved in invoking an immune response or it may represent immunoglobulin and its catabolic products. There is considerable evidence that antigen-antibody complexes play an important role in the induction of biological activities. It is suggested certain antibody fragments in combination with antigen lack the ability to induce complement fixing properties [7]. SJÖGREN et al. [15] propose that the blocking of the cell-mediated response is due to an antigen-antibody complex.

The biological function of each class of immunoglobulin associated with this tumor is unknown. The failure of the cell-mediated response may be attributable to the presence of one of these immunoglobulins, possibly IgA, in the inhibition or suppression of complement activation. That antigen-antibody complexes effect the metabolic production of a substance which inhibits the production of a complement activator or other immunosuppressive factor has been postulated [1]. The class of antibody involved has not been resolved.

The possibility exists that the immunoglobulins were extracted from the infiltrating mononuclear cells which are quite prevalent in melanoma tumors. This seems unlikely, since the IgG from the tumor was shown to be identical to serum IgG (fig. 3). The identity of IgA was not tested for.

Clearly, a great deal more needs to be done along these parameters. Clinically, the malignant melanoma appears to be the most immunologically responsive of human neoplasms. Elution, separation and testing of each immunoglobulin independently in tissue culture should clarify their biological roles with respect to malignant melanoma.

References

1 AMOS, D.B.; COHEN, I. and KLEIN, W.J., Jr.: Mechanisms of immunologic enhancement. Transpl. Proc. 2: 68–75 (1970).

2 ANTHONY, H. and PARSONS, M.: Globulin on cells of cancer patients. Nature, Lond. *206:* 275–276 (1965).
3 CROWLE, A.J.: Enhancement by cadmium of double-diffusion precipitin reactions. J. Immunol. *81:* 194 (1958).
4 HELLSTRÖM, I.; HELLSTRÖM, K.E.; EVANS, CHARLES A.; HEPPNER, GLORIA H.; PIERCE, GEORGE E. and YANG, JAMES P.S.: Serum-mediated protection of neoplastic cells from inhibition by lymphocytes immune to their tumor-specific antigens. Proc. nat. Acad. Sci., Wash. *62:* 362–368 (1969).
5 HELLSTRÖM, I.; HELLSTRÖM, K.E.; SJÖGREN, HANS O. and WARNER, G.A.: Demonstration of cell-mediated immunity to human neoplasms of various histological types. Int. J. Cancer *7:* 1–16 (1971).
6 HELLSTRÖM, I.; SJÖGREN, H.O.; WARNER, G. and HELLSTRÖM, K.E.: Blocking of cell-mediated tumor immunity by sera from patients with growing neoplasms. Int. J. Cancer *7:* 226–237 (1971).
7 ISHIZAKA, K.; ISHIZAKA, T. and SUGAHARA, T.: Biological activity of soluble antigen-antibody complexes. VII. Role of an antibody fragment in the induction of biological activities. J. Immunol. *88:* 690–701 (1962).
8 KOPF, A.W.; SILBERBERG, I. and COOPER, N.S.: Immunohistochemical study of human malignant melanoma for the presence of gamma globulin. J. Invest. Derm. *47:* 83–86 (1966).
9 LEWIS, M.G.; IKONOPISOV, R.L.; NAIRN, R.C.; PHILLIPS, T.M.; FAIRLEY, G.H.; BODENHAM, D.C. and ALEXANDER, P.: Tumour-specific antibodies in human malignant melanoma and their relationship to the extent of the disease. Brit. med. J. 547–552 (1969).
10 MCKENNA, J.M.; SANDERSON, R.P. and BLAKEMORE, W.S.: Extraction of distinctive antigens from neoplastic tissue. Science *135:* 370–371 (1962).
11 MORTON, D.L.; MALMGREN, R.H.; HOLMES, E.C. and KETCHAM, S.: Demonstration of antibodies against human malignant melanoma by immunofluorescence. Surgery *64:* 233–240 (1968).
12 MUNA, N.M.; MARCUS, S. and SMART, C.: Detection by immunofluorescence of antibodies specific for human malignant melanoma cells. Cancer *23:* 88–93 (1969).
13 RODRIGUEZ, J. and DEINHARDT, F.: Preparation of a semi-permanent mounting medium for fluorescent antibody slides. Virology *12:* 316 (1960).
14 ROMSDAHL, M.M. and COX, I. SEBASTIAN: Human malignant melanoma antibodies demonstrated by immunofluorescence. Arch. Surg. *100:* 491–497 (1970).
15 SJÖGREN, H.O.; HELLSTRÖM, I.; BANSAL, S.C. and HELLSTRÖM, K.E.: Suggestive evidence that 'blocking antibodies" of tumor-bearing individuals may be antigen-antibody complexes. Proc. nat. Acad. Sci., Wash. *68:* 1372–1375 (1971).
16 STACKPOLE, C.W.; AOKI, T.; BOYSE, E.A.; OLD, L.J.; LUMLEY-FRANK, J. and DE HARVEN, E.: Cell surface antigens: serial sectioning of single cells as an approach to topographical analysis. Science *172:* 472 (1971).
17 TAKAHASHI, T.: Possible examples of antigenic modulation affecting H-2 antigens and cell surface immunoglobulins. Transplant. Proc. *3:* 1217–1220 (1971).

Author's address: IRENE SEBASTIAN COX, M.D. Anderson Hospital and Tumor Institute, *Houston, TX 77025* (USA)

Tissue Culture Studies on Human Malignant Melanoma

R. H. WHITEHEAD and J. H. LITTLE

Queensland Institute of Medical Research and Department of Pathology, Princess Alexandra Hospital, Brisbane

Introduction

Early reports of tissue culture studies on malignant melanoma have been reviewed previously by COBB and WALKER [3] and ROHMSDAHL and HSU [12]. There has been only one detailed description of a cell line obtained from a melanoma and its subsequent use in melanoma antibody studies [8].

The high incidence of melanoma in Queensland [4] and recent suggestions of a common antigen in melanomas [5, 7] suggested that a detailed cultural study of melanoma was warranted in order to provide cell lines for antigenic study.

Attempts were therefore made to culture cells from most types of human melanoma. This paper reports the results and correlates tumour cell proliferation *in vitro* with clinical status of the patient and pathological classification [1, 2].

Materials and Methods

All available melanomas were collected from the Queensland Melanoma Project over a period of two years and a few were obtained elsewhere. Specimens included most pathological types of melanoma and varied considerably in size. Portion of the tumour was transported to the laboratory in tissue culture medium RPMI 1640 (Gibco Laboratories, Grand Island, N.Y.) containing penicillin (200 µg/ml) and streptomycin (200 µ/ml), washed and trimmed of normal tissue and teased with a scalpel blade to release tumour cells. The cells obtained were suspended in RPMI 1640 containing 20 % inactivated foetal calf serum, 0.05 % Bacto-tryptose broth (DIFCO Laboratories, Baltimore, Md.), penicillin (100 µg/ml) and streptomycin (100 µ/ml) at pH 7.1, and incubated at 37° in 60 mm Falcon petri dishes (Falcon Plastics, Los Angeles) in an atmosphere of 5 % CO_2 in air. The buffer used in early studies was NaHCO$_3$ (28 mM) but later was NaHCO$_3$ (18 mM) and Hepes (Calbiochem [Aust.] Carlingford (15 mM). Once cells were established in culture the serum level was reduced to 10%.

In live cultures 5 cell types (fibroblasts, epithelial cells, macrophages, lymphocytes and tumour cells) were present, but these could be distinguished on morphological grounds. The term 'persistence' is used to mean the attachment of viable tumour cells to the culture vessel. 'Proliferation' refers to an increase in the number of tumour cells often at a very slow rate.

Specimens for electron microscopy were trypsinized, fixed in glutaraldehyde and osmium tetroxide and embedded in araldite. Sections were cut on an LKB ultramicrotome and viewed on a Philips EM300 electron microscope.

Cells were grown to approximately 50% of confluence, the chemical to be tested for toxicity added to the growth medium and subsequent growth recorded at 72 h.

Chromosome preparations were made after incubating cultures in colchicine (1 μg/ml) for 4 h. Cells were washed and treated with hypotonic Hanks solution, fixed in acetic-alcohol and the air dried smears stained with Giemsa.

Results

Primary Melanomas

Thirty-two primary skin melanomas were cultured and the results are summarized in table I. The details of the 8 tumours showing proliferation are

Table I. Summary of tissue culture studies

Type of specimen	Classification	Tissue culture		
		No.	Persisted[1]	Proliferated[2]
Primary melanomas	lentigo malignant melanoma	9	4	2
	superficial spreading melanoma	11	4	1
	unclassified	7	4	1
	nodular	5	5	4
	Total	32	17	8
Secondary melanomas	lymph nodes	9	9	9
	skin nodules	18	15	9
	Total	27	24	18
Ocular melanomas		4	4	2
Other skin lesions		14	4	1
Total		77	49	29

1 Persistence of tumour cells in culture for 4 or more weeks.
2 Actual proliferation of tumour cells.

given in table II. The two lentigo malignant melanomas which proliferated were large, pedunculated tumours and both had high mitotic rates. The one superficial spreading melanoma showing tumour cell proliferation also had a high mitotic rate and was classified as papillary. The nodular melanomas showed better persistence and proliferation. Three had marked mitotic rates and one had a medium mitotic rate. All were larger than 12 mm in diameter. The association of high mitotic rates in tumours and proliferation in culture is shown in table III.

Fibroblast cultures were obtained from 24 of the 32 primary melanomas tested but there was no evidence that these were other than normal fibroblasts.

Fourteen skin tumours of other types were also cultured. Two of 9 naevi and both basal cell papillomas persisted for more than 4 weeks but

Table II. Details of primary melanomas proliferating *in vitro*

No.	Patient	Sex	Age	Type	Stage	Location	Size (mm)	Mitotic rate
MM1	KF	F	50	NA[1]	NA	Back	17 × 14	NA
MM69	MW	M	55	LMM	3A	Postauricular	45 × 50 × 13	Marked
MM90	SD	F	35	Nodular	3A	Knee	12 × 12 × 4	Marked
MM99	IH	F	87	LMM	3A	Neck	26 × 18 × 6	Marked
MM112	CB	M	36	SSM	3A	Wrist	15 × 12 × 3	Marked
MM158	RM	M	49	Nodular	3B	Face	22 × 17 × 10	Marked
MM187	AB	M	57	Nodular	3A	Thigh	22 × 20 × 12	Marked
MM191	MD	F	49	Nodular	4	Thigh	20 × 20 × 10	Medium

1 Not available.

Table III. Influence of mitotic rate of primary tumours on proliferation of tumour cells in culture

Mitotic rate[1]	Culture result	
	Persistence >4 weeks	Proliferation
Low	3/12	0/12
Moderate	3/6	1/6
Marked	6/6	6/6
Not available	5/8	1/8

1 As judged on examination of sections according to the criteria of BEARDMORE et al. [1]

only 1 naevus showed any evidence of proliferation. The cells persisting in these cultures could not be identified as naevus or papilloma cells but did not appear to be normal fibroblasts.

Secondary Melanomas

The results of cultures are summarized in table I. Three of the cultures from secondary nodules (MM127, MM138 and MM181) are still in culture, after 40, 20 and 10 passages respectively (table IV). Three of the cultures from metastatic lymph nodes are still growing, one (MM96) after 70 passages.

Lines of lymphoid cells were obtained from three metastatic lymph node deposits and from two subcutaneous nodules. Tumour cells and lymphoid cells grew together in the same dish. One lymphocyte culture was found to contain EPSTEIN-BARR virus complement fixing antigen [13].

Description of Cell Lines

The MM96 cells are pleomorphic with lightly pigmented epitheloid cells predominant. Approximately 10 % of cells are lightly pigmented dendritic cells with very long processes. At the 65th passage the cells were aneuploid with a mode of 49 and a minute chromosomal element in 10 % of cells. The intensity of pigmentation varies inversely with growth rate. The cells have a doubling time of 50 hours. The cells show no contact inhibition and although they normally grow attached, they may pile up or grow in suspension.

The MM127 cells are small non-pigmented spindle shaped cells growing attached to the culture vessel. Approximately half of the cells were diploid

Table IV. Details of cell strains established.

No.	Patient	Sex	Age	Specimen		Pigmentation	Growth	Passages
MM96	DG	F	66	lymph node		+	Attached	70
MM127	TM	M		Subcutaneous nodule		−	Attached	40
MM138	WK	F	32	Subcutaneous nodule	a b	+ −	Floating Attached	20
MM170	JW	M		Lymph node		+	Attached	10
MM173	EG	F	62	Lymph node		+	Attached	10
MM181	JA	F	16	Nodule		−	Attached	10

while the remainder were aneuploid with a mode of 48 at the 30th passage. The cell line developed from a patch of heavily pigmented cells which had persisted in culture for 3 months after an initial period of growth. Although the cells are non-pigmented a few melanosomes were found on electron microscopy. These cells have a doubling time of 80 hours.

The MM138 cells originally grew partly in suspension and partly attached. The attached cells were large, slightly pigmented, and spindle shaped. The floating cells were more heavily pigmented. This cell line changed at the 17th passage level and a cell-type emerged that was non-pigmented and grew attached to the culture vessel. Two sub-lines were then separated, one predominantly a suspension culture and the other predominantly attached. The cells are aneuploid with modes of 66 for the suspension sub-line and 128 for the attached sub-line at the 20th passage.

The MM170 cells are pleomorphic with both small and large spindle-shaped cells and epitheloid cells present in equal proportion, and occasional dendritic cells. The cells are slightly pigmented and are mainly diploid with 20% of cells having a mode of 49.

The other two cell lines, MM173 and MM181, are growing too slowly to enable further study at the present time.

Experimental Study of Cell Lines

A number of cytotoxic agents were tested using MM96 and MM127 and foetal fibroblasts. Greater sensitivity of tumour cells compared with fibroblast cultures were observed with 5-fluorodeoxyuridine and p-methoxyphenol but not with mitomycin C, actinomycin D, iododeoxyuridine, nitrogen mustard, hydroxyurea, or o-methoxyphenol.

Increasing the glucose concentration of the medium to 3% did not increase the pigmentation of MM96 or MM 127 but it did inhibit growth to some extent and altered the morphology of MM127 cells markedly. The addition of ACTH (1 U/ml) had no enhancing effect on growth or pigmentation.

Electron Microscopy

Four of the cell lines (MM96, MM127, MM138 and MM170) were examined and all contained melanosomes and premelanosomes. The number varied from 10 per cell in MM127 cells to more than 50 per cell in MM96 cells. The laminar structure characteristic of melanosomes could readily be detected in those that were less heavily pigmented. No virus particles were detected in any of the cultures examined at several different passage levels.

EB Virus

MM96 and MM127 cells were tested for Epstein-Barr virus antigen by complement fixation tests [13] but both were negative.

Virus Susceptibility

MM96 cells were found to be susceptible to infection with a number of viruses. Poliovirus type 2, reovirus type 3, *Herpes virus hominis,* adenovirus type 2 and Murray Valley encephalitis viruses all produced cytopathogenic effect (CPE). Neither mumps virus nor a serum rich in Australia antigen produced CPE.

Discussion

The specimens described in this paper were a representative sample of all melanomas seen by the Queensland Melanoma Project, and included many very small primary tumours presumably due to the general awareness of the public and the medical profession of Queensland [4]. The better growth of elevated or nodular primaries may have been due to a larger number of tumour cells. However, Beardmore et al. [1] previously reported that nodular melanomas have a poorer prognosis, even when not deeply invasive, and these two findings may indicate the presence of a cell type which renders the tumour both more invasive and more capable of adapting to tissue culture.

The relationship noted between mitotic rate and growth in culture is not surprising and melanomas with a high mitotic rate have been found to have a poor prognosis [J. H. Little, unpublished results].

Secondary melanoma deposits grew better in culture than primaries and the more disseminated the tumour the better the growth. All three subcutaneous nodules that proliferated were from terminal cases. It is possible that a series of changes in cell type occur as the tumour progresses, each change rendering the cell more likely to grow *in vitro.*

The change in growth pattern of two of the cell lines (MM127 and MM138) from slowly proliferating pigmented cells to small, rapidly dividing, non-pigmented, spindle-shaped cells in the space of one passage is of interest. The two events were unrelated and occurred 9 months apart. The differences in chromosome number between the cell lines helps to rule out the possibility of cross-contamination. Concurrent with this transformation the chromosome number of the MM138 cells increased from a mode of 66 to a mode of 128.

The culture of lymphocytes from subcutaneous nodules indicates a high degree of lymphocyte infiltration into the tumour. However, the concurrent proliferation of lymphocytes and tumour cells in the same culture vessel suggests that these lymphocytes were not cytotoxic for the tumour cells.

The effect of p-methoxyphenol on human melanoma cells *in vitro* parallels the results described by RILEY [10, 11] using guinea-pig melanocytes. He found that the effect of p-methoxyphenol was a function of the state of pigmentation and there was some support for this in the present work.

The establishment of six melanoma cell lines will enable more detailed immunological and biochemical studies of this tumour.

Acknowledgements

We wish to acknowledge the assistance of Dr. N. C. DAVIS and his colleagues of the Queensland Melanoma Project for providing the specimens for this project.

The electron microscopy was kindly performed by Dr. D. MORTON, CSIRO Meat Research Laboratory, Cannon Hill and Dr. RUTH MITCHELL, Department of Pathology, University of Tasmania.

Miss M. STEEN provided invaluable technical assistance in the maintenance of the cultures.

References

1 BEARDMORE, G. L.; QUINN, R. L. and LITTLE. J.H.: Malignant melanoma in Queensland: pathology of 105 fatal cutaneous melanomas. Pathology 2: 277–286 (1970).
2 CLARK, W. H., Jr.: A classification of malignant melanoma in man correlated with histogenesis and biologic behavior. Adv. Biol. Skin 8: 621–647 (1967).
3 COBB, J. P. and WALKER, D.G.: Studies on human melanoma cells in tissue culture. I. Growth characteristics and cytology. Cancer Res. 20: 858–867 (1960).
4 DAVIS, N. C.; HERRON, J.J. and MCLEOD, G. R.: Malignant melanomas in Queensland: analysis of 400 skin lesions. Lancet ii: 407–410 (1966).
5 LEWIS, M. G.; IKONOPISOV, R. L.; NAIRN, R. C.; PHILLIPS, T. M.; FAIRLEY, G. H.; BODENHAM, D.C. and ALEXANDER, P.: Tumour-specific antibodies in human malignant melanoma and their relationship to the extent of the disease. Brit. med. J. *iii*: 547–522 (1969).
6 MIODUSZEWSKA, O.: The influence of ACTH and MSH on human malignant melanoma cells *in vitro*; in PORTA and MUHLBOCK Symposium on structure and control of the melanocyte, Proc. 6th Int. Pigment Cell Conf., pp. 70–73 (Springer, New York 1966).
7 MORTON, D. L.; MALGREN, R.A.; HOLMES, E. C. and KETCHAM, A.S.: Demonstra.

tion of antibodies against human malignant melanoma by immunofluorescence. Surgery 64: 233–240 (1968).
8 OETTGEN, H. F.; AOKI, T.; OLD, L. J.; BOYSE, E. A.; DE HARVEN, E. and MILLS, G. M.: Suspension culture of a pigment-producing cell line derived from a human malignant melanoma. J. nat. Cancer Inst. 41: 827–843 (1968).
9 PONOMARYOVA, V. N.: Cytological and cytophysiological changes in the melanocytes of the human iris in tissue culture; in PORTA and MUHLBOCK Symposium on structure and control of the melanocyte Proc. VIth Int. Pigment Cell Conf., pp. 73–78 (Springer, New York 1966).
10 RILEY, P. A.: Hydroxyanisole depigmentation: in-vitro studies. J. Path. Bact. 97: 193–206 (1969).
11 RILEY, P. A.: Mechanism of pigment-cell toxicity produced by hydroxyanisole. J. Path. Bact. 101: 163–169 (1970).
12 ROMSDAHL, M. M. and HSU, T. C.: Establishment and biologic properties of human malignant melanoma cell lines grown in vitro. Surg. Forum 18: 78–79 (1967).
13 WALTERS, M. K. and POPE, J. H.: Studies of the EB virus-related antigens of human leukocyte cell lines. Int. J. Cancer 8: 32–40 (1971).

Author's address: Dr. R. H. WHITEHEAD, Queensland Institute of Medical Research, *Brisbane, Queensland* (Australia)

Genetic and Immunologic Approaches to Transplantable Mouse Melanomas[1]

M. Foster, J. Herman and L. Thomson

Mammalian Genetics Center, Department of Zoology, The University of Michigan, Ann Arbor, Mich.

Introduction

Many tumors have been demonstrated to possess, on their cell surfaces, tumor-specific transplantation antigens (TSTA), [5, 8, 13, 14]. Moreover, the use of immunosuppressive agents to improve the 'takes' of histoincompatible organ grafts or to suppress autoimmunity [15] has led to a significant increase in the occurrence of spontaneous tumors [8]. In addition, patients with primary immune deficiency exhibit more frequent spontaneous malignancies when contrasted with the frequency of tumor occurrence in the general population [8].

On the positive side, recent reports indicate successful attempts to induce in guinea pigs strong antitumor immune attack against weakly antigenic tumors, either by injecting BCG mixed with tumor cells [16], or by injecting BCG into established intradermal tumors [17]. However, this procedure was much less effective when BCG and tumor were injected contralaterally [4]. In this laboratory [6], a strong antigenic stimulus in the form of a histoincompatible tumor graft (i.e., invariably rejected B16 melanoma) implanted subcutaneously onto one flank of BALB/c mice has resulted in the rejection of another tumor subsequently implanted onto the other flank (i.e., Harding-Passey melanoma, to which the BALB/c host is regularly susceptible). Finally, Alexander and associates have reported that syngeneic, allogeneic and even xenogeneic immune lymphocytes transferred to tumor-bearing hosts have resulted in growth retardation or disappearance of these tumors [1–3].

[1] Supported in part by grants from the University of Michigan Cancer Research Institute and Project FRR-967, and by PHS grant 5-TO1-GM-00071-13.

These lines of evidence, together with the frustrating phenomenon of 'immunological enhancement', involving tumor-protecting blocking antibodies [12], implicate the host's immune system in both effective and unsuccessful responses to the challenge of tumors.

Our goal in this investigation is to develop immunologic procedures for provoking an effective antitumor attack by an otherwise susceptible individual. The underlying hypothesis is that a tumor capable of growing in a susceptible individual can differ from the host only by one or more 'weak' histocompatibility antigens. The problem is to induce the host to mount a strong immune attack against weakly antigenic tumor cells.

In working toward this goal we have used an animal model involving genetically standardized tumor-susceptible or resistant inbred strains of mice, as well as tumors of the same general type which consistently differ in their histocompatibility relations (mouse melanomas B16, S91A, and Harding-Passey). Such a system provides a basis for repeated testing of procedures designed to intensify a host's antitumor attack.

Materials and Methods

Three mouse melanomas were propagated in the susceptible host strains originally obtained for that purpose, i.e., B16 in C57BL/6, S91A in DBA/2 and Harding-Passey in BALB/c.

Tumor grafts involved subcutaneous flank implants of about 0.03 ml of minced B16 melanoma, or about 0.06 ml S91A or Harding-Passey. Occasional doubtful cases of rejection were resolved by an additional challenge. Such methods have led to our ability to classify a specific inbred strain as either 100% resistant or 100% susceptible to a given tumor.

F_1, F_2, and backcross progeny of crosses between susceptible and resistant parents were tested in the same ways.

In the spleen lymphocyte transfer experiments, spleens were removed from immunized, tumor-rejecting, or nonimmunized resistant mice, and placed in sterile Petri dishes with Tissue Culture Medium 199 (Difco). Each spleen was minced, drawn into a syringe, and injected intraperitoneally into a tumor-bearing susceptible recipient mouse.

Double challenge experiments involved two melanoma grafts: the normally rejected tumor on one flank and, generally about a week later, the normally 'taking' tumor on the other flank. A tumor size estimate was computed as the product of externally measured greatest and least diameters. Such size estimates are plotted in the growth curves of figures 1-6.

Finally, in crosses between susceptible and resistant strains, the genetic expectations based upon models involving independently assorting genes for 'strong' histocompatibility antigens, are as follows: all F_1 progeny should be susceptible, while $(3/4)^n$ of F_2 progeny and $(1/2)^n$ of backcross progeny (F_1's backcrossed to the resistant parental strain) should be susceptible (where n is the number of gene differences between the susceptible and resistant parental strains).

Results and Discussion

Histocompatibility Relations

The first phase of our investigation was directed toward a screening of various inbred strains for susceptibility or resistance to subcutaneous tumor graft challenge, followed by similarly testing the progeny of F_1, F_2, and backcross matings involving susceptible and resistant parental strains. Our findings are most extensive for the B16 melanoma, and some of these findings were recently reported [6, 7].

In the following listings of melanoma histocompatibility relations, each susceptible or resistant strain is followed in parentheses both by the allele of the major histocompatibility *locus* (*H-2*) fixed in that strain and by the recording, as a fraction, of the number of individuals succumbing to tumor challenge (numerator) out of the total number of individuals tested (denominator).

B16 Melanoma

This tumor exhibits a very limited host range, with subcutaneous implants growing in all tested individuals from the closely related strains C57BL/6J ($H-2^b$, 143/143) and C57BL/10ScSn ($H-2^b$, 10/10). The 100 % resistant strains are A/J ($H-2^a$, 0/5), A.BY/SnSf ($H-2^b$, 0/3), B10.A ($H-2^a$, 0/3), B10.D2 ($H-2^d$, 0/3), BALB/c ($H-2^d$, 0/12), C3H ($H-2^k$, 0/6), DBA/2J ($H-2^d$, 0/5), SEC/Re ($H-2^?$, 0/6), ST/bJ ($H-2^k$, 0/10), STOLI ($H-2^e$, 0/2), and YBR/a ($H-2^d$, 0/3). F_1 progeny of susceptible X resistant strain crosses are all susceptible. F_2 and backcross data are consistent with genetic models based upon differences between particular susceptible and resistant strains at 1, 2 or 3 gene *loci*. The fact that the B16-susceptible strains are both homozygous for allele $H-2^b$ does not implicate a difference at this gene *locus* as the sole factor concerned with susceptibility and resistance. For example, the congenic resistant strain A.BY is also homozygous for $H-2^b$, yet it is B16 resistant. Thus, one or more histocompatibility genes, other than the $H-2$ *locus*, must be involved in these differences.

S19A Melanoma

The following strains, as recently briefly reported [9], are 100 % susceptible: DBA/2J ($H-2^d$, 94/94), DBA/1J ($H-2^d$, 8/8), BALB/c ($H-2^d$, 5/5), NZB

(H-$2^{d?}$, 3/3) and SEC/Re (H-$2^?$, 8/8). The following strains are 100% resistant: A/J (H-2^a, 0/4), A.BY/SnSf (H-2^b, 0/20), B10.A (H-2^a, 0/25), B10.D2 (H-2^d, 0/28), C3H (H-2^k, 0/7) C57BL/6J (H-2^b, 0/10), C57BL/10Gn-Os (H-2^b, 0/4), C57BL/10ScSn (H-2^b, 0/4), and ST/bJ (H-2^k, 0/5). All tested F_1 progeny of susceptible X resistant matings are susceptible. Preliminary F_2 and backcross data are so far consistent with predictions based upon differences at one or two histocompatibility gene *loci* between the mated susceptible and resistant strains.

Harding-Passey Melanoma

This tumor has the widest host range of the three melanomas tested. All strains previously indicated are susceptible to this tumor. It is, nevertheless, rejected as a xenogeneic graft by the deer mouse, *Peromyscus maniculatus*.

Thus, a given strain can be susceptible to one melanoma but resistant to another, e.g., strain C57/BL6, or C57/BL10, is susceptible to B16 but resistant to S91A; or strain BALB/c is susceptible to Harding-Passey but resistant to B16. Such patterns provide the basis for the prophylaxis and immunotherapy experiments described later.

Double Challenge Experiments

The test system used for the double challenge experiments comprised BALB/c hosts challenged first on one flank by invariably rejected B16 melanoma, followed by a second challenge on the opposite flank by Harding-Passey melanoma, to which the strain is regularly susceptible. In a few cases the time interval between graft challenges was only a few hours, while in most cases the interval was about a week. Several different results are depicted in figures 1–4. Figure 1 illustrates a typical control Harding-Passey growth curve in the absence of a B16 challenge. Figure 2 illustrates a significant Harding-Passey growth retardation when this tumor challenge followed the strongly antigenic challenge of a B16 implant. Figures 3 and 4 illustrate multiple Harding-Passey rejections resulting from successive double challenges.

At this point the results of the double challenge experiments can be summarized as follows: in 41 control animals, singly challenged by Harding-Passey, all hosts succumbed. In 74 doubly-challenged experimental group hosts, only 22 tumors (30%) grew as rapidly as they grew in control hosts, 23 (31%) exhibited Harding-Passey growth retardation, while 29 (40%) exhibited apparently complete rejection of their Harding-Passey implants, including the two cases of multiple rejections previously described. Thus,

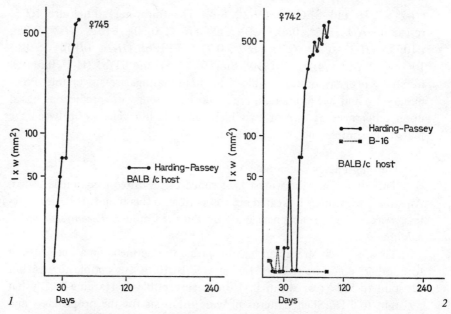

Fig. 1. Characteristic control growth curve of Harding-Passey melanoma in BALB/c host.

Fig. 2. Double challenge experiment, involving subcutaneous grafts of normally rejected B16 melanoma on one flank and normally growing Harding-Passey melanoma on the other. In this case growth retardation of the Harding-Passey melanoma included a short period of apparent disappearance.

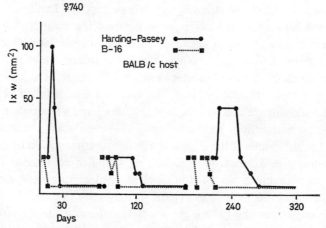

Fig. 3. Repeated double challenges (B16 followed by Harding-Passey 7–8 days later). Both melanomas were repeatedly rejected.

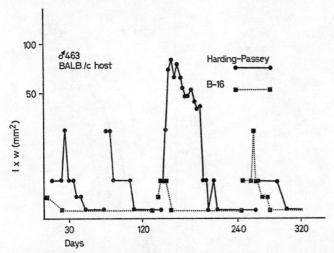

Fig. 4. Four consecutive Harding-Passey challenges, three of which were preceded by B16 challenges. All were rejected. This host rejected the second Harding-Passey graft, despite the absence of a B16 challenge a week before.

a large majority of the double-challenge experiments resulted in a significantly strengthened host antitumor attack against a melanoma to which these hosts are regularly susceptible.

In these cases of induced Harding-Passey melanoma graft rejections, it is not yet clear whether such rejections are due to a strong but non-specific immunogenic stimulus provided by the first (B16) tumor or whether both tumors possess weak common TSTA (perhaps virus-associated), so that the first allogeneic graft challenge provokes a strong enough cross reaction against the usually weakly antigenic second tumor (Harding-Passey) to cause the latter's rejection.

Spleen Lymphocyte Transfers

In the first of two similar experiments (fig. 5), three susceptible C57BL/10 litter mates were challenged by B16 melanoma. One of these hosts (♀171), showing earliest rapid tumor growth, was chosen as the recipient of spleen cell suspensions obtained from B16-challenged donors of resistant strains, or from resistant F_2 or backcross progeny. Her dosage regime was as follows (with tumor implantation on day zero): 2 or 3 spleens/day, days 2–11. Treatment was suspended, days 12–19, and then resumed with the same spleen dosages, days 20–29. In this case, tumor size decreased during the earlier treatments, but resumed growth after treatment was stopped. Upon

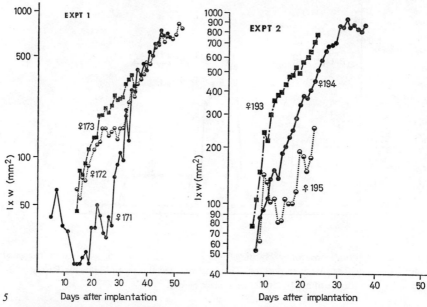

Fig. 5. First spleen lymphocyte transfer experiment. C57BL/10 littermates challenged by B16 melanoma. Note temporary tumor growth retardation for ♀ 171, treated with sensitized, resistant spleen cells. Also ♀ 172 exhibited brief tumor growth retardation during treatment with unimmunized, resistant spleen lymphocytes. For ♀ 173, similar brief treatment yielded no apparent effect on tumor growth. See text for additional details.

Fig. 6. Second spleen lymphocyte transfer experiment. C57BL/10 littermates challenged by B16 melanoma. ♀ 194 served as untreated control. ♀ 193 received unimmunized, resistant spleen cells, with no apparent tumor growth retardation. ♀ 195 received both sensitized and unimmunized resistant spleen cells, resulting in tumor growth retardation during days 11–19. See text for additional details.

resumption of spleen cell transfers another temporary decrease in tumor size occurred, followed by uncontrolled tumor growth.

Female 172 received 2 or 3 spleens/day from unimmunized resistant BALB/c donors, days 22-29. During this treatment period, the B16 tumor exhibited temporary cessation of growth. Tumor growth then resumed after treatment was suspended. This unexpected treatment effect needs additional verification. Female 173 received 3 sensitized spleens/day, days 22–25, and then 2 or 3 unimmunized spleens/day, days 26-29. No effect on tumor growth was apparent.

In the second experiment (fig. 6) three C57BL/10 littermates were given B16 implants on day zero, and then subjected to the following spleen cell dosage regimes: a) Female 194 served as untreated control. b) Female 193 received 3 unimmunized BALB/c spleens/day on day 3 followed by 1 unimmunized BALB/c spleen/day, days 4–17. No tumor growth retardation was observed. c) Female 195 received 3 immunized spleens on day 3; 1 immunized spleen/day, days 4 and 5; and then 1–4 unimmunized spleens/day from resistant donors of various strains, days 6–24. Decreased tumor size or growth retardation was observed during days 11–19. In this case it is not clear whether this antitumor attack was due to the earlier administration of sensitized spleen cells or to the later administration of nonimmunized cells, or to both. If the allogeneic sensitized spleen lymphocytes were rejected by the tumor host soon after administration, with no lasting antitumor effect, then the treatment effect could be attributed primarily to the unimmunized spleen cells. In any event, these preliminary observations corroborate the results of the first experiment, in that spleen lymphocytes from resistant donors can confer at least a temporary adoptive antitumor immunity.

These adoptive immunity experiments, while not particularly successful immunotherapeutically, are nevertheless instructive in that a) they implicate the immune system in these cases of tumor growth retardation, and b) suggest, in the as yet unverified results with nonimmune lymphocytes, that unchallenged resistant individuals might contain a small number of lymphocytes capable of immediate attack against allogeneic tumor histocompatibility antigens [10, 11].

Conclusion

Our results suggest the possibility of developing immunologic antitumor procedures which might be useful prophylactically or therapeutically, especially as adjuncts to surgical excision of primary tumors. An appropriate immunologic approach might be particularly useful in providing sensitized lymphocytes capable of specifically seeking out and destroying relatively small numbers of unnoticed or inaccessible metastatic cells. An immunologic approach might also serve as an adjunct to other antitumor treatments, such as chemotherapy or radiation, provided these treatments would not themselves be significantly immunosuppressive. Success in this approach would depend upon accurate histocompatibility typing of individual hosts (actual or potential) and of tumors, as well as the preparation of safe methods for administering antigenic challenges whether by means of lethally irradiated histoincompatible tumor cells or of cell surface antigen fractions coupled with other nonspecific agents capable of heightening the antitumor immune response.

References

1 ALEXANDER, P.: Immunotherapy of cancer. Experience with primary sarcoma in the rat. Proc. roy. Soc. Med. *60:* 1181–1182 (1967).
2 ALEXANDER, P.: Immunotherapy of leukemia: the use of different classes of immune lymphocytes. Cancer Res. *27:* 2521–2526 (1967).
3 ALEXANDER, P.; BENSTED, J.; DELORME, E.J.; HALL, J.G.; HAMILTON, L.D.G. and HODGETT, J.: Treatment of primary sarcomas by enhancing host defense with immune lymphocytes or their RNA; in The proliferation and spread of neoplastic cells, 21st Ann. Symp. on Fundamental Cancer Research, Univ. of Texas M.D. Anderson Hosp. pp. 693–710 (Williams and Wilkins, Baltimore, Md. 1967).
4 BARTLETT, G.L.; ZBAR, B. and RAPP, H.J.: Suppression of murine tumor growth by immune reaction to the bacillus Calmette-Guerin strain of *Mycobacterium bovis*. J. nat. Cancer Inst. *48:* 245–257 (1972).
5 BOYSE, E.A.: in R.T. SMITH and M. LANDY, Immune surveillance, 5–30 (Academic Press, New York 1970).
6 FOSTER, M.; HERMAN, J. and THOMSON, L.: Histocompatibility relations and immunotherapeutic role of B16 mouse melanoma. Genetics *68:* s20–s21 (1971).
7 FOSTER, M. and THOMSON, L.: Histocompatibility relations of mouse melanoma B16. Genetics *64:* s21 (1970).
8 GOOD, R.A.: in R.T. SMITH and M. LANDY, Immune surveillance, pp. 439–451 (Academic Press, Press, New York 1970).
9 HERMAN, J. and FOSTER, M.: Histocompatibility relations of mouse melanomas S91A and Harding-Passey. Genetics *68:* s27 (1971).
10 JERNE, N.K.: in R.T. SMITH and M. LANDY, Immune surveillance, pp. 345–362 (Academic Press, New York 1970).
11 JERNE, N.K.: The somatic generation of immune recognition. Europ. J. Immunol. *1:* 1–19 (1971).
12 KALISS, N.: Dynamics of immunological enhancement. Transplant. Proc. *2:* 59–67 (1970).
13 LAW, L.: Studies of tumor antigens and tumor-specific immune mechanisms in experimental systems. Transplant. Proc. *2:* 117–132 (1970).
14 MITCHISON, N.A.: Immunologic approach to cancer. Transplant. Proc. *2:* 92–103 (1970).
15 WALKER, S.E. and BOLE, G.G.: Augmented incidence of neoplasia in female NZB/NZW mice treated with long-term cyclophosphamide. J. Lab. Clin. Med. *78:* 978–979 (1971).
16 ZBAR, B.; BERNSTEIN, I.; TANAKA, T. and RAPP, H.J.: Tumor immunity produced by the intradermal inoculation of living tumor cells and living *Mycobacterium bovis* (strain BCG). Science *170:* 1217–1218 (1970).
17 ZBAR, B. and TANAKA, T.: Immunotherapy of cancer: regression of tumors after intralesional injection of living *Mycobacterium bovis*. Science *172:* 271–273 (1971).

Author's address: Dr. MORRIS FOSTER, Mammalian Genetics Center, Department of Zoology, The University of Michigan, *Ann Arbor, MI 48104* (USA)

Direct Heterologous Transplantation of a Malignant Human Melanoma and its Ultrastructure
Preliminary Report

J. M. JADIN and G. VAN DER SCHUEREN

Tumour Center, University of Leuven, Leuven

Introduction

Malignant melanoma has been successfully transplanted into the anterior chamber of the rabbit's eye [1]. More recently, MANUELIDIS [2] inoculated human malignant melanoma from a 4-month culture into conditioned hamsters and on the 11th day, portions of the subcutaneously growing tumour were successfully re-transplanted into the cerebrum of guineapigs.

In the present study, human malignant melanoma cells were injected directly into hairless mice, without immunosuppression.

Materials and Methods

Hairless mice have the advantage that after a normal hair cycle they lose their hair permanently, and growth of tumours in the subcutaneous tissues can be observed directly, owing to the transparency of their thin pink skin.

The source of the material was an extensive primary melanoma on the face of a male patient 71 years of age. The clinical diagnosis of malignant melanoma arising in Dubreuilh's melanosis was confirmed histologically.

A grafting needle was inserted into the animal at the level of the hind leg and pushed subcutaneously as far as the shoulder region where portions of the tumour were deposited.

Similar to the original tumour, there was an irregular distribution of pigment in the various grafts. The original tumour and the grafts were all studied ultrastructurally.

Results

In the first passage, 4 mice were inoculated, all of which developed within 2 weeks a visible and palpable mass, which grew for a while and then

Table I. Four mouse-passages of a human malignant melanoma

Hairless mice (unconditioned)	9 weeks
Hairless mice (unconditioned)	13 weeks
Hairless mice (unconditioned)	23 weeks
Hairless mice (without thymus)	2 weeks
Original graft alive	47 weeks

remained stationary. Some degree of satellitosis was apparent in the exposed tumours but no metastases developed.

All the animals receiving the second transplant developed tumours, although some were very small.

Further transfers were made in week 13 and from these another transfer was made in week 23 into thymectomised mice. Some growth occurred but the animals, already in a delicate state, all died before any further transfer could be made.

Successful passage was maintained until the 47th week.

The similarity between the tumour in the animal grafts and the original human tumour at the time of removal is well illustrated electron-microscopically (fig. 2).

The hairless mouse presented itself as an excellent host for the study

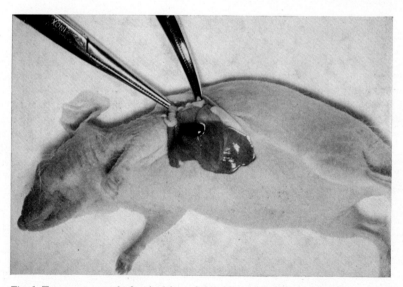

Fig. 1. Tumour exposed after incision of the skin and before grafting.

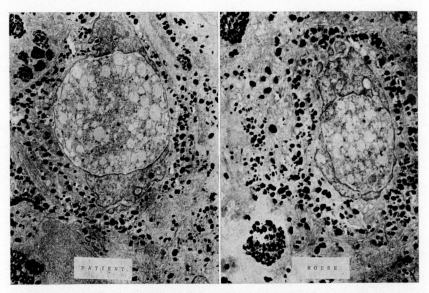

Fig. 2. Electron micrographs showing similarity between the patient's biopsy and the mouse transplant on the 329th day.

of subcutaneous melanoma, although the tumour growth was never abundant; better growth may perhaps be obtained by pretreatment.

Conclusion

Hairless mice can be recommended as experimental animals suitable for a close follow-up study of a subcutaneously grafted, human malignant melanoma. In non-conditioned hairless mice a malignant melanoma survived over four passages during 47 weeks. At the end of this period, the ultrastructural picture of the graft resembled the structure of the original biopsy.

References

1. GREEN, H.: The heterologous transplantation of human melanomas. Yale J. Biol. Med. *22:* 611–620 (1950).
2. MANUELIDIS, E. E.: Heterologous transplantation of a tissue culture line of a human melanoma (TC 491). Yale J. Biol. Med. *43:* 307–322 (1971).

Author's address: Dr. J. M. JADIN, Centrum Gezwelziekten, University of Leuven, *Leuven* (Belgium)

Morphologic Patterns of Spontaneously Regressing Malignant Melanoma in Relation to Host Immune Reactions

R. L. Ikonopisov

Oncological Research Institute, Sofia

Partial or complete spontaneous regression of a tumour is in fact, one of the most striking phenomena, witnessing an overpowering reaction in favour of the host. Spontaneous regression of malignant melanomas offers highly intriguing aspects in the study of host response to autologous tumours. Clinical and morphological evidence for immune response versus autologous malignant melanoma has been provided *inter alia* by observations on: a) 'good' primary melanomas remaining localized for a long time before they metastasize to regional lymph nodes or to remote visceral or cutaneous sites [12]; b) metastatic lesions remaining localized and stationary for a prolonged period [13]; c) spontaneous regression of a primary malignant melanoma in the face of regional lymph node metastases [6, 15, 9]; d) spontaneous regression in precancerous melanosis with multicentric melanomas [10]; e) spontaneous regression in lymph node metastases [2]; f) 'smouldering disease' – some melanoma lesions fade while new ones appear [4].

This is a report on the immunological and morphological data in four cases of spontaneously regressing malignant melanomas falling within the class of the abovementioned points c) and d) Although our study covers a rather small number of cases and our immunomorphological data inevitably bear shortcomings, we feel that as far as the phenomenon of spontaneous regression of malignant melanoma is concerned, any piece of evidence for the correlation of the morphology of the tumour and its surroundings with the integral immune response of the host (as detected by immunological techniques) would perhaps be of help in putting together the elements and the phases of the so far highly hypothetical mechanisms of the immune cytotoxic action.

Case Reports

Case I

J.M., a 24-year-old Australian (a joint patient of Professor G. W. MILTON and Mr. C. I. COOLING), was first seen in October 1969. Two years before that, the patient had discovered that a mole, present since birth on his left calf, grew larger and bled readily on contact. Ten months before presentation at the RMH-London, a fairly marked swelling occurred at the site of the mole, subsiding later, leaving a palish halo. Only weeks before being referred to the RMH, the patient discovered a lump in his left groin. On examination, a bluish-black lesion with a diameter of 0.5 cm surrounded by a wide depigmented halo was observed on the back of his left mid calf (fig. 1) An enlarged, firm but mobile lymph node was present in his left groin. On October 14th, 1969, a wide excision of the lesion was carried out as well as block dissection of the left inguinal and iliac lymph nodes. Both histological and immunological studies were done on the tumour. Histology (Dr. A. LEVENE, RMH-London) showed no melanocytes at the dermoepidermal junction. The pigmentation visible clinically was entirely of dermal origin with macrophages laden with pigment (fig. 2). In one field only there was a collection of small lymphoid cells. No evidence of residual tumour was seen in this representative section. Histology of the lymph node showed a highly pleomorphic malignant melanoma with focal degeneration and cytophagocytosis. Some of the melanin was contained in cells which were probably not tumour cells. Reaction centres were present in some of the uninvolved lymphoid tissue. The patient's serum was tested against a cell suspension of his own metastatic melanoma. No cytotoxic antibodies were detected in the complement dependent RNA synthesis inhibition test. A skin test with 3×10^6 autologous irradiated melanoma cells revealed a positive cutaneous delayed hypersensitivity reaction.

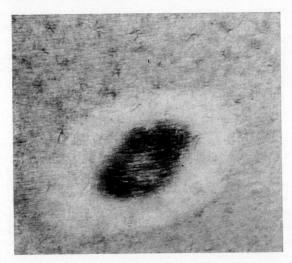

Fig. 1. Case I. Complete spontaneous regression of primary malignant melanoma. Residual bluish-black lesion surrounded by a wide depigmented halo.

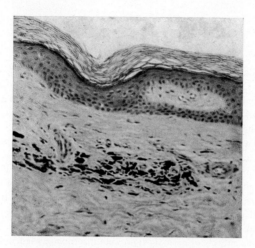

Fig. 2. Case I. Light-microscopic appearance. No evidence of residual tumour. No melanocytes in the dermo-epidermal junction. A collection of dermal melanocytes laden with pigment.

Case II

S. P., a 36-year-old male presented at the Dermatological Department of the Oncological Research Institute in Sofia in July 1971 with an enlargement of lymph nodes in the right axilla. Histology of these nodes revealed invasion with malignant melanoma (Prof. R. RAICHEV). In a detailed examination we detected a small depigmented halo surrounding a slightly sessile nevus in the middle of the patient's back. The patient stated that he had had a pigmented mole in this same region since birth. Two years prior to presentation to hospital the mole enlarged, ulcerated, then fell off. The remaining lesion was excised. Histology of this lesion (Prof. R. RAICHEV) showed a dermoepidermal nevus with almost no cellular reaction in the dermis, but abundant fibrosis. The epidermis immediately surrounding the nevus was devoid of melanocytes. Immunological studies revealed no serum cytotoxic antibodies, but a strong delayed hypersensitivity reaction to autologous irradiated melanoma cells. A BCG heaf gun inoculation resulted in a strongly positive reaction.

Case III

B. L., an 80-year-old woman was first seen in July 1969 at the RMH-Surrey Branch-UK. She had had a pigmented patch on her back for many years which, two months prior to hospital presentation, became larger and raised. At its lower end an elevated and infiltrating nodularity occurred, surrounded by a slightly erythematous and depigmented halo in the adjacent normal skin. No palpable nodes were discovered in the axillary and the inguinal regions. Histology (Dr. A. LEVENE, RMH-London) showed moderate malignant junction changes and moderate numbers of subepidermal heavily pigmented macrophages with a few scattered lymphocytes. A polypoid, poorly pigmented primary melanoma in the area of malignant lentigo was observed. Indirect immunofluorescence (T. PHILLIPS) revealed a positive antigen-antibody binding reaction. Delayed hypersensitivity to autologous irradiated melanoma cells was positive as well.

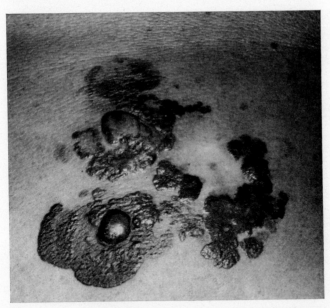

Fig. 3. Case IV. Multicentric malignant melanoma within a malignant lentigo with regions of spontaneous regression and depigmentation.

Case IV

K. D., a 66-year-old male seen at the Dermatology Department of the Oncologica Research Institute in Sofia in August 1971 stated that he had noticed a bluish black patch on his back for the last 7 years. Two months prior to medical examination, the lesion enlarged and formed a few nodularities, some of which bled. On examination, a large, irregularly shaped malignant lentigo with several nodular melanomas within the lesion were seen (fig. 3). There were a few regions exhibiting spontaneous regression and depigmentation. No enlarged lymph nodes were found in the axillary and the inguinal regions. The lesion on the back was excised and the defect was covered with a free skin graft. Multiple serial histologic sections revealed various areas of dermoepidermal activity and malignant transformation to fully invasive melanoma, along with regions of spontaneous regression and macrophages laden with pigment. These areas were surrounded by a variety of cells of the lymphoid series and macrophages (Fig. 4). Immunological studies revealed the presence of serum cytotoxic antibodies and a strong delayed hypersensitivity reaction to autologous irradiated melanoma cells. BCG heaf gun inoculation resulted in a positive skin reaction (table I).

Discussion

The morphological picture at the site of a spontaneously regressing tumour involving the action of cells of the lymphoid series is accepted as

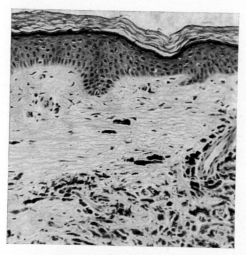

Fig. 4. Case IV. Nodular malignant melanoma with abundant stromal reaction.

Table I. Immune reactions in spontaneously regressed malignant melanoma

Patient	Sex	Age	Clinical stage	Immunity Humoral	Cellular	Immunostimulation BCG
J. M.	M	22	$T_0N_1M_0$	–	+	nt
S. P.	M	36	$T_0N_2M_0$	–	+	+ +
B. L.	F	80	$T_1N_0M_0$	+	+	nt
K. D.	M	66	$T_2N_0M_0$	+	+	+ + +

circumstantial evidence for cellular-mediated immunity. As knowledge on the effector mechnisms of an immune response is now accumulating [1, 7, 8], it becomes clear that cellular and humoral immunity may be initiated by the presence of an antigenic stimulus: the tumour specific membrane antigenic determinants. The latter are thought to be released through the afferent arm of the immune response and the antigenic information transferred by cell mediators to the regional lymph nodes. The strength of the reaction at this end would depend largely upon the physiological properties of the in-transit tissues providing a certain degree of accessibility to the lymph node. Although detailed information on the cellular reactions within the efferent arm of cell-mediated immune response *in vivo* remains unelucidated, specific cytotoxic lymphocytes have been detected in *in vitro* studies on animal tumours [16]. It is assumed that an antigenically stimulated lymph node reacts with the production of immunoblasts, small cytotoxic lymphocytes and plasma cells, part of which may remain in the lymph node comprising the armamentarium of the phenomenon of immunologic memory, while the rest of the immunocompetent cells may be transferred through the efferent arm of the immune response to the melanoma target cell. It is very likely that this effector response is the result of a joint and correlated action of cellular elements competent to respond to the antigenic stimulation. On their way to the target or in its vicinity the macrophages may be armed [6] with cytotoxic antibodies produced and carried by the immune lymphocytes. Again, the action at the melanoma cell target would depend upon the accessibility through the tissues of the efferent territories of the immune response.

Reminding of the above hypothetical assumptions becomes necessary when we come to interpret the presence of peritumoural infiltrates of cellular elements of variable numbers and type. It could be suggested that the cells involved in the ultimate end of the cell-mediated effector mechanisms would perhaps differ in type and numbers depending upon the presence of whole melanoma cells bearing tumour-specific membrane antigens. As long as the latter persist at the tumour bed the regional lymph nodes and the entire immune system will be specifically stimulated and induced into releasing antibody producing and armed immunocompetent cells. The lymphocytes and plasma cells would initially appear and increase in number but will subsequently diminish in number with the continuous destruction of the target melanoma cells. On the other hand, macrophages would be increasing in number as they would be involved in engulfing the dying cells, breaking them down into digestible parts and digesting them through phagocytosis.

The situation with a variable lymphocyte infiltrate from sparse, moderate, to abundant, is illustrated in cases III and IV with nodular melanomas arising in a malignant lentigo and displaying partial spontaneous regression. If we try to assemble separate instances of these processes of growth and regression within the entire lesion, it does give us an idea of the gradual build-up of the mass of round cell infiltrate combatting the existing malignant melanoma cells. The immunological tests carried out in these patients revealed a corresponding intactness of their humoral and cellular immunity.

In cases I and II where the primary melanomas have been subjected to complete spontaneous regression in the face of regional lymph node metastases, histology reveals masses of macrophages laden with pigment but no lymphocyte, plasma cell or histiocyte infiltrates. Regression here is complete, no tumour-specific membrane antigens are present which could have provoked the influx of immune 'combat' cells. According to YOUNG and COWAN [17], cellular infiltration was much less in tumours which have regressed for several months. BLACK [3] reports on two cases of complete spontaneous regression with no inflammatory infiltrate in the surroundings of the regressed tumour. Our patients' humoral immunity was negative while a positive delayed hypersensitivity reaction to autologous melanoma cells was present (table I). The latter provides evidence for the capacity of effective cell-mediated immunity which presumably has brought about the spontaneous regression of the primary tumour. It could be that the lymph node metastases in cases I and II have arisen and grown as immunologically competent cells have been released in greater numbers and drained in the direction of the efferent arm of the immune response for the destruction of an antigenically 'highly attractive' cell target. Drainage of lymph has been known to enhance lymph node metastases [14].

Finally, the presence of serum cytotoxic antibodies may not correlate with the cytotoxic capacity of lymphoid cells. The discrepancy between a positive delayed hypersensitivity reaction and the lack of detectable serum antimelanoma autoantibodies could be explained either by the presumption that humoral and cellular immunity are of different origin and different factors contribute to their selective blockage, or by the simple absorption of circulating antibodies by the regressing primary or the growing metastases, and, last but not least, in the process of arming the macrophages.

Surely, any discussion of the problem of spontaneous regression of tumours puts forth more questions than it answers, but a further study of the immunomorphological aspects of this phenomenon may prove to be highly rewarding.

Acknowledgements

This work was supported by a Wellcome Foundation (UK) grant and was started during the period December 1970–March 1971 at The Royal Marsden Hospital, Surrey Branch, Sutton, Surrey, UK. The author wishes to thank Dr. P. E. THOMPSON HANCOCK, Director, Department of Clinical Research, The Royal Marsden Hospital, London for his kindness and help. This study was made possible mainly through the wholehearted cooperation and encouragement of Mr. C. I. COOLING, Consultant Surgeon at the RMH, the histopathology interpretations of Dr. ARNOLD LEVENE, Consultant Pathologist at the RMH, the assistance of Miss MILLER and her staff at Medical Records, RMH, Surrey Branch, and Mrs. S. F. B. BELL, Follow-up Dept., Metropolitan South-East Cancer Registry.

It has been completed at the Dermatology Department, assisted in histopathology interpretation by Professor R. RAICHEV and Dr. KONSTANTIN TSANEV, Oncological Research Institute (Director: Professor Dr. GERASSIM MITROV) Sofia, Bulgaria.

References

1 ALEXANDER, P. and HALL, J. G.: The role of immunoblasts in host resistance and immunotherapy of primary sarcomas. Adv. Cancer Res. *13:* 1–37 (1970).
2 BESWICK, I. P. and QUIST, G.: Brit. med. J. *ii:* 930 (1963).
3 BLACK, M. M.: Discussion; in L. SEVERI Immunity and tolerance in oncogenesis, vol. 2, p. 1164 (Perugia 1970).
4 BODENHAM, D. C.: A study of 650 observed cases of malignant melanoma in the South-West region. Ann. roy. Coll. Surg. Engl. *43:* 218–239 (1968).
5 DAS GUPTA, T.; BOWDEN, L. and BERG, J. W.: Malignant melanoma of unknown primary origin. Surg. Gyner. Obstet. *117:* 341–345 (1963).
6 EVANS, R. and ALEXANDER, P.: Cooperation of immune lymphoid cells with macrophages in tumour immunity. Nature, Lond. *228:* 620–622 (1970).
7 HALL, J. G.: Effector mechanisms in immunity. Lancet *iv:* 25–28 (1969).
8 HUMPHREY, L. J.; BARKER, C.; BOKESCH, C.; FETER, P.; AMERSON, J. R. and BOEHM, O. R.: Immunologic competence of regional lymph nodes in patients with mammary cancer. Ann. Surg. *174:* 383–389 (1971).
9 IKONOPISOV, R. L.: Malignant melanoma with unknown primary. Sbornik Trudove (Scientific Papers), Canc. Res. Inst., Sofia *9:* 101–105 (1964).
10 IKONOPISOV, R. L. and RAICHEV, R.: Host-tumour relationship in malignant melanoma arising in melanosis circumscripta precancerosa Dubreuilh. Sbornik Trudove (Scientific Papers), Canc. Res. Inst., Sofia *14:* (1972)
11 IKONOPISOV, R. L.; LEWIS, M. G.; HUNTER-CRAIG, I. D.; BODENHAM, D. C.; PHILLIPS, T.; COOLING, C. I.; PROCTOR, J.; HAMILTON FAIRLEY, G. and ALEXANDER, P.: Autoimmunization with irradiated tumour cells in human malignant melanoma. Brit. med. J. *ii:* 752–754 (1970).
12 LEWIS, M. G.: Possible immunological factors in human malignant melanoma in Uganda. Lancet *ii:* 921–922 (1967).
13 MILTON, G. W.; LANE BROWN, M. M. and GILDER, M.: Malignant melanoma with an occult primary lesion. Brit. J. Surg. *54:* 651–658 (1967).
14 RUDENSTAM, C. M.: Experimental studies on trauma and metastasis formation. Acta chir. scand. Suppl. *391* (1968).
15 SMITH, J. L. and STEHLIN, J. S.: Spontaneous regression of primary malignant melanoma with regional metastases. Cancer *18:* 1399–1415 (1965).
16 TAGASUGI, M. and KLEIN, E.: Transplantation *9:* 219 (1970).
17 YOUNG, S. and COWAN, D. M.: Nature, Lond. *205:* 713 (1965).

Author's address: Dr. R. L. IKONOPISOV, Department of Dermatology and Clinical Immunology of Tumours Laboratory, Oncological Research Institute, *Sofia-Darvenitza 56* (Bulgaria)

Subject Index

Albinism,
    cancer 236
    incidence in Africa 236
    melanocytes of jungle fowl 47
    melanoma 236
Acid phosphatase in skin 66
ACTH in melanocyte differentiation 1
Actinomycin D inhibition of melanocyte responses 188
Adenosine 3,5-monophosphate (cyclic AMP) 1,188
Adrenergic nerve – melanophore system 195
Antimelanoma agents
    BCG 270, 350, 390
    chlorpromazine compound 215
    interferon 208
    4-isopropyl catechol 202
    neutron capture 215
    polyinosinic-polycytidylic acid 208
    thymectomy-irradiation 208
APUD system of cells 55

Cancer of skin in Africa 236
Celticity 229
Circadian colour change 180
Cyclic AMP (C-AMP) 1, 188
Cyclic nucleotide phosphodiesterase 188
Cytochalasin B 1, 27
Cytochrome b_5 134

Dendritic cells (indeterminate) in hair matrix 20
Dopa (3,4-dihydroxyphenylalanine)
    administration associated with melanoma recurrences 300
    blood levels 312
    cysteinyl- 171
    fluorimetry of dopa conjugates 171
    incorporation in melanosomes 39
    in Dubreuilh's melanosis 292
    melanocytes 55
    metabolism in humans 308
    metabolites in urine 317
    pagetoid melanoma 292
Dubreuilh's precancerous melanosis 292

Electron microscopic studies of,
    dendritic cells of hair matrix 20
    Golgi during melanogenesis 47
    melanocyte-keratinocyte interaction 27
    melanoma, recurrent 300
    melanosomes 165, 166, 366
    melanoma tissue culture 382
    pagetoid melanoma 292
    skin colour 66
    transplanted human melanoma 399
    tyrosine oxidation 98
Endocrine polypeptide series of cells (APUD) 55

Epidemiology of melanoma 222, 229, 246, 273
Epidermal melanin unit 27, 66

Fluorescence
  APUD series 55
  dopa conjugates 171
  formaldehyde-induced 55, 171
  marsupial pigments 142
  melanoma 171
Fluorimetry
  dopa conjugates 171
  marsupial pigments 142
  melanoma 55, 171

Genetics of melanoma
  human 229, 261
  murine transplantable 390
Golgi system 47

Haemopexin 346
Hair, human, melanocytes 20
  dendritic cells 20
Hormones
  ACTH 1
  MSH 1, 188, 195
  norepinephrine 195

Imidazole (cyclic nucleotide phosphodiesterase activator) 188
Immunity
  lymphocytotoxicity 350, 382
  melanoma, human 261, 285, 350, 360, 366, 372, 402
  melanoma, mouse 390

Keratinocyte
  epidermal-melanin unit 27, 66
  interaction with melanocytes 27
  microfilaments 27

Langerhans cells 20
Lymphocytic infiltrates in association with melanoma 285

Marsupial pigments 142
Melanin
  aggregation 66, 195
  cutaneous 66, 236
  epidermal 66
  fungal 158
  plant 158
  squid 158
  structure 151, 158
  synthesis, in culture 90
    peroxidase-induced 98
    ultra-violet-light-induced 195
  x-ray diffraction 158
Melanocyte
  albino-like of jungle fowl 47
  cytochemistry 55
  differentiation 1
    effect of ACTH 1
    effect of C-AMP 1
  dopa 55
  hair 20
  in APUD system 55
  interaction with keratinocytes 27
  negro 66
  New Guinean 6
  patterns of distribution in relation to naevi 246
Melanocyte-stimulating hormone (MSH) 1, 188, 195
Melanocytogenesis 1
Melanocytoma 292
Melanogenesis
  cytochrome b_5 134
  dopa incorporation 39
  electron transfer system 134
  Golgi system 47
  inhibition 118
  in vitro 39, 90
  jungle fowl 47
  melanosomal 125, 134
  peroxidase induction 98
  tyrosine 82, 98
Melanogens, urinary 312, 317, 323
Melanoma, human
  africa 236
  albino 236
  amino acids of plasma and urine 331
  antigens 360, 366

Subject Index

APUD series relationship 55
associated primary tumours 255
BCG therapy 270, 350, 390
catechols in urine 308, 312, 317, 323
Celtic susceptibility 229
dopa in blood 312 [323
dopa metabolites in urine 308, 312, 317,
Dubreuilh's precancerous melanosis 292
epidemiology 222, 229, 246, 261, 273
familial 222
fluorescence, formaldehyde-induced 55
haemopexin 346
host resistance 255
immune reactions 55, 270, 285, 350, 360, 366, 372, 402
immunoglobulins 222
immunofluorescence 55, 366
Japan 246
lymphocytic infiltrates 285
melanogens, urinary 312, 317, 323
mortality related to incidence 222
neutron-capture therapy 215
ocular 222
pagetoid 292
plasma amino acids 331
prognosis 285
recurrences after dopa administration 300
regression, spontaneous 402
research at M. D. Anderson Hospital and Tumor institute 261
solar circulating factor 222
tissue culture
    lymphocytotoxicity 350, 382
    tyrosinase inhibition 111
    virus susceptibility 382
transplantation to mice 399
urinary amino acid 331

Melanoma, mouse
  anti-melanoma agents
    interferon 208
    4 isopropyl catechol 202
    polyinosinic-polycytidylic acid 208
    spleen lymphocyte transfers 390
    thymectomy-irradiation 208
  electron transfer system 134
  genetic aspect 390
  histocompatibility 390
  immunology 390
  lymphocytotoxicity 382
  melanogenesis, peroxidase-induced 98
  microsomal cytochrome b_5 134
  plasma amino acid 331
  urinary amino acids 331

Melanophores
  adrenergic nerve-melanophore system 195
  response to
    actinomycin D 188
    C-AMP 188
    cytochalasin B 27
    imidazole 188
    isoproterenol 188
    MSH 188, 195
    norepinephrine 195
    procaine 188
    puramycin 188
    tetracaine 188
    tetracyclines 188
    UVL 195

Melanosomes
  aggregation 195
  chemical composition 165
  distribution in keratinocytes 66
  dopa incorporation 39
  melanisation 39, 66
  membrane permeability 125
  responses to
    aromatic amino acids 125
    chemical agents 125
    irradiation 125
    metabolic inhibitors 125
    UVL 195
  size 66
  structure 125
  tyrosinase activity 39

Melatonin
  assay 180

Subject Index

circadian rhythm 180
Microfilaments of keratinocytes 27
Mouse hairless, suitability for transplants 399

Naevi of childhood 276
Naevocytoma 292
Neutron-capture therapy of melanoma 215
New Guinea 'red skins' 6

Organ culture of xanthic fish fins 1

Peroxidase-induced melanogenesis 98
Piebaldism
   canine 14
   hypopigmentation 14
   Waardenburg's syndrome 14, 236
Pigment
   chemical structure 151, 158
   donation 27, 66
   fluorescence 55, 142, 171
   in Africans 236
   in New Guineans 6
   marsupial 142
   relationship with cancer susceptibility 236
   retinal, in New Guineans 6
   skin colour 66, 236
   tryptophane-derived in marsupials 142
Pigment donation inhibition by cytochalasin B 27
Polyinosinic-polycytidylic acid, anti melanoma effect 208
Psoralens 66
Pterinosomes 1
Puromycin 188

Red skins in Africa 236
   in New Guinea 6
Retinal pigment in New Guineans 6

Skin colour
   African 236
   Celtic 229
   classes of 66
   dihidroxyacetone 66
   negroid 66
   New Guinean 6
   ultrastructure 66
Spectrophotometry of melanoma extracts 171
Spleen lymphocyte transfers 390

Tyrosinase activity
   biochemistry 82
   effect of cyanide 125
   Golgi system 47
   inhibition 1, 111, 118, 125
   in vitro 90, 111
   irradiation 125
   localisation in vertebrates 82
   1-phenylalanine inhibition 125
   1-tryptophane inhibition 125
   melanosomal 39
Tyrosine metabolism 90, 98, 125

Ultraviolet light
   melanogenesis, relationship 66
   melanoma epidemiology relationship 222
   melanophore response 195

Vitiligo in Africa 236

Waardenburg syndrome 14, 236

Xanthophores 1